Challenges and Opportunities for Transforming From STEM to STEAM Education

Kelli Thomas
University of Kansas, USA

Douglas Huffman
University of Kansas, USA

A volume in the Advances in Educational Technologies and Instructional Design (AETID) Book Series

IGI Global
DISSEMINATOR OF KNOWLEDGE

Published in the United States of America by
IGI Global
Information Science Reference (an imprint of IGI Global)
701 E. Chocolate Avenue
Hershey PA, USA 17033
Tel: 717-533-8845
Fax: 717-533-8661
E-mail: cust@igi-global.com
Web site: http://www.igi-global.com

Library of Congress Cataloging-in-Publication Data

Names: Thomas, Kelli, 1969- editor. | Huffman, Douglas, editor.
Title: Challenges and opportunities for transforming from STEM to STEAM education / Kelli Thomas and Douglas Huffman, editors.
Description: Hershey, PA : Information Science Reference, 2020. | Includes bibliographical references and index. | Summary: "This book explores new approaches and advances to science, technology, engineering, and mathematics (STEAM) education"-- Provided by publisher.
Identifiers: LCCN 2019040439 (print) | LCCN 2019040440 (ebook) | ISBN 9781799825173 (hardcover) | ISBN 9781799825180 (paperback) | ISBN 9781799825197 (ebook)
Subjects: LCSH: Interdisciplinary approach in education. | Science--Study and teaching. | Technology--Study and teaching. | Engineering--Study and teaching. | Arts--Study and teaching. | Mathematics--Study and teaching.
Classification: LCC LB1027 .C434 2020 (print) | LCC LB1027 (ebook) | DDC 372.35/044--dc23
LC record available at https://lccn.loc.gov/2019040439
LC ebook record available at https://lccn.loc.gov/2019040440

This book is published in the IGI Global book series Advances in Educational Technologies and Instructional Design (AETID) (ISSN: 2326-8905; eISSN: 2326-8913)

British Cataloguing in Publication Data
A Cataloguing in Publication record for this book is available from the British Library.

For electronic access to this publication, please contact: eresources@igi-global.com.

Advances in Educational Technologies and Instructional Design (AETID) Book Series

ISSN:2326-8905
EISSN:2326-8913

MISSION

Education has undergone, and continues to undergo, immense changes in the way it is enacted and distributed to both child and adult learners. In modern education, the traditional classroom learning experience has evolved to include technological resources and to provide online classroom opportunities to students of all ages regardless of their geographical locations. From distance education, Massive-Open-Online-Courses (MOOCs), and electronic tablets in the classroom, technology is now an integral part of learning and is also affecting the way educators communicate information to students.

The **Advances in Educational Technologies & Instructional Design (AETID) Book Series** explores new research and theories for facilitating learning and improving educational performance utilizing technological processes and resources. The series examines technologies that can be integrated into K-12 classrooms to improve skills and learning abilities in all subjects including STEM education and language learning. Additionally, it studies the emergence of fully online classrooms for young and adult learners alike, and the communication and accountability challenges that can arise. Trending topics that are covered include adaptive learning, game-based learning, virtual school environments, and social media effects. School administrators, educators, academicians, researchers, and students will find this series to be an excellent resource for the effective design and implementation of learning technologies in their classes.

COVERAGE

- Digital Divide in Education
- Educational Telecommunications
- Web 2.0 and Education
- Virtual School Environments
- Classroom Response Systems
- Game-Based Learning
- Adaptive Learning
- Curriculum Development
- Online Media in Classrooms
- Bring-Your-Own-Device

The Advances in Educational Technologies and Instructional Design (AETID) Book Series (ISSN 2326-8905) is published by IGI Global, 701 E. Chocolate Avenue, Hershey, PA 17033-1240, USA, www.igi-global.com. This series is composed of titles available for purchase individually; each title is edited to be contextually exclusive from any other title within the series. For pricing and ordering information please visit http://www.igi-global.com/book-series/advances-educational-technologies-instructional-design/73678. Postmaster: Send all address changes to above address.

Titles in this Series

For a list of additional titles in this series, please visit:
http://www.igi-global.com/book-series/advances-educational-technologies-instructional-design/73678

Transforming Music Education in P-12 Schools and the Community
Taichi Akutsu (Okayama Prefectural University, Japan & Seisa University, Japan)
Information Science Reference • ©2020 • 230pp • H/C (ISBN: 9781799820635) • US $175.00

UXD and UCD Approaches for Accessible Education
Ricardo Mendoza-González (Tecnológico Nacional de México, Instituto Tecnológico de Aguascalientes, Mexico) Huizilopoztli Luna-García (Universidad Autónoma de Zacatecas, Mexico) and Alfredo Mendoza-González (Universidad Autónoma de Zacatecas, Mexico)
Information Science Reference • ©2020 • 290pp • H/C (ISBN: 9781799823254) • US $185.00

Preparing 21st Century Teachers for Teach Less, Learn More (TLLM) Pedagogies
Pradeep Kumar (Taylor's University, Malaysia) Michael James Keppell (Taylor's University, Malaysia) and Chee Leong Lim (Taylor's University, Malaysia)
Information Science Reference • ©2020 • 332pp • H/C (ISBN: 9781799814351) • US $195.00

Developing Technology Mediation in Learning Environments
Filomena Soares (Polytechnic of Porto, Portugal) Ana Paula Lopes (Polytechnic of Porto, Portugal) Ken Brown (Letterkenny Institute of Technology, Ireland) and Anne Uukkivi (TTK University of Applied Sciences, Estonia)
Information Science Reference • ©2020 • 330pp • H/C (ISBN: 9781799815914) • US $195.00

Global Perspectives on Gameful and Playful Teaching and Learning
Matthew Farber (University of Northern Colorado, USA)
Information Science Reference • ©2020 • 330pp • H/C (ISBN: 9781799820154) • US $195.00

Handbook of Research on Integrating Computer Science and Computational Thinking in K-12 Education
Jared Keengwe (University of North Dakota, USA) and Patrick Wachira (Cleveland State University, USA)
Information Science Reference • ©2020 • 385pp • H/C (ISBN: 9781799814795) • US $265.00

For an entire list of titles in this series, please visit:
http://www.igi-global.com/book-series/advances-educational-technologies-instructional-design/73678

IGI Global
DISSEMINATOR OF KNOWLEDGE

701 East Chocolate Avenue, Hershey, PA 17033, USA
Tel: 717-533-8845 x100 • Fax: 717-533-8661
E-Mail: cust@igi-global.com • www.igi-global.com

Editorial Advisory Board

List of Reviewers

Table of Contents

Section 4
STEAM in Teacher Education

Detailed Table of Contents

In some high school learning environments, hundreds of students engage in collaborative, term-long, project-based learning STEM or STEAM units of work. In this chapter, the authors report on an ongoing design-based research project in which researchers and teachers collaborate to design, teach, and assess STEAM units of work. They draw on research on project-based learning and interdisciplinary collaboration to inform the analysis drawing on Jonassen's typology of problem solving. The purpose of this chapter is to present two tools that functioned as boundary objects mediating learning and self-assessment: (1) a 'STEAM slider' used by students in groups to reach consensus on the use of the tools, knowledge, and processes employed in their project and (2) a criteria sheet developed to mediate students' and teachers' engagement in self- and teacher-assessment. The authors use these results to make recommendations for the next iteration of the project.

Biotechnology, the use of organisms or parts of organisms as tools to support and innovate human wants or needs, plays an extensive role in our daily lives. However, biotechnology is minimally addressed in K-12 educational environments. An issue with the lack of biotechnology in curricula is that we, as end-users, are becoming increasingly defined and dependent on biotechnological innovations. We should be able to think critically and form educated decisions about our medical care, the food we eat, and the biotechnological products we do or do not use. The end-user's acceptance of biotechnological solutions to solve a human want or need also may hinge on the ability of the problem-solver to be artistically creative. This chapter introduces artistic biotechnology and provides three exemplars of K-12 classroom lessons anchored in the Design Thinking Model (DTM) with purposeful art integration. The DTM guides instructors and students through solving ill-defined problems for end-users with artistic biotechnological solutions.

The 21st century offers many technological advancements resulting from cutting-edge research being conducted by scientists and engineers at higher education institutions, industries, and national labs. As exciting scientific research is being conducted, the need to address the challenges of inspiring and engaging new learners in science, technology, engineering, and mathematics (STEM) fields continues to grow. The authors present the Science Cartoons (SciToons) program as a model for making

Section 3
STEAM Education Beyond the Classroom

Section 4
STEAM in Teacher Education

This chapter investigates how teacher candidates' experiences in STEAM field studies with community partners can inform work in teacher education within an integrated practicum based on curriculum of place. The overall goal of the inquiry is to better understand and articulate the particular ways in which people value place-based knowledge. Through relationships with Indigenous communities, the team of educators has a deeply held conviction that sustained deliberations on the connections between Indigenous knowledge systems and place-based thinking can provide significant opportunities for reframing education. Learning from place emphasizes a relationship with the land, something deeply respected in Indigenous communities and something absent from much of place-based education. The research explores this tension as we come to a deeper and shared understanding of co-responsibility within Treaty 7 relationships. The project seeks to close this gap by considering varying perspectives of place as it informs STEAM teacher education pedagogy.

Preface

The ways in which people around the world interact and integrate their lives has changed dramatically in the past two decades with the convergence of technological innovations and the opening of international borders. Thomas Friedman highlighted this notion of globalization in his compelling book, *The World is Flat* (Friedman, 2005). Friedman argued that changes brought on by communication technologies were essentially leveling the economic playing field around the world into an interconnected, interdependent, global flat structure. The integration of globalization into our lives is affecting how people learn, work, and play. The interconnectedness of national economies through increased trade and exchanging of resources magnifies economic globalization. The ease of sharing ideas and information among people in different countries is unprecedented and digital technologies continue to support social globalization. The rapid expansion of digital technologies signaled what Klaus Schwab conceptualizes as the Fourth Industrial Revolution. The Fourth Industrial Revolution is "characterized by a range of new technologies that are fusing the physical, digital, and biological worlds, impacting all disciplines, economies and industries, and even challenging ideas about what it means to be human," (World Economic Forum, 2019).

Education systems across the world recognize the proliferation of changes in globalization. Concerns for how to educate a productive citizenry are increasingly focused on education reforms in science, technology, engineering and mathematics and specifically the modern transdisciplinary nature of the individual fields, commonly referred to as STEM. The term STEM was first coined in the United States by the National Science Foundation in the 1990's and represented a strategic decision by scientists, technologists, engineers, and mathematicians to build on the interconnected nature of the disciplines. Although there has been widespread attention to STEM integration within education, there is often uneven representation of each discipline in STEM education (English and Krishner, 2016; Hoachlander, 2014/2015; Honey, Pearson, & Schweingruber, 2014). Hoachlander (2014/2015) summarized the concern in the United State that "despite more than a decade of strong advocacy by practitioners, employers, and policymakers, STEM education in US schools leaves a great deal to be desired. In too many schools, science and math are still taught mostly in isolation from each other, and engineering is absent," (p. 74).

Recently, there has been a movement towards more technology, engineering, and arts integrated with science and mathematics education. This movement has been supported by the development of standards for education such as those established by the fields of science and mathematics. The initial standards in mathematics published by the National Council for Teachers of Mathematics in the late 1980's and the National Science Education Standards in the mid-1990's supported the notion of science technology engineering and mathematics (STEM) education and a focus on these multiple disciplines in education initiatives gained traction. Many individual states across the United States adopted new standards, as they re-think the goals of PreK-20 education. The Next Generation Science Standards (NGSS) and the Common Core State Standards for Mathematics (CCSSM) have increased emphasis on technology and engineering. The new emphasis on technology and engineering has changed education programs from both a theoretical and practical point of view. The CCSSM and NGSS documents also include standards on the practices that are essential for doing mathematics, science, and engineering.

Recognizing the critical need to attract students to STEM education and considering how to promote integration across the STEM disciplines the notion of adding arts to STEM emerged from a group at the Americas for the Arts National Policy Roundtable in 2007 (Perignat & Katz-Buonincontro, 2019). Around the same time, the Rhode Island School of Design began using the term STEAM to promote the role of the arts and design in STEM K-20 education. In 2015, the National Academies of Science, Engineering, and Medicine convened the National Academies Keck *Future Initiative* Conference on Art, Design and Science, Engineering and Medicine Frontier Collaborations: Ideation, Translation, and Realization. One seed group at that Futures Initiative conference focused their discussion and recommendations on Creating a Learning Educational System to Identify the Benefits of STEM to STEAM (National Academies of Science, Engineering, and Medicine [NASEM], 2016). The STEM to STEAM seed group noted the need to improve STEM educational systems at "every stage in society: formal primary, secondary, and higher education; life-long and evergreen learning; and education of the general public," (NASEM, 2016, p. 33) The group explored the idea that "art can be essential to conveying more effectively the critical messages of science, technology, AND medicine to our citizens; to enabling more informed choices about our future; and to building an innovative workforce" (p. 33). The group argued for better integration and "cross-talk" among disciplines to enhance solutions to "interdisciplinary global problems—issues like climate change and food shortages" (p. 34) and proposed a model for teachers to experience cross-discipline professional development through what the group called the Undisciplined Classroom. The group's notion of the Undisciplined Classroom for teacher professional development aligned with Vasquez, Sneider, and Comer's (2013) definition of transdisciplinary integration where knowledge and skills from multiple disciplines are applied to real-world problems and projects. The group believed the Undisciplined Classroom model would encourage educators to implement teaching and learning beyond individual disciplines in their classrooms.

The inclusion of the Arts within STEM education expands discipline specific conceptions of science, technology, engineering, and mathematics to include creativity, innovation, and design from the visual and performing arts as well as communication skills from the communication arts. Use of the term STEAM (an acronym for science, technology, engineering, arts, and mathematics) has become more prevalent in education and the shift from STEM to STEAM is changing what it means for students to learn within and across these disciplines. This new conceptualization of science, technology, engineering, art and mathematics, known as STEAM, and has raised many issues about how these different disciplines inter-relate. There is considerable discussion about the degree of integration versus separation that this conceptualization might imply (Moore, Stohlmann, Wang, Tank, & Roehrig, 2014; Roehrig, Moore, Wang, & Park, 2012). Besides being more inclusive of technology, engineering, and the arts, the field of science and mathematics education has moved in several new directions. There is increased consideration of broader entities such as nations, states, schools, and communities (Fleischman, Hopstock, Pelczar, & Shelley, 2010; Provasnik et al., 2012) as mediating variables for science and mathematics education outcomes.

Science and mathematics learning is also being recognized as a lifelong endeavor (Faulk, Dieking, & Storksdieck, 2005) and as something that can occur outside of formal settings (NRC, 2012a). There is more emphasis on science and mathematics learning in higher education and in disciplinary specific settings (NRC, 2012b), as well as in consideration of workforce development and for college and career readiness (CCSS, 2010; NGSS Lead States, 2013; NRC, 2012c). There is increased emphasis on individualized, self-motivated learning rather than more group-delivered education (NRC, 2011). Finally, science and mathematics education is becoming more global. Science and mathematics education professional associations, journals, and conferences are generally considered to speak to international audiences.

This edited book on science, technology, engineering, arts, and mathematics (STEAM) will help the field of education understand the movement from STEM towards a broader vision of STEAM. The intertwined nature of STEAM creates a challenging environment for educators. It is important to understand the increasingly complex and integrated nature of the field. It requires expertise in multiple areas, multiple methods, and the integration of multiple perspectives. The movement toward more technology, engineering, and the arts as part of science and mathematics raises many questions about what we mean by STEAM, and what are the best ways to create

STEAM programs. Integrating arts, technology, and engineering goes beyond the current traditional view of science and mathematics education. It is important for scholars and practicing educators to understand the underlying assumptions of such a priority and the implications for both research and practice. The chapters in the book explore this new vision of STEAM, present new approaches to STEAM, and advance new notions for science, technology, engineering, arts, and mathematics education. The book also includes state-of-the-art examples of STEAM from experts in the field that will help readers see how these new approaches to STEAM can be applied in practice.

The book is organized into four sections including "STEAM Education Learning Environments," "STEAM Education in Schools and School District Initiatives," "STEAM Education Beyond the Classroom," and "STEAM in Teacher Education."

SECTION 1: STEAM EDUCATION LEARNING ENVIRONMENTS

The first section of the book includes four chapters on designing STEAM environments with specific examples of how to design STEAM environments. The first chapter is on how to use Universal Design for Learning (UDL) to design flexible and useable STEAM learning spaces. The emphasis of the chapter is on thinking about and planning for learner variability as a key component towards designing inclusive educational experiences for all students, including students with disabilities.

The second chapter is on tools students and teachers can use to mediate STEAM learning environments. Two tools are used to mediate learning through consensus: (1) a 'STEAM slider' that students in groups used to reach consensus on the knowledge and processes employed in their project; and (2) a criteria sheet developed to mediate students' and teachers' engagement in self- and teacher-assessment. The chapter provides specific tools and advice on ways teachers can mediate engagement in STEAM environments.

The third chapter is on artistic biotechnology. This chapter focuses on the use of biotechnology as a basis for design thinking in STEAM. The authors define *artistic biotechnology* as a pedagogical approach to purposefully integrate art through the natural intersections of design thinking and biotechnology education. Learning biotechnology with purposeful art integration can provide a platform for educators to motivate students to collaborate with one another, utilize science, technology, engineering, and mathematics to enhance their art-making processes, and purposefully strive for critical and creative problem-solving solutions.

The fourth chapter is on the development of science animation videos to increase accessibility in STEAM. The purpose of this chapter is to describe an innovative and inclusive approach for integrating the Arts with STEM through the development and dissemination of science animation videos called SciToons. The animations are designed to be engaging and accessible to audiences with diverse academic backgrounds.

SECTION 2: STEAM EDUCATION IN SCHOOLS AND SCHOOL DISTRICT INITIATIVES

The second section of the book includes two chapters on school and district initiatives to enhance STEAM. The first chapter focuses on two case descriptions drawn from the perspectives of leaders in two school districts in which schools adopted a STEAM focus. The chapter describes the challenges and opportunities school districts faced with implementation of a STEAM initiative, and proposes five features to consider when implementing models to becoming a STEAM-focused school or school district. The five features drawn from analysis of the two cases are: buy-in from teachers and school administrators; adequate time for teacher learning and planning; community connections, real-world and problem-based or project-based learning; mutual decision-making and support between teachers and administrators; and budget planning and allocation.

The second chapter focuses on a school district integration of Maker Space Movement. The Maker Space Movement brought innovation and creativity into education. Maker Space has helped students and educators question beliefs about activities, lessons, and classes, pushing them to embrace a more student and experience driven learning environment. "Making Success" was a two-year research project to investigate and describe the integration of making into a school district. The research focused on describing the critical characteristics and capacities that allow maker space to integrate into secondary schools.

SECTION 3: STEAM EDUCATION BEYOND THE CLASSROOM

This section provides examples of STEAM education outside the classroom. Included is a chapter on the use of STEAM in a local arts center. The focus is on a laboratory for creativity that engages students on ways of integrating art with science, technology, engineering, and mathematics.

The second chapter is on the use of dance as a means to integrate the arts in science, technology, engineering and mathematics. Black girls "STEAMing" through dance provides a motivating means of engaging under represented students in STEAM.

SECTION 4: STEAM IN TEACHER EDUCATION

Finally, the last section of the book focuses on teacher education and ways of helping beginning and experienced teachers integrated the arts into science, technology, engineering, and mathematics. The first chapter presents a transdisciplinary view of STEAM and various ways to conceptualize the integration of STEAM into a STEAM instructional methods course.

The second chapter is centered on the use of modeling as a means of integrating STEAM. A key component of the science and engineering practices in the Next Generation Science Standards is modeling. The author uses modeling as a means of fully integrating STEAM.

The last chapter describes place-based methods of learning STEAM. This chapter investigates how teacher candidate experiences with community partners in STEAM can inform work in teacher education. As educators who teach within an integrated STEAM practicum semester, they are interested in linking STEAM courses, Indigenous knowledge systems, and place-based education. This chapter seeks to foster new understandings around a curriculum of place among Indigenous communities, not-for-profit organizations, public schools, and a Canadian post-secondary institution.

Overall, together these chapters about STEAM education provide a comprehensive glimpse into different ways and different places to integrate the Arts into Science, Technology, Engineering, and Mathematics. This edited book can inform the work of educators and education administrators from Kindergarten through post-secondary settings, as well as educational researchers and education services providers. It is important for educators to not only think about new definitions and new theoretical approaches to STEAM integration, but to also consider different practical ways the Arts can be integrated into STEM. Through comprehensive thinking about STEAM we can envision a future of STEAM education that includes interactive learning environments that are designed to insure all students are included in STEAM.

REFERENCES

Common Core State Standards Initiative. (2010). *Common Core State Standards for mathematics.* Retrieved from http://www.corestandards.org/assets/CCSSI_Math%20 Standards.pdf

English, L. D., & Kirshner, D. (2016). Changing agendas in international research in mathematics education. In L. D. English & D. Kirshner (Eds.), *Handbook of international research in mathematics education* (3rd ed.; pp. 3–18). New York: Taylor & Francis.

Falk, J. H., Dierking, L. D., & Storksdieck, M. (2005). *A review of research on lifelong science learning*. Washington, DC: Board on Science Education, The National Academies.

Fleischman, H. L., Hopstock, P. J., Pelczar, M. P., & Shelley, B. E. (2010). *Highlights From PISA 2009: Performance of U.S. 15-Year-Old Students in Reading, Mathematics, and Science Literacy in an International Context (NCES 2011- 004). Washington, DC: U.S. Department of Education.* Retrieved from http://nces.ed.gov/pubsearch/pubsinfo.asp?pubid=2011004

Friedman, T. (2005). *The world is flat: A brief history of the twenty-first century.* New York, NY: Farrar, Straus, and Giroux.

Hoachlander, G. (2014/2015). Integrating SET&M. *Educational Leadership*, 74–78.

Honey, M., Pearson, G., & Schweingruber, A. (2014). *STEM integration in K-12 education: status, prospects, and an agenda for research.* Washington, DC: National Academies Press.

Mazuzan, G. T. (1994). *The National Science Foundation: A brief history.* Office of Administrative and Public Affairs, the National Science Foundation. Retrieved from https://www.nsf.gov/about/history/nsf50/nsf8816.jsp

Moore, T. J., Stohlmann, M. S., Wang, H. H., Tank, K. M., & Roehrig, G. H. (2014). Implementation and integration of engineering in K-12 STEM education. In J. Strobel, S. Purzer, & M. Cardella (Eds.), *Engineering in PreCollege Settings: Research into Practice*. Rotterdam, The Netherlands: Sense Publishers. doi:10.2307/j.ctt6wq7bh.7

National Academies of Sciences, Engineering, and Medicine. (2016). *Art, Design and Science, Engineering and Medicine Frontier Collaborations: Ideation, Translation, and Realization: Seed Idea Group Summaries*. Washington, DC: The National Academies Press. doi:10.17226/23528

NGSS Lead States. (2013). *Next generation science standards: For states, by states.* Washington, DC: The National Academies Press.

National Research Council (NRC). (2011). *Learning science through computer games and simulations*. Washington, DC: The National Academies Press. Retrieved from http://www.nap.edu/catalog.php?record_id=13078

National Research Council (NRC). (2012a). *A framework for K-12 science education: Practices, crosscutting concepts, and core ideas*. Washington, DC: The National Academies Press. Retrieved from http://www.nap.edu/catalog.php?record_id=13165

National Research Council (NRC). (2012b). *Discipline-based education research: Understanding and improving learning in undergraduate science and engineering*. Washington, DC: The National Academies Press. Retrieved from http://www.nap.edu/catalog.php?record_id=13362

National Research Council (NRC). (2012c). *Education for life and work: Developing transferable knowledge and skills in the 21st Century*. Washington, DC: The National Academies Press. Retrieved from http://www.nap.edu/catalog.php?record_id=13398

Perignat, E., & Katz-Buonincontro, J. (2019). STEAM in practice and research: An integrative literature review. *Thinking Skills and Creativity*, *31*, 31–43. doi:10.1016/j.tsc.2018.10.002

Provasnik, S., Kastberg, D., Ferraro, D., Lemanski, N., Roey, S., & Jenkins, F. (2012). *Highlights From TIMSS 2011: Mathematics and Science Achievement of U.S. Fourth- and Eighth-Grade Students in an International Context (NCES 2013- 009)*. Washington, DC: National Center for Education Statistics, Institute of Education Sciences, U.S. Department of Education. Retrieved from http://nces.ed.gov/pubsearch/pubsinfo.asp?pubid=2013009

Roehrig, G. H., Moore, T. J., Wang, H. H., & Park, M. S. (2012). Is adding the E enough?: Investigating the impact of K-12 engineering standards on the implementation of STEM integration. *School Science and Mathematics*, *112*(1), 31–44. doi:10.1111/j.1949-8594.2011.00112.x

Vasquez, J., Sneider, C., & Comer, M. (2013). STEM lesson essentials, grades 3–8: integrating science, technology, engineering, and mathematics. Portsmouth, NH: Heinemann.

World Economic Forum. (2019). *The fourth industrial revolution, by Klaus Schawb*. Retrieved at https://www.weforum.org/about/the-fourth-industrial-revolution-by-klaus-schwab

Section 1

STEAM Education Learning Environments

Chapter 1
Designing STEAM Learning Environments

Haidee A. Jackson
University of Kansas, USA

James D. Basham
University of Kansas, USA

Kelli Thomas
University of Kansas, USA

Cassandra L. Hunt
University of Kansas, USA

ABSTRACT

This chapter highlights some of the technological changes in society that have led to an increased need to consider instructional and design challenges in implementing STEAM education. Specifically, the chapter discusses how challenges related to designing learning environments in STEAM education can be mediated through application of the Universal Design for Learning (UDL) framework. Consideration is given towards designing for flexible and useable STEAM learning spaces by thinking about and planning for learner variability as a key component towards designing inclusive, humanistic educational experiences. In addition, STEAM learning spaces are discussed in terms of catalyzing learner creativity, providing for individualizing instruction, and empowering 21st century learners to develop collaborative, problem-solving tools, and soft skills.

DOI: 10.4018/978-1-7998-2517-3.ch001

INTRODUCTION

The increased focus on the crosscutting approaches of curriculum design and education systems emerged from the needs in producing a global citizenry that is prepared for modernity (Basham, Israel, & Maynard, 2010). There is little doubt that the world has changed and is continuing to change. As a global society, we are in the midst of disruption across all aspects of life, from the way we socialize, consume news and information, to buying goods, and work. The World Economic Forum (Schwab, 2016) has identified these changes as the beginning of the fourth industrial revolution.

The revolutions of the past have shaped the schools of today and the emergence of the fourth industrial revolution will reshape the education system of the future. As identified in Basham, Han, Zhang, and Yang (in press) the first industrial revolution (circa 1760s) brought machine based manufacturing and railroads to existence. These changes started to shift populations of people living and surviving in rural villages and towns to larger population centers driven by the new capitalist economy. During this time, the value of education moved informal apprenticeship models to support the rural area to more formalized education models for educating more people. Schools and universities started to develop specialized areas of preparation and training to prepare the workforce for survival in the new economy.

The second industrial revolution (circa 1870s) was about developing technological solutions for society's changing needs. With a more educated populace, came the development and adoption of technological solutions and early standards-based systems in manufacturing, production, health care, sanitation, and communications at scale. During this time, society witnessed the development of steel, expansion of railroads, internal combustible engine, electrification, assembly line factory models, and scientific management of manufacturing and business systems. The education system modeled the newly found need for standardization through forming standards and a level based education system. This increased focus on specializations, beginning the science of education, and a focus on the efficiency of student learning in the system.

The third industrial revolution (circa 1950s) emerged with a focus on globalization of systems through standardized processes and increased capabilities in communications. Today, this is represented in the adoption of digital systems across society in areas such as manufacturing, communications, commerce, and daily life. The education system reacted through development and acceptance of national and international standards, support for wide-scale achievement measures, the classification of learners with differences and disabilities, and increased scientific focus on education and learning. This all placed a greater focus on specializations across education.

Today, education is beginning another revolutionary shift in modernity. The fourth industrial revolution is being shaped by the global decentralization and increased acceptance of *interdisciplinary innovation* and the *rapid adoption of innovation* across society. Together, these core components are supporting a rapidly increasing and self-reinforcing process in iterative development, which supports continued change and ongoing disruption throughout society. The education system is in a chaotic state attempting to find a balance between the needs of the past while preparing learners for the future.

Recognizing a need for change, there has been an increased focus on the crosscutting approaches integrated across science, technology, engineering, and mathematics (STEM). While the terminology may be up for debate, the basic premise and rationale of STEM education has always integrated the transdisciplinary understandings of four subject areas to encompass a more holistic view of societal development (Basham et al, 2010). Thus, whether called STEM or science, technology, engineering, arts, and mathematics (STEAM), the focus should be to prepare a citizenry for the future that includes the associated knowledge and skills integrated into the arts and humanities.

As society moves into this next revolutionary cycle, there is a need to consider how the education system will move past the rigid structures associated with the previous revolutions, and move toward a dynamic system that rapidly evolves with the emergent pluralistic existence of the human experience. To provide for some level of stability in the midst of continual innovation decisions for designing such a system should be principled on a humanist understanding for how to support human variability and inclusion while developing lifelong learners (Basham et al, 2010; Schwab, 2016; WEF, 2016). Within STEAM education, this means there is a need to consider how we design both our learning experiences as well as our environments.

LEARNER VARIABILITY

Variability within the student population has traditionally been conceived of as being undesirable within the established education system (Basham, Hall, Carter, & Stahl, 2016, Zhao, 2016). This is evident in the traditional education system with the focus on "fixing" student variability to align with a conceptualized norm (Basham et al, 2016, Zhao, Wehmeyer, Basham, & Hansen, 2019). Within this traditional deficiency based model, students that fall outside of the norm are tracked and remediated to fit within the norm (Zhao, 2016). The educational system has been designed to provide the corrective action deemed necessary for students to become successful members of society. Yet, new societal and technological changes have helped to overhaul the previous educational system in light of a more modern conception of adopting and supporting an educational design in support of individual learner variability (Goldin & Katz, 2008; Zhao et al., 2019).

Consequently, an educational system that is designed to support individual learner variability not only ceases to continue efforts towards normalization of its greater student population, but strives towards the incorporation of individual inclusive educational practices. To this end, the societal and educational paradigm has shifted towards maximizing individual potential and fostering student growth through personalization of the educational process (Basham et al., 2016). The skill sets that have become essential to the modern age are markedly different from what was once valued within the educational system. Variability, which was once considered a liability to the educational system, and one that was to be normalized, has now become an essential factor in designing universally inclusive educational practices (Zhao, 2018).

DESIGNING FOR USABILITY AND FLEXIBILITY

The modern learning environment is designed in consideration of usability in function and form of tools (e.g., instructional technology), accessibility (part of usability), and flexibility of learning tools (Basham et al., 2010; Rubin & Chisnell, 2008). Learning tools should provide for flexibility of use among a diverse group of learners within a variety of learning environments. Today's learning environments are equally variable to that of diverse student populations in schools. For instance, students may now participate in online or blended learning formats alongside traditional classroom style learning spaces. These types of learning spaces have evolved beyond the conceptualization of learning occurring within a particular dimension of space or even at a particular point in time.

Learning spaces that may have been previously designed to provide access for students to develop a particular subset of skills may no longer be apt for the task. Basham, Israel, and Maynard (2010) describe an example of this in regards to the use of computer labs in schools. Computer labs were designed to support teacher-centered or computer-centered instruction in a technology rich-environment (Basham et al, 2010). Yet, by distributing the computers (e.g. laptops, tablets), such as in the use of digital backpacks, the model of learning moves away from teacher- or computer-centered to more learning goal and even learner-centered (Basham, Meyer, & Perry, 2010).

As we consider the need to STEAM learning experiences, there is a need to recognize how learning spaces themselves are no longer fixed and rigid mechanisms in a set place or time. The flexibility of instructional spaces can be re-fashioned in terms of creating learning environments that are conducive to facilitating inclusive personalized learning outcomes for all students. Consequently, while multiple chapters in this book consider the curricular content and learning activities associated with the STEAM experience, this chapter develops a foundational understanding for how the environment itself plays a role in the learning interactions and outcomes of these experiences.

TRADITIONAL CLASSROOMS

Conventional school learning spaces, with desks traditionally placed in rows, is viewed as representing a teacher-centered pedagogy that standardizes the learning process (Dovey & Fisher, 2014; Dumont et al., 2010). This design focuses all attention to a single source of information, the teacher. Many consider this top-down disseminating model of learning as ineffective for engaging the modern day 21st century learner (Benade, 2017; Dumont and Istance, 2010; Hattie & Donoghue, 2016; Theisens, Benavides, & Dumont, 2008). The traditional process of disseminating knowledge is itself, non-participatory by design, and consequently, has been thought to elicit only surface learning (Hattie & Donoghue, 2016; Theisens, Benavides, & Dumont, 2008). The non-collaborative and non-responsive nature of these educational practices are noted as being ineffective for the development of life-long and expert learners (Benade, 2017; Dumont & Istance, 2010; Theisens et al., 2008).

Consequently, learning spaces in schools have become reimagined in terms of producing a contemporary or creative model of an innovative learning environment (ILE) (Dovey & Fisher, 2014; Dumont, Istance, and Benavides, 2010; Mulcahy, Cleveland, & Aberton, 2015). Mulcahy Cleveland, and Aberton (2015) identify that the re-imagining of learning spaces as ILEs may help foster educators in developing lifelong learners. Specifically, the conventional or traditional school model and associated learning spaces restrict the ability to adopt student-centered and technologically savvy pedagogy (Alterator & Deed, 2013; Benade, 2016; Dovey & Fisher, 2014). Conversely, ILE's are characterized as being goal driven with the use of technology within a modernized, flexible learning environment (Mulachy et al., 2015; OECD, 2013; OECD, 2015).

The ILE or modern learning environment is premised on the idea that educational practices of teachers are shaped by the environmental space. ILEs are assumed to support greater learning opportunities for students than what a conventional learning space can provide to learners (Dovey & Fisher, 2014; Lippman, 2010; Thomas, 2010). Hence, educational policies in Australia and New Zealand support the creation of ILEs in schools (MCEETYA, 2008; New Zealand Ministry of Education, 2014; OECD, 2013). The policy roll-outs are, in large part, a result of how technological innovations are shaping and altering the landscape of the economy, and consequently changing the perception of the utility of conventional pedagogies (Alterator & Deed, 2013; Benade, 2016; Dovey & Fisher, 2014; Dumont et al., 2010).

TRADITIONAL CONTAINERS TO STEAM LEARNING SPACES

Conventional school structures are viewed as industrial schemes, encompassing rigid and ineffective environmental learning spaces for the 21st century learner (Alterator & Deed, 2013; Benade, 2016; Dovey & Fisher, 2014; Dumont et al., 2010). Within this process, the traditional classroom is critiqued on the grounds of being an architectural container of sorts, wherein the enactment of student-centered learning activities are hampered by the rigidity of the space (Fisher, 2006; Mulcahy, 2015; Tanner, 2008; Upitis, 2004).

A pragmatic pedagogy that engages student variability within a co-designed learning space places goal-driven learning interactions at the forefront of educational design and changes the power structure in the classroom. For instance, according to Basham et al. (2019), the capacity for a learning space to enable the development of relationship and community building becomes an important key design feature in the modern learning environment. This design consideration places collaborative STEAM based problem-solving as a central goal within these spaces. Additionally, Christensen, Horn, and Johnson (2008) note that success of an educational system depends on its measure of changeability to meet the demands of society. Thus, the accessibility of STEAM learning experiences that authentically engages all students in active problem-solving, collaboration, while using both STEAM content and associated soft skills, calls for educators to reconsider the measure of what constitutes a productive learning space.

One example of a school that has re-envisioned what constitutes productive learning spaces, is Prairie Elementary, a STEAM-focused school in the Midwest United States. Prairie Elementary is five years into the process of redesigning learning spaces within the context of STEAM education. The school community actively participates in co-designing both the learning spaces and learning experiences enacted in the school. Building administrators, teachers, students, students' families, and community members are all engaged in contributing to the learning functions of the school. Several changes in the learning environment of the school including aspects of time, space, and resources demonstrate a commitment to establishing an ILE that supports all students in becoming life-long learners who work collaboratively to solve real problems.

The notion of time for learning was changed by removing the rigid single-subject time schedules typical in many elementary classrooms. Instead of the more traditional use of time, where one might find fifty-five minutes is dedicated to mathematics followed by ninety minutes for language arts, at Prairie Elementary, time during the school day is used flexibly depending on the current STEAM project or problem for which students are seeking solutions and outcomes. The projects are used to integrate content and address curriculum standards from across STEAM subjects, where the A is interpreted broadly, to include visual and performing arts as well as communication and language arts. The school still designates time for lunch and time for recess, but now, all the other time in between is considered learning time.

While the school structure still has walls and separate classrooms, those boundaries around learning spaces have faded and space around the school property is used in a more fluid way to accommodate cross-class, cross-grade level project based learning. The school's library has been redesigned to be a multi-functional media center with furniture that is easily moved or modified, a Maker Space area with storage for supplies and resources that is easily accessible, an open floor plan in the center that is ready for project specific set up, and wall space with both electronic screens and whiteboards as well as convertible bulletin boards. Additionally, multiple areas on the school grounds outside the building are used as outdoor classroom spaces.

The acquisition, allocation, and use of learning resources have been reimagined at Prairie Elementary as well. The school transitioned from a computer lab model to a full one device to one student model. Students in the primary elementary grades are provided with personal tablets and students in the intermediate elementary grades are issued personal laptops. The school's budget for classroom supplies was restructured using a school-wide supply chain model rather than an individual teacher or grade level model. Before, each teacher was given a small budget for purchasing supplies at the beginning of the school year. The traditional approach promoted the individual teacher acquisition of supplies and resources that often went unused. The new model allows the school to better distribute, share, and optimize resources allocated for consumable and non-consumable supplies and promotes a more collaborative approach to planning.

HOW UNIVERSAL DESIGN FOR LEARNING SUPPORTS SPACE DESIGN

Emergence of Universal Design

In the 1970s, Ron Mace developed principles of architecture under what is now called Universal Design (UD) that supported more flexible and inclusive design for all users of the space (Center for Universal Design, 1997; Story, Mueller, & Mace, 1998). UD aimed to support the needs of persons with or without disabilities when inhabiting or navigating a building or space. Over time, these principles were increasingly utilized to create environments and eventually in designing products (e.g., Apple's iPhone) that were equally suitable for the goal-driven use of everyone. This resulted in little need to create separate designs or adaptations for environmental interactive competency or for trying to retrofit the environment to meet the needs of people with disabilities. The principles of UD emerged from designers considering how to create and incorporate design specifications that were inclusive enough to comprise the needs and abilities of all people (Story et al., 1998). Hence, designing for diversity in environmental spaces became a centralized practice through utilizing UD principles in spatial design and eventually product design. To this end, seven UD principles were developed to meet the goal of inclusive design: *Equitable Use, Flexibility in Use, Simple and intuitive use, Perceptible information, Tolerance for error, Low physical effort,* and *Size and space for approach and use* (Story, 2001). The incorporation of these seven UD principles into both space and product design enabled designers to create spaces and products in accordance with the needs of its diverse end users (Story et al., 1998).

UD space design supports the inclusivity for the diverse array of needs of people to enter a space. For instance, many schools and even stores now have widened doors that are automated, as well as ramps, to support individuals carrying packages, shopping carts, or even those in wheelchairs. Other design supports included the use of door handles instead of door knobs, so that a building can be accessible to more people. Eventually UD principles moved beyond architecture to broader design areas to include such things as closed captioning, enlarged text, manipulating font, color, and sound to make information accessible to diverse populations within instructional and digital environments (CAST, 2009). During the same time, students with disabilities began to become more included in the general education classroom. Innovative developments across disciplines in architectural and technological design increased the capacity for students to participate in these inclusive environments. Then, the United States of America, passed the Americans With Disabilities Act of 1990, Pub. L. No. 101-336, 104 Stat. 328 (1990) to support basic accessibility as part of the functional design of learning spaces. Additionally, the ADA opened doors for accessibility of educational content (e.g. textbooks, websites). The pioneering work that took place in UD supported a broader understanding of the foundational barriers within environmental design.

From UD to Universal Design for Learning

Founded in the late 1990s at CAST (see www.cast.org), Universal Design for Learning (UDL) was built upon the same premise as UD with a targeted focus on learning. Today, UDL is known as a research-based design framework for supporting the design of goal driven, inclusive learning environments and experiences for all learners. From a research stance, UDL integrates research findings on learning across fields associated with mind, brain, and education (Basham et al., 2016). In practice, UDL uses both proactive and iterative design (e.g., design thinking) to support all learners to meet the goal driven design of a learning experience. This is done through providing multiple means for students to engage within the experience, providing multiple means of representation, and multiple ways for students to demonstrate their understanding (Basham & Marino, 2013; CAST, 2018). Different from other strategies or practices in education, UDL is a design framework and not a single strategy but a framework for considering learner variability and a host of practices associated with proactive and iterative design.

Different from UD, the UDL framework is more narrowly focused on supporting learners within the process of learning. However, as one can quickly surmise, the design of learning moves well beyond providing a framework for designing K-12 lesson plans. UDL is focused on designing both learning environments (or places where learning takes place) and more broadly the learning experiences (the interactions that take place in the process of learning). Thus, UDL design considers the goals of the learning experience, the known and unknown learner variability, the measures of success and assessment, and finally the instructional strategies and tools used within the experience (Basham & Marino, 2013). Currently, UDL has been adopted across the globe in the design of a variety of learning environments and experiences across the lifespan (preK-12, higher education, workforce development; Basham & Blackorby, in press).

Within UDL, the UD principles of spatial design such as *Equitable use, Flexibility in use, Low physical effort, and Size and space for approach and use* (Connell et al., 1997) were translated into encompassing principles of inclusive cognitive instructional design. For example, UD focuses on the principle of *Equitable use* to ensure the space or product is designed to ensure that it is useful for diverse ability levels (Story et al., 1998). *Equitable use* within UDL can be translated by ensuring that all learners have the ability to access all aspects of the learning experience, they are supported in their ability to build the necessary knowledge and skills to demonstrate understanding, and that they are provided with the supports necessary to develop internalized abilities to regulate understanding, support continued comprehension, and executive functioning (CAST, 2018). *Flexibility in use* has been outlined in the UD guidelines to ensure that a design provides options and choices for an individual user's preferences in adapting a design to meet the needs of the user (Story et al., 1998). Within UDL, flexibility in use is supported through proactively designing learning experiences that anticipate both known and unknown learner variability through ensuring there are multiple ways for students to engage in the learning, supporting multiple ways for students to access and comprehend the necessary knowledge and skills, and providing various ways for students demonstrating understanding (Basham & Marino, 2013).

Within practice, STEAM teachers can adopt pedagogical practices that support designing instruction with student variability in mind by providing flexible options for student engagement and learning. Teachers can also use technological applications to ensure active student engagement and participation in learning through flexible options that facilitate student choice. Flexible instructional methods in pedagogy and in designing learning environments provide a greater scope for ensuring that teachers plan for and teach with student variability in mind. Consequently, UDL supports instructional design thinking. UDL aims to meet the instructional needs of diverse learners by designing instruction to address the variable cognitive needs of students as a whole within the learning environment (learners with or without disabilities). Educators as instructional designers can employ UDL principles in structuring lessons and delivering content in the classroom to accommodate the needs of diverse student populations. Moreover, instructional content can be designed in a way to ensure that all students, with or without disabilities, are provided with flexible methods, tools, and choices by which they can engage in the learning environment.

The UDL principle of providing for multiple means of representation in instructional content (illustrations, videos, podcasts) helps facilitate the UD architectural design principle of *Equitable use* in learning. Instead of content having to be modified or retrofitted to meet the needs of a new individual, content can be designed to be displayed and represented in multiple formats to students (Cook & Rao, 2018). For instance, with the use of new technological tools, instructional digital content can be modified by adjusting font type or magnifying text to meet the individual cognitive needs of students. Additionally, rather than only using written texts or teacher lecture to provide information about key concepts related to a STEAM problem, teachers can incorporate short videos with visualizations and closed captioning designed to present complex ideas in short segments. Videos created and available through programs like SciToons at Brown University, Crash Course, or Kurzgesagt on YouTube are resources that teachers can integrate and use to incorporate student choice into instruction. As another example, the SciToons, *Data Visualization and Representation* demonstrates how an instructional video can be used within a STEAM problem or project to support basic learning. The video highlights a key big idea in both sharing and consuming data and cautions viewers about the potential misuse of data representation. To support multiple means and flexible use, viewers have the option to select closed captioning of the video script, which includes visual graphics to illustrate the content, and the playback speed can be adjusted.

The UDL principles can thereby maximize learning for all students through supporting the instructional experience and equity across all learners. The emergence of UD and UDL principles in architectural design, partnered with the use of technological tools, and innovative practices to support the needs of diverse learners in education, have allowed us to re-imagine STEAM learning spaces. One way to accomplish this is for educators to use the UDL framework when developing STEAM problem-based units by considering the goal of the unit, the barriers that exist in the curriculum, and how to remove barriers by adding flexibility to the curriculum. The goals of a STEAM problem-based unit can be effectively established through essential questions that guide the learning experiences. Knowledge and skills from the content disciplines can be embedded in the STEAM unit in a purposeful way, driven by the essential questions rather than by the content. Using the essential questions as an anchor for instructional design of STEAM units can support flexibility in instructional design. For example, at Prairie Elementary, a sixth grade STEAM unit was designed to address the following essential questions: *How do environmental factors affect the landscape around the school including outdoor classroom spaces, playground facilities, and the creek that flows beside the school property? How can we mitigate negative impacts of environmental factors on the outdoor school landscape?* The goals embedded within the essential questions allowed students to work in teams on innovative solutions for improving the school property in ways that were personalized for each student on the team. The unit was designed to address student variability because students chose what aspects of the school property to investigate and because teachers provided multiple means of engagement for students, multiple means of representing information, and multiple means of action and expression for students throughout the unit. The students were motivated to engage in learning with a real purpose to improve their school property. For more information on the UDL framework, including implementation, readers are encouraged to visit www.LearningDesigned.org.

Designing for STEAM Learning

The predominant student skills that are held in high esteem in the literature with regards to modern learning comprise learning objectives related to developing student problem solving, collaboration, and soft skills (Basham et al., 2010). In particular, problem solving is portrayed as being at the confluence of critical thinking processes aimed at investigating and interpreting information (Morrison, 2015). Meaningful interpretations and solutions of problems are viewed as arising out of student inquiry practices that revolve around deeper reflective practices to fully engage students' critical thinking skills. Consequently, the STEAM curriculum is designed to enrich students' problem-solving skills through engaging students in higher-order thinking practices aimed at conceptualizing the underlying principles of problem-solving

(Dixon, 2012; Pintrich & De Groot, 1990). To this end, students are equipped with collaborative team problem-solving skills.

The collaborative orientation of both STEM and STEAM problem-solving is described as encouraging the development of teamwork through the completion of group challenges. The completion of a problem objective that arises out of collaborative efforts is predominantly stressed above individual effort alone (Zhao et al., 2019). The STEM and STEAM transdisciplinary curriculum are oriented around producing problem situations that ensure collaborative skills are instilled in students, as collaborative skills are highly prized within the modern world (Zhao, 2018). Students are instructed in the development of collaborative problem-solving skills which focus on how individual student roles contribute towards task completion in a whole group effort. Consequently, students learn to collaborate with a diverse array of students with various interests, abilities, and backgrounds. The diverse problem-solving collaborative efforts of individual students striving towards meeting a learning objective contributes towards fostering a creative atmosphere for student learning and engagement exercises (Householder, 2012).

The development of soft skills are highly valued among institutions of higher learning as well as within future ready careers (Bancino, 2007; Harris, 2008; Oleson, 2014). Soft skills such as a positive attitude, solid work ethic, openness, friendliness, and an ability to communicate well, among others have been outlined in the literature as important skills (Harris, 2008). Overall, communication skills are especially stressed throughout the literature as being important for the development of students' problem-solving and collaborative skills. The ability for students to work with others, manage conflict, and communicate well are all viewed as intrinsic soft skills necessary for collaborative problem-solving in the modern world.

These skills are born from social interactions. As students are empowered to determine that nature of their own work space in a flexible learning environment, peer interaction and teamwork is pushed towards maximizing soft skill development. The opportunity to develop soft skills in a shared learning space culminates in providing a learning space that ensures optimal levels of STEAM instruction take place. Students that are given the flexibility and freedom of choice to move around in a classroom environment and determine the components of their own learning space are better able to self-regulate and self-manage their own workload. The measurable result is that learners are able to realize greater expression in creativity and problem-solving. The development of problem-solving skills through positive social interactions in the classroom could facilitate greater student understanding and learning of content. Design thinking, as proposed by the UDL framework, could be an essential tool for teachers to provide the necessary background and foundation for students to actively participate in the space with content. The potential for learners to experience a richer and more engaging experience through social interactions in their environment could impact students' ability to enter into the problem-solving skills necessary to master content.

Designing Space for STEAM

The idea that STEM fields require more rigid problem-solving skills that rely on more mathematical, deductive types of logic is losing ground in the new paradigmatic awakening of STEAM based instruction (Catterall, 2002; Henriksen, 2014). The creative aspect of the arts, it is argued, is always present in innovative problem-solvers (Caper, 1996; Dail, 2013; Eger, 2013; Root-Bernstein, 2003). The shift in instructional strategies to facilitate the initiative towards integrating the Arts into STEM calls for a radical shift in the types of traditionally practiced pedagogies in the STEM field. Educational initiatives that aim towards dissolving separations between science, technology, engineering, arts, and mathematics involve creating a conceptually different educational structure-one that focuses on integrating STEM and the arts into a STEAM model (Yakman, 2008).

Within the engineering field, conversations pertaining to changing instructional strategies revolve around guiding training efforts towards the development of greater student understanding in problem-solving, which is defined as being illustrated by students' display of skill, competence, and the realization of contextualized knowledge on a deeper level (Connor, Karmokar, & Whittington, 2015). Instructional practices such as lecturing are not found to be conducive for developing student problem-solving skills, and moreover, have been found to be ineffective within college settings for preparing engineers. Such practices have been observed to be bereft of the opportunity for students to engage in creative and/or critical thinking, which are viewed as being essential for a student's capacity to think on a deeper level, and promote higher order reasoning skills (Connor et al., 2015; Johnson, 1999). Consequently, engineer preparation efforts call for alternative educational strategies that focus on student-centered learning (Connor et al., 2015).

Arts-based teaching methods like creating visual works of art that display science concepts are an example of STEAM type instructional design methods. A key aspect of STEAM design involves employing students in artistic and creative visual designs of traditional science-minded concepts to facilitate greater student understanding and engagement. Students are encouraged to write songs or develop poems that express science ideas. The body is kinesthetically engaged to imitate science ideas in the same way that students would typically engage their bodies in theatre. The underlying idea in STEAM educational practices is that cross-disciplinary type engagement between science and art educational practices produces more creative and innovative thinkers (Henriksen, 2014).

The learning space not only becomes re-imagined in terms of providing the means to create problem-solving situations to encourage collaborative efforts and the ability for students to develop soft-skills, but also re-imagined in terms of inspiring student creative impetus within the environment. The emergence of the A in Arts in STEM towards a new STEAM based instructional design necessitates a new consideration of the role that space design plays in facilitating the addition of more artistic and overall humanities focused instructional practices in the environment.

Space Design and Pedagogy

The interdisciplinary emergence of architecture and education calls for an integration of architectural design knowledge and consideration of student learning and developmental needs in a learning space (Taylor, 2009). The result that emerges is a new educational design paradigm-one that puts educators at the center of reconfiguring pedagogy and educational spaces to meet the needs of student-centered educational practices that support a STEAM design model. Educators must design to meet diverse student learning needs, and so direct their efforts towards designing cognitively accessible instruction, but must also learn to think about the environmental context of how instruction is occurring within the learning space. Indeed, the learning space emerges as a unification between the physical and the non-physical environment. This calls for rethinking and establishing new pedagogies in terms of instructional design and instructional space.

The establishment of a new pedagogy built on thinking about how to design instruction in a newly designed and modern student-centered learning space akin to an ILE calls for a philosophy of education that considers a learning space as a holistic construct (Taylor, 2009). Educators must learn how to design learning experiences in these flexible spaces, meeting the goals of STEAM curriculum, in an inclusive environment. In short, it is not enough to architecturally design an ILE and expect that student-centered pedagogy will automatically arise in the educator without proper professional development of the educator. Within the literature, the ILE or learning space has not been found to be sufficient in and of itself for establishing automatic student-centered pedagogical practices among educators placed in such environments.

The idea that ILEs will stimulate teachers to change their practices is not supported within the literature. A few researchers have asserted that any changes in teaching practices from teachers transitioning from conventional to ILE learning spaces is due to the teachers' own knowledge base, flexibility, understanding of content, and technical proficiency (Alterator & Deed, 2013, 2016; Deed & Lesko, 2015). Lackney (2008) regards teachers' knowledge base and technical skill to be pre-existing as a form of "environmental competency." Lackney (2008) pointed out that teachers who are ill-equipped in environmental competency may run into quite a few challenges when attempting to teach within an ILE, without proper preparation.

For instance, the Linking Pedagogy, Technology, and Space (LPTS) measurement tool was used in quasi-experimental studies that explored how different types of learning spaces may affect teachers and students in various classroom environments (Byers, Hartnell-Young, & Imms, 2018; Byers & Imms, 2014, 2016; Byers, Imms, & Hartnell-Young, 2014; Imms & Byers, 2016). The LPTS is a behavioral analysis tool implemented by researchers observing student behavior and engagement in the ILE classroom.

Within these studies, the ILEs were identified as being linked to fostering positive student attitudes towards learning, which in turn lead to an increase in positive academic outcomes. However, these studies suggested that the impact of learning spaces on pedagogical practices was not a cause and effect relationship (Byers, 2016). The research illustrated that teacher practices do impact student learning. Professional development of teachers or pedagogical practices of teachers in the STEAM classroom becomes important for student engagement.

The results suggested that the technical skills of the teacher played a role in determining the effect of learning spaces on students (Byers, 2016). Specifically, it was found that teacher's technical and technological skills, as well as the teacher's willingness to utilize the ILE in alignment towards incorporating digital technology factored into students' attitudes and learning outcomes (Byers et al., 2014; Imms & Byers, 2016). These studies revealed that the technical skill and willingness of the teachers to utilize the learning spaces had a significant effect on how pedagogy was implemented in the classroom. The resulting pedagogical practices that impacted student learning in a statistically significant way (student engagement, learning outcomes) was a result of pedagogical practices by the teachers within the ILEs and not simply a matter of inhabiting an ILE (Byers, 2016; Byers et al., 2014, 2016; Imms & Byers, 2016). The ILE demands teacher flexibility and technical proficiency in facilitating learning. Teachers that may lack environmental competency or training into how to navigate and facilitate learning in an ILE may revert to conventional educational practices (Byers et al., 2018b).

Consequently, considering how to re-fashion teacher professional development in terms of considering how to utilize student-centered pedagogies in a modernized learning space becomes of utmost importance in education. Designed learning spaces are holistically considered in terms of student movement, mobility of furniture, and flexibility (Taylor, 2009). Consequently, the instructional methods that facilitate project-based learning, and teacher as coach and facilitator within a UDL framework, necessitates the need for increased teacher professional development in design thinking (Taylor, 2009). As educators begin to consider learning spaces anew, and in terms of design, new interdisciplinary practices can bring about a change in pedagogy that supports thinking about designing learning environments and experiences in a more holistic and inclusive manner.

From UDL and STEM to UDL and STEAM Ahead

UDL utilizes multiple means of representation, expression and action, and engagement within its framework to design curriculum to meet the needs of diverse learners (CAST, 2011b). In the STEM instructional setting, the use of technology can be incorporated into the curriculum to provide multiple formats by which information can be conveyed to students through the use of virtual graphical displays replete with videos and sound (Curry, Cohen, & Lightbody, 2006). STEAM instructional practices are equally conducive to incorporating multiple means by which content can be accessed and displayed by a learner. For example, in revisiting some of the STEAM educational practices that were aforementioned, learners were encouraged to engage through artistic expression of content. This could mean that a learner engages with scientific concepts through creating possible artistic visual designs that display said concepts. The STEAM educational practices encourage multiple means of action and expression of content, and are therefore, in perfect alignment with the UDL framework.

The UDL framework defines the curriculum in terms of a learner's encountered experience in a learning space. This includes the curriculum goals and standards by which a learner is assessed, and the instructional content and its materials or tools by which instruction is carried out in a learning space. UDL as a framework utilizes pedagogy and instructional tools or technological innovations to create a learning environment that engages diverse learners (Basham & Marino, 2013). STEAM educational practices encourage a wide array of social interactive experiences that engage diverse learners through individualized and artistic self-expression. Consequently, STEAM practices foster the ability for students to engage in diverse educational experiences as a matter of choice in the act of creative self-expression in learning. UDL is a design-based framework that provides an established research-based foundation for the design of both STEAM learning environments and experiences that meets the needs of all learners.

CONCLUSION

Societal changes will continue to make great impacts upon the educational needs of learners. These changes require rethinking how we are designing both learning environments and experiences. Educational practices which aim towards supporting a humanistic paradigm expressive of inclusivity and consideration of learner variability are very much needed (Basham et al, 2010; Schwab, 2016; WEF, 2016). In light of considering STEAM education, thinking about how to design instructional practices and learning spaces becomes a pivotal consideration to meeting the needs of the 21st century learner.

The principles of UD space design coupled with an instructional design framework like UDL can provide opportunities for STEAM educators to begin envisioning new inclusive practices in education. STEAM based instructional practices that provide opportunities for student choice may help to facilitate a learning environment that inspires students towards creative thinking and self-directed learning. Pedagogy that fosters STEAM centered practices within a UDL design framework can play a significant role in determining how learning spaces are re-imagined in the 21st century. A STEAM curriculum which utilizes principles of UDL in support of providing multiple means of engagement, representation, and action and expression can be augmented by considering the design of a learning space.

In STEAM based education, collaborative problem-solving becomes an essential element within instructional design. Educators can consider the capacity of a learning space for fostering positive social interactions that are conducive for community building as well as for learners in developing soft skills (Basham et al., 2019). Learning spaces that are designed to be flexible for student movement can create a holistic infrastructure in support of STEAM goals. Flexible learning spaces may also provide greater opportunities for learners to be provided with personalized learning experiences. The design of educational practices and learning spaces in STEAM provides a key opportunity for educators as designers to consider the role that space plays in supporting goal-driven learning activities. In conclusion, learning environments need no longer be considered as fixed or rigid spaces, but instead as flexible, malleable spaces in support of inclusivity and learner variability.

REFERENCES

Alterator, S., & Deed, C. (2013). Teacher adaptation to open learning spaces. *Issues in Educational Research*, *23*(3), 315.

Alterator, S., & Deed, C. (2016). Reacting to 'irregular' learning environments in a senior secondary school. *Curriculum and Teaching*, *31*(2), 47–69. doi:10.7459/ct/31.2.04

Americans With Disabilities Act of 1990, Pub. L. No. 101-336, 104 Stat. 328 (1990).

Bancino, R., & Zevalkink, C. (2007). Soft Skills: The New Curriculum for Hard-Core Technical Professionals. *Techniques: Connecting Education and Careers (J1), 82*(5), 20-22.

Basham, J. D., & Blackorby, J. (in press). UDL next: The future of the framework. In K. Lowrey (Ed.), *Critical Issues in Universal Design for Learning*. Oviedo, FL: Knowledge by Design.

Basham, J. D., Hall, T. E., Carter, R. A. Jr, & Stahl, W. M. (2016). An operationalized understanding of personalized learning. *Journal of Special Education Technology, 31*(3), 126–136. doi:10.1177/0162643416660835

Basham, J. D., Israel, M., & Maynard, K. (2010). An ecological model of STEM education: Operationalizing STEM for all. *Journal of Special Education Technology, 25*(3), 9–19. doi:10.1177/016264341002500303

Basham, J. D., & Marino, M. T. (2013). Understanding STEM education and supporting students through universal design for learning. *Teaching Exceptional Children, 45*(4), 8–15. doi:10.1177/004005991304500401

Basham, J. D., Meyer, H., & Perry, E. (2010). The design and application of the digital backpack. *Journal of Research on Technology in Education, 42*(4), 339–359. doi:10.1080/15391523.2010.10782555

Benade, L. (2017). Is the classroom obsolete in the twenty-first century? *Educational Philosophy and Theory, 49*(8), 796–807. doi:10.1080/00131857.2016.1269631

Byers, T. (2016). *Development of an observation metric for linking pedagogy, technology and space*. Academic Press.

Byers, T., Hartnell-Young, E., & Imms, W. (2018). Empirical evaluation of different classroom spaces on students' perceptions of the use and effectiveness of 1-to-1 technology. *British Journal of Educational Technology, 49*(1), 153–164. doi:10.1111/bjet.12518

Byers, T., & Imms, W. (2014). *Making the space for space: The effect of the classroom layout on teacher and student usage and perception of one-to-one technology*. Paper presented at the 26th Australian Computers in Education Conference, Adelaide, Australia.

Byers, T., & Imms, W. (2016). Evaluating the change in space in a technology-enabled primary years setting. In K. Fisher (Ed.), *The translation design of schools: An evidence based approach to aligning pedagogy and learning environment design* (pp. 215–236). Sense. doi:10.1007/978-94-6300-364-3_10

Byers, T., Imms, W., & Hartnell-Young, E. (2014). Making the case for space: The effect of learning spaces on teaching and learning. *Curriculum and Teaching*, *29*(1), 5–19. doi:10.7459/ct/29.1.02

Byers, T., Imms, W., & Hartnell-Young, E. (2018). Evaluating teacher and student spatial transition from a traditional classroom to an innovative learning environment. *Studies in Educational Evaluation*, *58*, 156–166. doi:10.1016/j.stueduc.2018.07.004

Caper, R. (1996). Play, experimentation and creativity. *The International Journal of Psycho-Analysis*, *77*, 859–869. PMID:8933214

CAST. (2018). *Universal Design for Learning Guidelines version 2.2*. Retrieved from http://udlguidelines.cast.org

Catterall, J. S. (2002). The arts and the transfer of learning. In R. J. Deasy (Ed.), *Critical Links: Learning in the arts and student academic and social development*. Washington, DC: Arts Education Partnership.

Christensen, C. M., Horn, M. B., & Johnson, C. W. (2008). How' disruptive innovation' will change the way we learn. *Education Week*, *27*(39), 25–36.

Connor, A., Karmokar, S., & Whittington, C. (2015). From STEM to STEAM: Strategies for enhancing engineering & technology education. *International Journal of Engineering Pedagogy*, *5*(2), 37–47. doi:10.3991/ijep.v5i2.4458

Cook, S. C., & Rao, K. (2018). Systematically applying UDL to effective practices for students with learning disabilities. *Learning Disability Quarterly*, *41*(3), 179–191. doi:10.1177/0731948717749936

Curry, C., Cohen, L., & Lightbody, N. (2006). Universal design in science learning. *Science Teacher (Normal, Ill.)*, *73*(3), 32.

Dail, W. (2013). On cultural polymathy: How visual thinking, culture, and community create a platform for progress. *The STEAM Journal*, *1*(1). Available at: http://scholarship.claremont.edu/steam/vol1/iss1/7

Deed, C., & Lesko, T. (2015). 'Unwalling' the classroom: Teacher reaction and adaptation. *Learning Environments Research*, *18*(2), 217–231. doi:10.100710984-015-9181-6

Dixon, R. A., & Brown, R. A. (2012). Transfer of Learning: Connecting Concepts during Problem Solving. *Journal of Technology Education*, *24*(1), 2–17. doi:10.21061/jte.v24i1.a.1

Dovey, K., & Fisher, K. (2014). Designing for adaptation: The school as socio-spatial assemblage. *The Journal of Architecture, 19*(1), 43–63. doi:10.1080/13602365.2014.882376

Dumont, H., Instance, D., & Benavides, F. (Eds.). (2010). The nature of learning: Using research to inspire practice. Organization for Economic Cooperation and Development.

Dumont, H., & Istance, D. (2010). Analyzing and designing learning environments for the 21st century. *The nature of learning: Using research to inspire practice*, 19-34.

Eger, J. (2013). STEAM...Now! *The STEAM Journal, 1*(1). Available at: http://scholarship.claremont.edu/steam/vol1/iss1/8

Fisher, K. (2006). The new learning environment: Hybrid designs for hybrid learning. *Education Futures Public, 2.*

Goldin, C., & Katz, L. F. (2008). *The Race between Education and Technology.* Cambridge, MA: Harvard University Press.

Harris, K. S., & Rogers, G. E. (2008). Soft Skills in the Technology Education Classroom: What Do Students Need? *Technology Teacher, 68*(3), 19–24.

Hattie, J. A., & Donoghue, G. M. (2016). Learning strategies: A synthesis and conceptual model. *NPJ Science of Learning, 1*, 16013.

Henriksen, D. (2014). Full STEAM ahead: Creativity in excellent STEM teaching practices. *The STEAM Journal, 1*(2), 15.

Householder, D. L., & Hailey, C. E. (2012). *Incorporating Engineering Design Challenges into STEM Courses.* National Center for Engineering and Technology Education. Retrieved from http://eric.ed.gov/?id=ED537386

Imms, W., & Byers, T. (2016). Impact of classroom design on teacher pedagogy and student engagement and performance in mathematics. *Learning Environments Research, 19*(2), 1–14.

Johnson, P. A. (1999). Problem-based, cooperative learning in the engineering classroom. *Journal of Professional Issues in Engineering Education and Practice, 125*(1), 8–11. doi:10.1061/(ASCE)1052-3928(1999)125:1(8)

Lackney, J. A. (2008). Teacher environmental competence in elementary school environments. *Children, Youth and Environments, 18*(2), 133–159.

Lippman, P. C. (2010). *Can the physical environment have an impact on the learning environment?* Academic Press.

Morrison, J. (2006). *Attributes of STEM Education.* TIES STEM Education Monograph Series. Retrieved from http://www. tiesteach.org/monographs.aspx

Morrison, J., Roth McDuffie, A., & French, B. (2015). Identifying key components of teaching and learning in a STEM school. *School Science and Mathematics, 115*(5), 244–255. doi:10.1111sm.12126

Mulcahy, D., Cleveland, B., & Aberton, H. (2015). Learning spaces and pedagogic change: Envisioned, enacted and experienced. *Pedagogy, Culture & Society, 23*(4), 575–595. doi:10.1080/14681366.2015.1055128

MYCEETYA. (2008). *Melbourne Declaration on Education Goals for Young Australians.* Melbourne, Australia: Ministerial Council on Education, Employment, Training and Youth Affairs.

New Zealand Ministry of Education. (2014). *The ministry of education's statement of intent 2014-2018.* Wellington, New Zealand: New Zealand Ministry of Education.

Oblinger, D. G. (2006). Space as a change agent. *Learning Spaces, 1*, 1–2.

Kasza, P., & Slater, T. F. (2017). A Survey of Best Practices and Key Learning Objectives for Successful Secondary School STEM Academy Settings. *Contemporary Issues in Education Research, 10*(1), 53–66.

OECD. (2013). *Education at a Glance 2013: OECD Indicators.* OECD Publishing. doi:10.1787/eag-2013-

OECD. (2015). *Education at a Glance 2015: OECD Indicators.* Paris: OECD Publishing. doi:10.1787/eag-2015-

Oleson, A. K., Hora, M. T., & Benbow, R. J. (2014). *STEM: How a Poorly Defined Acronym Is Shaping Education and Workforce Development Policy in the United States.* WCER Working Paper No. 2014-2. Wisconsin Center for Education Research. Retrieved from http://eric.ed.gov/?id=ED556481

Pintrich, P. R., & De Groot, E. V. (1990). Motivational and Self-Regulated Learning Components of Classroom Academic Performance. *Journal of Educational Psychology, 82*(1), 33–40. doi:10.1037/0022-0663.82.1.33

Root-Bernstein, R., & Root-Bernstein, M. (2003). Intuitive tools for innovative thinking. *The international handbook on innovation*, 377-387.

Rubin, J., & Chisnell, D. (2008). *Handbook of usability testing: How to plan, design, and conduct effective tests* (2nd ed.). Indianapolis, IN: Wiley.

Schwab, K. (2016). *The fourth industrial revolution.* New York, NY: Crown Business.

Steele, F. (1980). Defining and developing environmental competence. *Advances in Experiential Social Processes, 2*, 225-244.

Story, M. F. (2001). Principles of universal design. *Universal Design Handbook.*

Story, M. F., Mueller, J. L., & Mace, R. L. (1998). *The universal design file: Designing for people of all ages and abilities*. Academic Press.

Tanner, C. K. (2008). Explaining relationships among student outcomes and the school's physical environment. *Journal of Advanced Academics, 19*(3), 444–471. doi:10.4219/jaa-2008-812

Taylor, A. (2009). *Linking architecture and education: Sustainable design for learning environments*. UNM Press.

Theisens, H., Benavides, F., & Dumont, H. (2008). *OECD work on future educational environments*. Academic Press.

Thomas, H. (2010). Learning spaces, learning environments and the dis 'placement' of learning. *British Journal of Educational Technology, 41*(3), 502–511. doi:10.1111/j.1467-8535.2009.00974.x

Upitis, R. (2004). School architecture and complexity. *Complicity: An International Journal of Complexity and Education, 1*(1).

Zhao, Y. (2016). From deficiency to strength: Shifting the mindset about education inequality. *The Journal of Social Issues, 72*(4), 720–739. doi:10.1111/josi.12191

Zhao, Y., Wehmeyer, M., Basham, J., & Hansen, D. (2019). Tackling the wicked problem of measuring what matters: Framing the questions. *ECNU Review of Education, 2*(3), 262–278. doi:10.1177/2096531119878965

ADDITIONAL READING

Benyon, D. (2014). Spaces of interaction, places for experience. *Synthesis Lectures on Human-Centered Information, 7*(2), 1–129. doi:10.2200/S00595ED1V01Y201409HCI022

Bracken, S., & Novak, K. (Eds.). (2019). *Transforming Higher Education Through Universal Design for Learning: An International Perspective*. Routledge. doi:10.4324/9781351132077

Garrett, J. J. (2010). *The elements of user experience: user-centered design for the web and beyond*. Pearson Education.

Glass, D., Meyer, A., & Rose, D. (2013). Universal design for learning and the arts. *Harvard Educational Review, 83*(1), 98–119. doi:10.17763/haer.83.1.33102p26478p54pw

Hall, T. E., Meyer, A., & Rose, D. H. (Eds.). (2012). *Universal design for learning in the classroom: Practical applications*. Guilford Press.

Johnson, J. (2013). *Designing with the mind in mind: simple guide to understanding user interface design guidelines*. Elsevier.

Meyer, A., Rose, D. H., & Gordon, D. (2014). *Universal design for learning: Theory and practice*. Wakefield, MA: CAST.

Schwab, K. (2017). *The fourth industrial revolution*. Currency.

Zhao, Y. (2017). What works may hurt: Side effects in education. *Journal of Educational Change, 18*(1), 1–19. doi:10.100710833-016-9294-4

KEY TERMS AND DEFINITIONS

Innovative Learning Environment (ILE): A learning environment that uses technology to implement flexible learning and educational design practices in education.

Learning Experience: The series of events and interactions within the comprised with a goal-driven learning process.

Soft Skills: Social skills conducive for effective communication and human interaction.

Traditional Classroom: These spaces are designed as an arrangement of similar desks placed in rows which are often difficult to rearrange and move about the room.

Universal Design (UD): Design principles originally developed to provide for accessibility, and inclusion in the design of buildings, products, and tools.

Universal Design for Learning (UDL): A scientifically based framework focused on supporting the variability of every learner through proactive and iterative design that integrates multiple means of engagement, representation of information, and action and expression of understanding.

Chapter 2
Tools to Mediate Learning and Self-Assessment in a STEAM Unit of Work

Kate Thompson
Queensland University of Technology, Australia

Susan Narelle Chapman
Queensland University of Technology, Australia

Harry Kanasa
https://orcid.org/0000-0002-9718-4074
Griffith University, Australia

ABSTRACT

In some high school learning environments, hundreds of students engage in collaborative, term-long, project-based learning STEM or STEAM units of work. In this chapter, the authors report on an ongoing design-based research project in which researchers and teachers collaborate to design, teach, and assess STEAM units of work. They draw on research on project-based learning and interdisciplinary collaboration to inform the analysis drawing on Jonassen's typology of problem solving. The purpose of this chapter is to present two tools that functioned as boundary objects mediating learning and self-assessment: (1) a 'STEAM slider' used by students in groups to reach consensus on the use of the tools, knowledge, and processes employed in their project and (2) a criteria sheet developed to mediate students' and teachers' engagement in self- and teacher-assessment. The authors use these results to make recommendations for the next iteration of the project.

DOI: 10.4018/978-1-7998-2517-3.ch002

INTRODUCTION

STEM education is a priority nationally and internationally, as countries identify areas of investment for future workforce planning (Reider, Knestis, & Malyn-Smith, 2016; Timms, Weldon, & Mitchell, 2018). STEAM education includes an "A" to represent the Arts, and much of the literature which advocates for this approach argues that scientific progress will be impoverished without the Arts or creative thinking (e.g. Herro & Quigley, 2017). The use of STEM or STEAM education can indicate units of work that address any single discipline areas (Science, Technology, Engineering, the Arts or Mathematics), or those that require students to make connections between discipline areas (Costin, Thompson & Chapman, 2018). The latter is the focus of this paper. In some high school learning environments, hundreds of students engage in collaborative, project-based learning STEM or STEAM units of work, using their own devices, connecting to school systems, over several weeks. Such learning environments can be described as complex (Goodyear & Carvalho, 2013), and provide challenges both for researchers to understand the interplay of teacher and learner activity, as well as for the design of assessment by teachers (Costin et al., 2018).

In this paper, we present a STEAM unit of work developed over several years, as part of ongoing design-based research project in which researchers and school teachers work closely to collaboratively design, teach and assess STEAM units. The study was conducted with Year 7 and 8 students (aged 12-13 years) as part of a STEM excellence program in an Australian High School. Year 7 is the first year of high school in Australia. A range of scaffolds were provided to students to support the development, implementation and reporting of their collaborative STEAM projects using a portfolio available in a shared, online folder. In this paper, we examine two tools that mediated learning through consensus: (1) a 'STEAM slider' that students in groups used to reach consensus on the use of the tools, knowledge and processes employed in their project; and (2) a criteria sheet developed to mediate students' and teachers' engagement in self- and teacher-assessment. We initially compare students' and researchers' identification of the connection of STEAM disciplines in the projects, and then students' self-assessment with teachers' assessment of the final product and report. We provide some examination of the differences in particular elements of the rubric that may help in the provision of feedback in the next iteration of the project. We draw on research about collaborative learning (e.g. Goodyear, Jones, & Thompson, 2014) and particularly in terms of interdisciplinary teams (e.g. Flowers, Mertens, & Mulhall., 2000), as well as project based learning (e.g. Jonassen, 2000) and self-assessment and evaluation (Boud, 1995) to investigate key elements of the design of this collaborative, project-based STEAM unit of work, and provide recommendations for the next iteration of the project.

BACKGROUND

Project-Based Learning

In project-based learning students work in small groups where a question or problem organizes or drives activities, and these activities produce artefacts, products or solutions which address the driving question (Blumenfeld, Soloway, Marx, Krajcik, Guzdial, & Palincsar, 1991). These authors observe that project-based learning provides a framework in which, "students pursue solutions to nontrivial problems by asking and refining questions, debating ideas, making predictions, designing plans and/or experiments, collecting and analyzing data, drawing conclusions, communicating their ideas and findings to others, asking new questions, and creating artefacts" (Blumenfeld et al., 1991, p. 371). They explain that, "learners construct knowledge by solving complex problems in situations in which they use cognitive tools, multiple sources of information, and other individuals as resources" (Blumenfeld et al., 1991, p. 371). Though Jonassen's work is generally in the area of technology, the principles he presents have relevance across the curriculum as they resonate with issues common to a wide range of disciplines. The research conducted at the STEAM workshop sites holds a constructivist view of education which is based on three principles: that knowledge is constructed by the individual learner who creates meaning through a cumulative history of experiences, that knowledge construction is produced through activity and meaning making created through personal interactions, and that, "meaning is in the mind of the knower and can be shared and expressed through discussion or disagreement but can never be totally experienced by another person, thus creating a world of multiple perspectives" (Jonassen, Peck, & Wilson, 2001, p. 46). In addition, the previous experience of the student as well as the learning context will influence the process of problem solving (Jonassen, 2000). Each type of problem-solving requires different forms of instruction to scaffold learning.

As Jonassen (2003) discusses, one of the implications of constructivism for teachers and students is that thinking and problem-solving should be valued at least as much as memorising. Jonassen (2000) provides a description of eleven problem types: logical, algorithmic, story, rule-using, decision making, trouble-shooting, diagnosis-solution, strategic performance, case analysis, design, and dilemmas. The implications for teachers centre on their willingness to relinquish forms of authority and, in Jonassen's view, to build the skills required to be technology literate. This includes a need for teachers to loosen their hold on intellectual and management authority. It follows that for students to become intentional, constructive learners they need to assume some authority and take responsibility for their learning. This can be done by setting goals and working in collaboration with others (Jonassen, 2003). The interdisciplinary STEAM pedagogy at the research sites often drew on technology components and consistently focused on problem-solving.

Collaborative Learning and Interdisciplinary Teams

Collaborative learning is distinct from cooperative learning (Dillenbourg, 1999) as collaborative learning requires the group to complete the task together through negotiation, rather than working individually on components that are then brought together. Several studies identify convergence as an important aspect of collaboration (e.g. Raca & Dillenbourg, 2015; Jeong & Chi, 2007). In considering STEAM units of work, models can include both cooperative and collaborative learning, where students take on the role of disciplinary 'experts' and then work together in the group to complete the task. Research that explores how individuals learn to work in interdisciplinary teams in graduate or professional settings (e.g. Pennington et al., 2015) focuses on the integration of disciplinary perspectives to create a shared model of a given system to then represent and mediate conversations about the phenomenon in study. Pennington et al. (2015) identified five indicators of productive collaboration: (1) identification of an appropriate question; (2) development of a shared vocabulary; (3) the co-creation of boundary negotiating objects; (4) the use of tools for visualizing and combining data; and (5) a new, more connected understanding of the question. Boundary negotiating objects are found in situations that require the negotiation of the research question or the conceptualization of a problem (Akkerman & Bakker, 2011; Lee, 2007). In this paper, the focus is on the support provided to students to reach consensus on the disciplines included in their project through the use of two boundary negotiating objects (the STEAM sliders and the criteria sheets) to support the development of their ability to assess their resulting understanding of the question.

Self-Assessment

Recent research in assessment has concentrated on the importance of formative assessment. Black and Wiliam (2009) describe this as when "evidence about student achievement is elicited, interpreted, and used by teachers, learners, or their peers, to make decisions about the next steps in instruction that are likely to be better, or better founded, than the decisions they would have taken in the absence of the evidence that was elicited." (p.9). Panadero, Andrade, and Brookhart (2018) identify the following as key components of formative assessment: Learning progressions should clearly articulate the sub-goals of the ultimate learning goal; learning goals and criteria for success should be clearly identified and communicated to students; evidence of learning is elicited during instruction; students should be provided with evidence-based feedback that is linked to the intended instructional outcomes and criteria for success; both self- and peer assessments are important for providing students an opportunity to think metacognitively about their learning; and a classroom culture in which teachers and students are partners in learning should be established. Black and Wiliam (2009) identify the core role of peers in formative assessment (through peer teaching and collaborative learning) and the learner themselves (providing opportunities to develop metacognition, motivation, interest, attribution, and self-assessment).

Boud and Falchikov (1989) state that "Self-assessment refers to the involvement of learners in making judgements about their own learning, particularly about their achievements and the outcomes of their learning" (p. 529). Self-assessment, in addition to the cognitive purposes, aims to increase students' responsibility for their own learning (Panadero, Andrade, & Brookhart, 2018). Self-assessment has been found to improve interest in criteria and feedback rather than in a grade (Klenowski, 1995). And the case for using it has been made at all levels of education (Towler & Broadfoot, 1992; Andrade, Du & Mycek, 2010). Self-evaluation refers to approaches in which students grade their own work (Andrade & Valtcheva, 2009). There have been several studies indicating a discrepancy between teacher or tutor evaluation of student work, and self-evaluation, even using criteria co-developed with students (Orsmond, Merry & Reiling, 2000; Tousignant & DesMarchais, 2002). Boud (1989) states that when self-evaluation of effort is removed, that there is little difference between teachers' and learners' assessment marks. In the study described in the following, students identified their own learning objectives within criteria related to the Australian Curriculum, and co-constructed with the teachers and one of the researchers.

METHODS

This study adopted an inquiry approach to design (Thompson et al., 2016; Alhadad & Thompson, 2017). The inquiry approach (Thompson et al., 2016) connects two frameworks: Carvalho and Goodyear's (2014) Activity Centred Analysis and Design (ACAD) framework, and Sandoval's (2014) conjecture mapping. Design-based research methods connect theory to the design of a particular activity, which is then implemented and evaluated (Sandoval and Bell, 2004). Often, the design component of design-based research is underemphasized in favor of describing the implementation. ACAD provides the structure for the design of the unit, including tasks (epistemic), role and rules of interactions (social) and digital and physical learning environment and tools (set); (2) learner activity (observable learner behavior); and (3) the learning outcomes (measurable changes over time). Conjecture mapping (Sandoval, 2014) aids in the identification of design and research conjectures. In this approach to design-based research, the design of the unit is mapped to the ACAD framework and the connection between the elements of design and the learner activity that the researchers are interested in are Sandoval's design conjectures – what is it about the design that provided the opportunity (or not) for particular learner activity. Observable learner behavior is identified for data collection to help to answer that question. The intended learning outcomes are then identified, and the connection between learner and teacher activity and learning outcomes are addressed by Sandoval's research conjectures – what is it about learner activity that results in learning outcomes? The answers to these questions are then used to inform the subsequent iterations of the design, and new questions are identified. In this section we will describe the context of the study, outline the key elements of the design of the STEAM unit of work, (having already been implemented and evaluated through iterations over several years) and then identify the design and research questions that guided our inquiry.

The unit of work was collaboratively designed by the researchers and the participating teachers. Initial discussions were held to ensure the overarching goals of the project were appropriate to meet the curriculum and assessment needs of the school, a suitable time frame for the project was established and ensuring any pedagogical or assessment priorities for the school were built into the design of the unit. Roles and responsibilities of the planning team were also established. In the second meeting, feedback was sought from teachers regarding the planned activities in terms of pacing and sequencing. The researchers presented draft versions of the unit's activities, scaffolds and assessment tools, were reviewed by the teachers and feedback was given. The third meeting presented the final version of the unit plan and the roles of the researcher and teachers during the implementation phase were clearly defined. The unit was then taught and assessed by the third author and the collaborating teachers. A number of different teaching modes were used. In some lessons, the teacher exposition (e.g. how to write a hypothesis) was delivered solely by the researcher with the teacher acting as an observer. These types of lessons also doubled as a professional development for the teacher. Both the researcher and teacher would then circulate the room as students undertook independent work. Other lessons were team taught where the exposition was shared equally between the researcher and teacher. All lessons were concluded with informal verbal reflection between the researcher and teacher to assess the lesson and make plans for the direction of future lessons.

This study was conducted in an Australian high school, with year 7 and 8 students. The curriculum for all schools until Year 10 is mandated by a national body, the Australian Curriculum and Assessment Authority (ACARA). While the curricula have been implemented by each State in a largely unaltered fashion, assessment and reporting has been inconsistent between States, and the way assessment is conducted also varies between subject areas. Given the different expectations of working for students in the STEAM areas, designing integrated assessment is challenging.

Participants for the study were drawn from three Year 7 classes (students were typically 12 years old, $n = 83$) and two Year 8 classes (students were typically 13 years old, $n = 52$). Students were required to apply for and be accepted into the STEM excellence program at this school. Selection was based upon their marks from primary school, an interview, and performance on an assessment task addressing basic mathematical and scientific concepts and skills. The program ran for eight weeks with the aim of scaffolding independent STEAM inquiry projects in preparation for a local, regional science fair. Students were allowed to work in small groups (2-3) or individually. The design of the program is outlined in Table 1.

Table 1. Design of STEAM program

Week	Tasks	Social arrangements	Resources
1	Introduction to the concept of STEAM and the supporting disciplines	Whole group	Presentation
	Brainstorm project ideas	Individuals/ pairs/ small group	Paper, post-its, pens
2	The criteria for evaluating project ideas were co-created by students, researcher and teachers	Whole group	Decision making matrix
	Identify topic for project	Individuals/ pairs/ small group	Paper, post-its, pens
3	Project planning	Individuals/ pairs/ small group	Field specific (i.e., either science, technology/engineering or mathematics) template to scaffold their inquiry ***STEAM Slider***
4	Students conduct projects	Individuals/ pairs/ small group	Materials specific to student projects Students used template to scaffold their inquiry
5	Students conduct projects		
6	Students conduct projects		
7	Student presentations	Individuals/ pairs/ small group	Students used the template to structure their presentation to the class
8	In-class STEAM fair	Pairs/individual/small group	Students presented their work to their peers in a STEAM fair context ***Criteria Sheet***

Learners received scaffolded support which began with presenting a range of relevant topic examples to stimulate the formation of ideas (week 1, Table 1). Students were asked to identify a problem, two features of problem solving were assumed: that the answer to the problem was unknown by the students, and that the problem had some, "social, cultural, or intellectual value" (Jonassen, 2000, p. 65) to the students. The researchers guided the students through an inductive process which began with brainstorming three questions: (1) What questions do you want to answer? (2) What real world problems do you want to solve? (3) What products do you want to create? The questions, problems and products identified by students are included in the results section. After identifying their problem (week 2, Table 1), students identified the types of knowledge needed for their project in order to solve the problem using the STEAM slider (week 3, Table 1). Students were also provided with a template PowerPoint file that further scaffolded their inquiry. Each week they were asked to add a part of their investigation to the template, saved on the shared drive at school, and to provide a reflection about their learning. One PowerPoint file was created for each group, and consensus was needed before adding to the slides (weeks 4-6, Table 1). Students then presented to the class using the template (week 7, Table 1) and in a STEAM fair style (week 8, Table 1).

The STEAM slider in Figure 1 was developed to help students conceptualize the contribution of the various STEAM fields to their projects. Each field is represented on an individual slider and the student is asked to consider their project in terms of the relative contributions of each of the fields to their project. The STEAM slider acknowledges that student projects may not sit neatly within a single field and allows students to see the synergistic nature of the fields. The lowest level of contribution from a field is the use of the physical tools typically associated with that field. For example, using a thermometer to measure temperature, a ruler to measure distance or a paint brush to paint. A higher contribution would be a student uses knowledge from that field. For example, applying Newton's Law to calculate force, using triangles to a structure to provide support, or selecting colors from a color wheel. The highest level of use of a field is the employment of the processes, or methods, of that field. For example, using the scientific method to design and conduct an experiment, Polya's method to solve mathematical problems or the artistic process to create art. It is assumed that a student indicates using the processes of a field then they are necessarily using both the tools and knowledge from that field.

In week 3, once the use of the sliders was explained to students they were asked to consider their projects in terms of the tools, knowledge and processes from each of the fields, and move the sliders to indicate the contribution of each field to the project. Moving the slider higher indicates a greater engagement with that field

Figure 1. The STEAM Slider

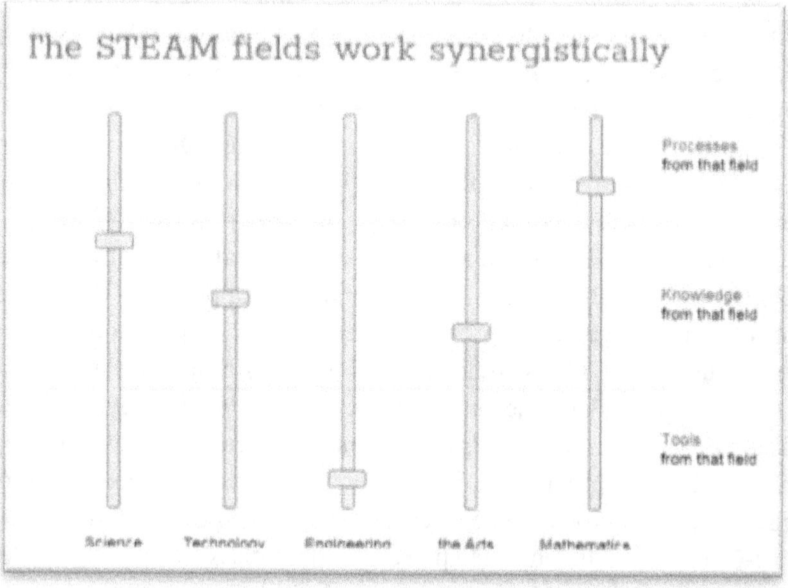

while at the same time acknowledging real-world projects can only be completed by using tools, knowledge and processes from more than one field. Students were asked the following questions to scaffold the process:

1. What tools did you use? Were they scientific, technological, engineering, artistic or mathematical tools?
2. What knowledge/ideas/concepts did you use? Did they come from science, technology, engineering, the arts or mathematics?
3. What inquiry process did you use? The scientific method, the design and engineering process, artistic exploration or Polya's problem solving method?

Once students had decided whether they had used to the tools, knowledge or processes from a particular field, they were then asked to gauge whether their tool use, for example, was low, medium or high, and then adjust the slider within that level. Students were also required to be able to verbally justify the positioning of the slider to their peers or to a teacher, thus functioning as a conversation tool to promote discussion and as a cognitive tool to guide their thinking process (Jonassen, Peck & Wilson, 2001). Each student project was also classified according to the STEAM slider by researchers, and the comparison is provided in the results section.

The student STEAM projects were classified by the researchers according to Jonassen's (2000) typology of problem-solving (discussed in the background section), and underpinned by a constructivist educational perspective (Howland, Jonassen, & Marra, 2012; Jonassen, 2003). Jonassens' (2000) typography of eleven different categories of problem-solving are based on the following aspects: learning activity, inputs, success criteria, context, structuredness and abstractness. The classification of these is provided in the results section.

The criteria sheets used by teachers and students in Figure 2 were based upon standardized versions currently in use in the school. The use of criteria sheets is standard procedure in Australian high schools as a means of assessment. Primarily designed for teacher use, in this study it was used by students as a formative assessment, reflective tool for the purposes of identifying gaps in their work so they can addressed, at the end of the project as a means of self-summative assessment, and in the longer term increasing their assessment literacy. Regardless of the intended audience, in alignment with criterion-based assessment principals, students and teachers read each of the statements (criteria) along the continuum and a determination was made as to which statement best described the work. Researchers and teachers worked collaboratively to frame the statements in the first person and to significantly simplify the language to word the statements in 'kid speak'. There is a state mandated requirement that levels of achievement be reported using the levels A (highest) to E lowest (C representing the expected performance for that age-level).

Figure 2. Criteria sheet example

Science investigation criteria sheet

	A	B	C	D	E
Scientific Processes	I've used books, websites and videos in my research. Resources are listed in my bibliography.	I've used 2 out of the three (books, websites and videos) sources for my research. Resources are listed in my bibliography.	I've used 1 out of the three (books, websites and videos) sources for my research. Resources are listed in my bibliography.	I have used resources but I don't have a bibliography.	I didn't used any resources and I don't have a bibliography.
	I have identified 5 causes and 5 effects. I have highlighted my IV and DV.	I have identified 4 causes and 4 effects. I have highlighted my IV and DV.	I have identified 3 causes and 3 effects. I have highlighted my IV and DV.	I have identified 2 causes and 2 effects. I have highlighted my IV and DV.	I have identified 1 cause and 1 effect. I didn't highlight my IV and DV.
	In my hypothesis, the IF segment states the direction of the effect, the THEN segment makes a clear prediction and the BECAUSE segment gives a scientific reason.	In my hypothesis, the IF segment states the effect, the THEN segment makes a clear prediction and the BECAUSE segment gives a scientific reason.	In my hypothesis, the IF segment states the effect, the THEN segment makes a prediction and the BECAUSE segment gives a reason.	My hypothesis is missing the BECAUSE segment.	My hypothesis is an aim OR my hypothesis is missing.
	I have listed ALL materials and equipment (brand, amount, and concentration) AND I have completed 3 rows of the safety audit.	I have listed ALL materials and equipment (brand, amount, OR concentration) AND I have completed 2 rows of the safety audit.	I have listed MOST materials and equipment (brand, amount, OR concentration) AND I have completed 1 row of the safety audit.	I have listed SOME materials and equipment. I have not completed the safety audit.	I have not listed any materials or equipment. I have not completed the safety audit.
	My method is written in the passive voice, past tense. I have included pictures and video of the method. I have conducted a minimum of 5 trials for each condition.	My method is written in the passive voice, past tense. I have included pictures OR video of the method. I have conducted a minimum of 4 trials for each condition..	My method is written in the passive voice, past tense. I have conducted a minimum of 3 trials for each condition.	The method is written in the active voice, future tense. I have conducted a minimum of 2 trials for each condition.	My method is missing. I have conducted only 1 trial for each of my conditions.
	I presented my data as tables and/or graphs. I have followed ALL tabulating and graphing conventions. My raw data is in an appendix.	I presented my data as tables and/or graphs. I have followed MOST tabulating and graphing conventions.	I presented my data as tables and/or graphs. I have followed SOME tabulating and graphing conventions.	I have presented raw data in the results section. I have followed SOME tabulating and graphing conventions.	I have presented raw data in the results section. I have followed NO tabulating and graphing conventions.
	My poster has ALL key elements of a science poster. There is a mix of words, images and graphs.	My poster has ALL key elements of a science poster. There is a predominance of words, images or graphs.	My poster has SOME key elements of a science poster. There is a predominance of words, images or graphs.	My poster has SOME key elements of a science poster. My poster only has words.	My poster is the pages of journal printed out and stuck on cardboard.
Knowledge and Conceptual Understanding	In my conclusion, I clearly state whether my hypothesis was supported or rejected. I justify my conclusion in 2 paragraphs.	In my conclusion, I clearly state whether my hypothesis was supported or rejected. I justify my conclusion in 1 paragraph.	In my conclusion, I clearly state whether my hypothesis was supported or rejected. I justify my conclusion.	In my conclusion, I clearly state whether my hypothesis was supported or rejected	In my conclusion, I state my hypothesis was proved or disproved.
Evaluating and Concluding	In my discussion, I have described (in paragraph form) 3 ways to improve my experiment OR suggested future experiments, and justified.	In my discussion, I have described (in paragraph form) 2 ways to improve my experiment OR suggested future experiments, and justified.	In my discussion, I have described (in paragraph form) 2 ways to improve my experiment OR suggested future experiments.	In my discussion, I have described (in paragraph form) 1 way to improve my experiment OR suggested future experiments.	In my discussion, I have described (in paragraph form) 1 way to improve my experiment.

To encourage reflection and assessment literacy on the part of the student, a self-assessment rubric was included in the template. A homework task in Week 2 was for students to read the rubric and bring questions to the next class and in the last week, students were asked to self-assess their work as part of the submission process. In the results section we compare the overall score allocated by students and the teacher, and investigate the elements of the rubric in which there was the greatest discrepancy in assessment.

FINDINGS

Students compiled an extensive list of ideas in three categories (Questions to be answered, problems to be solved, products to be created) which were written on sticky notes and attached to a large poster. Broadly speaking, students' questions, problems and products represented aspects of decision-making and design. Many of the questions (Year 7 brainstormed: 21 Questions, 31 products, 47 problems; Questions; Year 8 brainstormed: 51 Questions, 36 products, 28 problems) and problems took the form of a dilemma such as large-scale environmental, political and social issues, or intimate and complex personal concerns. They were also usually focused within one subject area, rather than extending beyond. Table 2 shows examples of research questions from the initial stage of generating ideas which emerged from all groups.

Table 2. Examples of student questions (Year 7 and Year 8)

Discipline	Year 7	Year 8
Science	What type of microorganisms live in water from different sources? How does slime work? How deep is the deepest part of the ocean? Which fruit has the best vitamins for you? What different sports require different muscles?	How do plants provide and transfer oxygen to our world? Where is the most bacteria in a normal household? If gravity and friction controlled a ball, what would happen to the ball? What is a black hole made from?
Technology	Can we create an app that detects if students are on their mobiles?	
Engineering	How much solar energy is needed to build a solar plane?	Is it possible to make a machine that effects a human by influencing time?
Arts	Does music improve your concentration?	Does dancing effect your way of learning?
Mathematics		What are the Fibonacci numbers?

In Table 2, examples of research questions demonstrate the range of research questions identified by students. In year 7, students did not identify mathematics research questions and in year 8 no students identified a technology question. The majority of questions were classified as science, with a range of sciences represented including physics and anatomy. The arts questions that were identified were linked to learning sciences and cognition research.

The second brainstorming topic was the identification of real-world problems that students wanted to solve. The results for the two year levels are shown in Table 3, below.

Year 7 students identified 71 problems during this session, and thirty problems were identified by year 8 students. When focusing on problems, no students in either year level identified arts problems, one student who identified a mathematics problem focused on learning mathematics (Table 3). Some of the identified problems were phrased as questions. Common across both year levels of students was the identification of large-scale problems such as climate change/global warming or curing cancer. In both groups, many of the identified problems were social problems, rather than any of the other discipline areas – including large-scale such as hunger, war, homelessness, as well as personal problems such as too much homework and chores.

Table 3. Examples of student questions (Year 7 and Year 8)

Discipline	Year 7	Year 8
Science	To stop global warming Why can't we find a cure for cancer? Malaria	Are there ways to reduce the amount of carbon dioxide levels in the air? Water quality Pollution and global warming
Technology	How can you stop people learning your password? How to stop watching YouTube and not get distracted	
Engineering	Recycling straws-machine What can't plastic disintegrate? How do we get faster wifi?	Making a fan without using electricity
Arts		
Mathematics		How can learning subjects such as maths be made easier for those who don't put effort in learning!
Social Sciences	Stress Chores Too much homework Bad traffic How to help end world hunger?	To solve the problem of the homeless How to solve addiction (DMA, smoking, gaming) To solve the problem of domestic violence Poverty and inequality

The final brainstorming topic was the identification of products that the students wanted to make. Examples of these are shown in Table 4 below.

Year 7 students generated 114 product ideas and, 24 products were identified by year 8 students. Most products were categorized as technology, and many of those that were technology were also for another purpose (e.g. an app that stops stress is technology, social science and science, Table 4). Year 7 students generated many more product ideas than year 8 students. Year 7 students' ideas were mostly related to their own immediate life (e.g. something to clean their room), although some did address larger-scale problems (e.g. a cure for Ebola). The Year 8 students identified mostly technology and engineering products, and the science products were more related to research questions. The final groups and their topics and year level are presented in Table 5 below. Figure 3, Figure 4, Figure 5, and Figure 6 show samples of the posters.

Table 4. Examples of student questions (Year 7 and Year 8)

Discipline	Year 7	Year 8
Science	A replacement for plastic, petrol and sugar cure for ebola Medicine for cancer	How to control mold growth? Does Carbon dioxide raise the temperature of the atmosphere
Technology	A mobile app detecting if mobile phones are being used in the classroom	A machine that blocks ads on computers or spams from affecting your laptop
Engineering	Flying cars Cleans your room when it senses items on the floor I want to be able to create an engine on my surf board so you don't have to paddle A digital whiteboard that records your ideas without the need to speak	An automatic feeder that gives out the food at a certain time for pets Waterproof submarine car Time machine
Arts	A new OS designed for film/art/ photography with inbuilt applications like modified (paint tool SAI, photoshop, AE, PR, SVP) more storage	A device that knits things for you
Mathematics		
Social Sciences	A lockdown system if there are intruders coming An app that stops stress Something that turns off your computer if you get angry/annoyed	

Table 5. Projects and group details

Project name	Guiding question	Social arrangement	Year level
Bacteria resistance	Which band aid is the most resistant to bacteria?	2 females	7
Microwave radiation	Do different periods of microwave radiation affect radish seeds' growth?	1 female, 2 males	7
Bacteria growth	Will there be a difference in growth between two samples of bacteria when one is fed a drop of sugar water?	2 females	7
Atmospheric pressure	How can dry ice affect carbon dioxide concentration and therefore affect atmospheric pressure?	1 female, 2 males	7
Air pressure and distance	Will the air pressure inside the ball affect the distance it will travel?	2 males	7
Crushing garbage	Can crushing garbage help protect the environment?	1 male	7
iSock WR	Design a device (iSock WR) to prevent hypothermia when a person is submerged in cold water.	3 males	7
Potato battery	How many white washed potatoes would be required to equal the same voltage to that of a double A battery?	3 females	8
Water level alarm	Design a water level alarm.	3 males	8

Figure 3. Sample poster

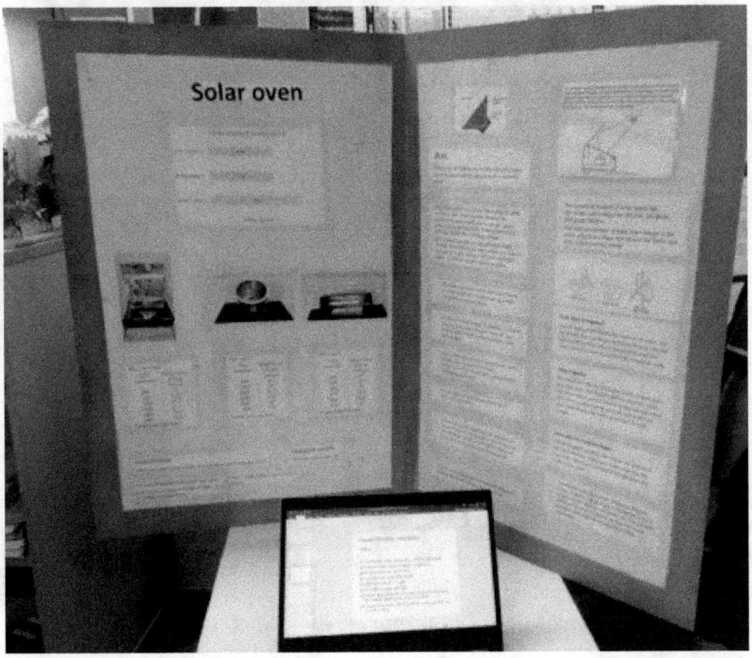

Figure 4. Sample poster

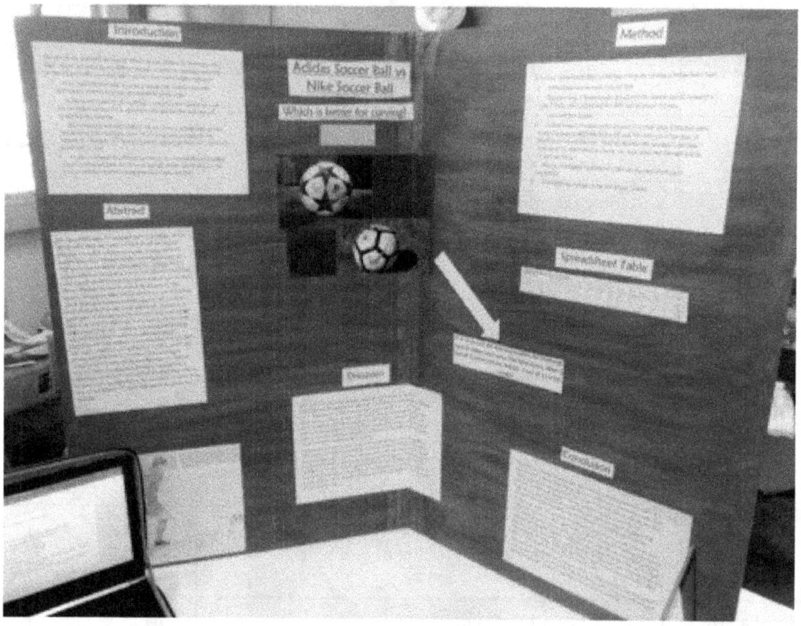

Figure 5. Sample poster

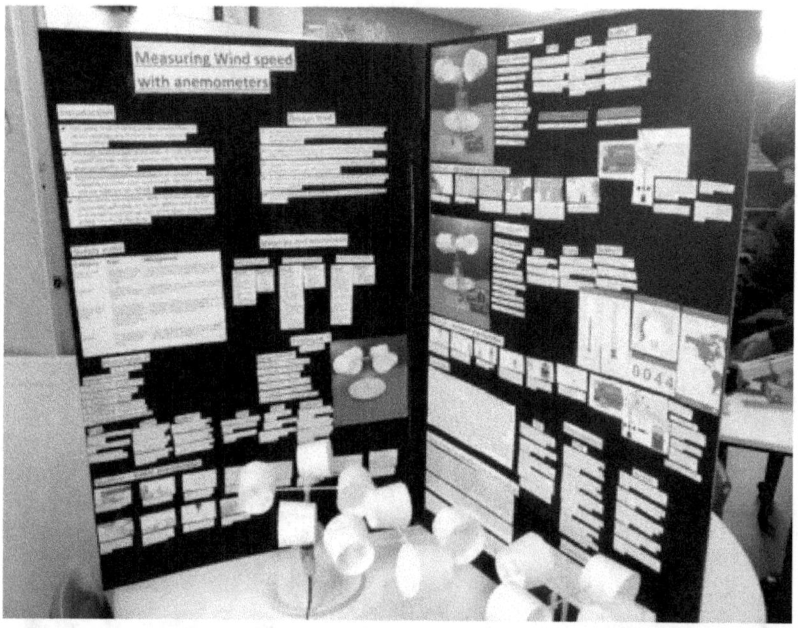

Figure 6. Sample poster

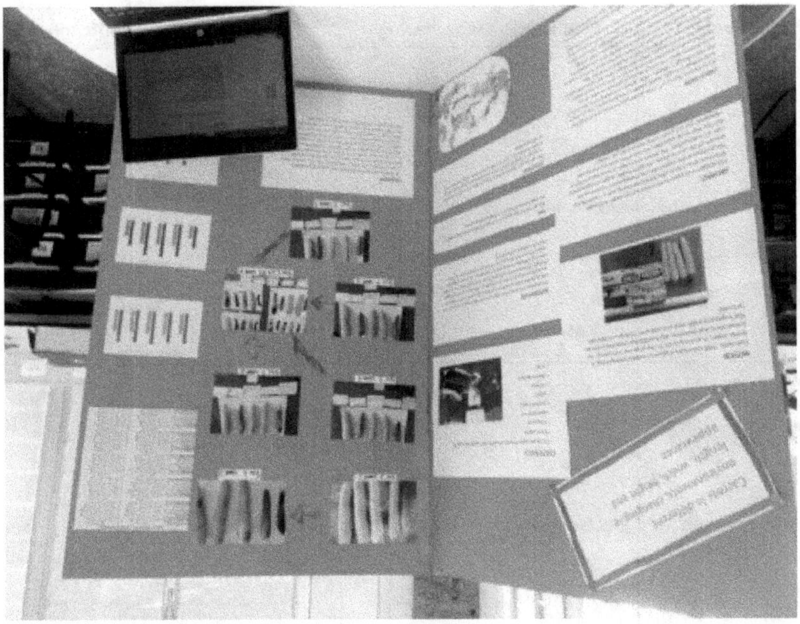

Seven groups of year 7 students and two groups of year 8 students participated in this project (Table 5). Of these, there were eight girls and 14 boys, and groups were of mixed gender and number of students. Students identified a topic for their project in groups. Each project was different, and addressed either a science or a technology topic. The researchers rated each project according to Jonasson's typology as shown in Table 6. Both students (solid blue line) and researchers (dotted orange and black lines) rated the relative contribution of each of the inquiry fields using the STEAM Slider (Tables 6). Note that the information contained in the STEAM sliders has been re-represented as a polar area chart with the three levels (tools, knowledge and processes) of the STEAM sliders indicated the nested set of pentagons. A dot placed at the very centre of the chart indicates that no tools, knowledge or processes were used from that field.

Table 6. Project identified according to problem-solving category and STEAM Slider

<table>
<tr><th>Project name</th><th>Problem-solving category</th><th>STEAM classification</th></tr>
<tr><td>Bacteria resistance</td><td>Decision-making</td><td>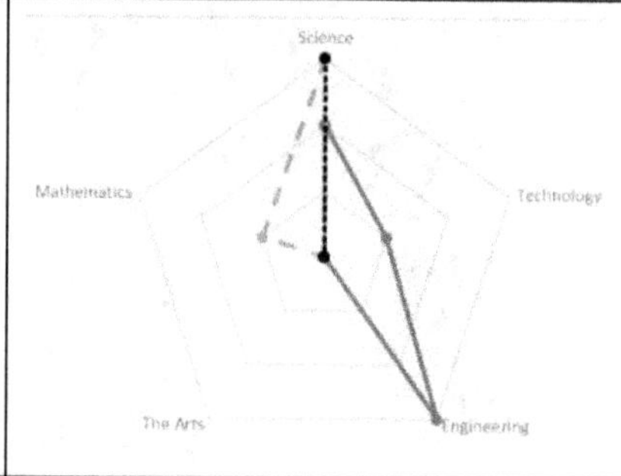
</td></tr>
<tr><td>Microwave radiation</td><td>Decision-making</td><td>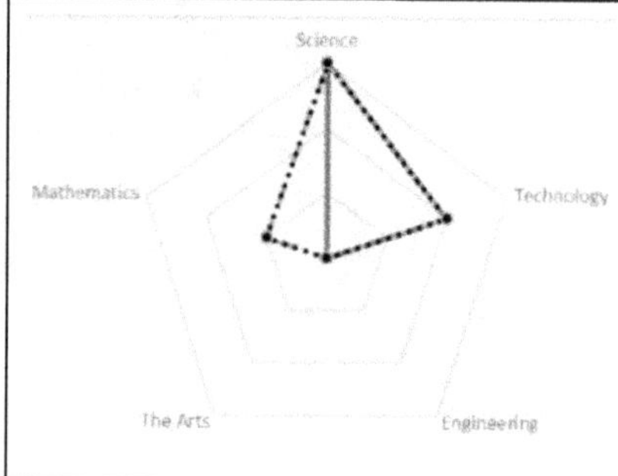
</td></tr>
<tr><td>Bacteria growth</td><td>Decision-making</td><td>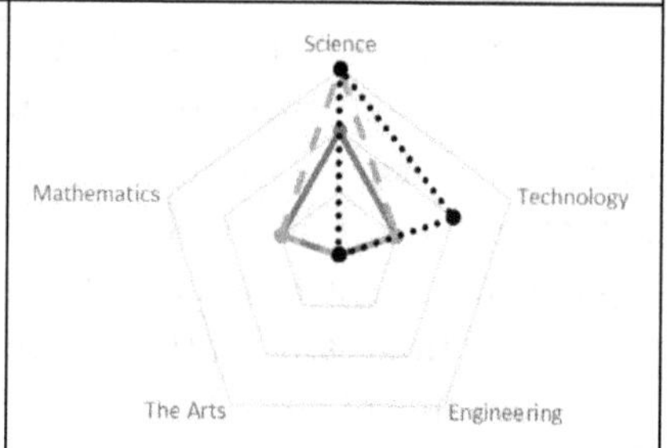
</td></tr>
<tr><td>Atmospheric pressure</td><td>Decision-making</td><td>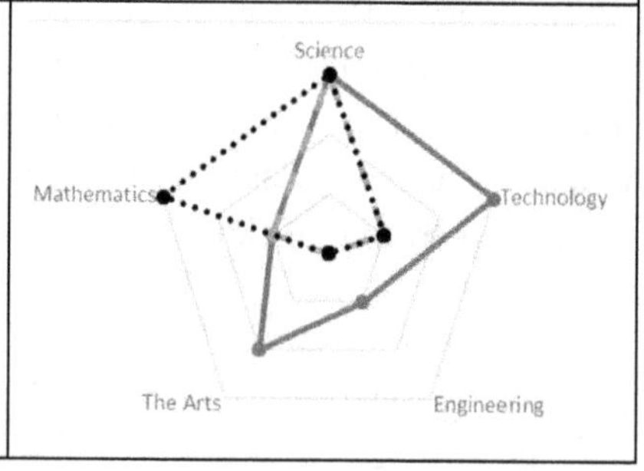
</td></tr>
</table>

continues on following page

Table 6. Continued

Project name	Problem-solving category	STEAM classification
Air pressure and distance	Decision-making	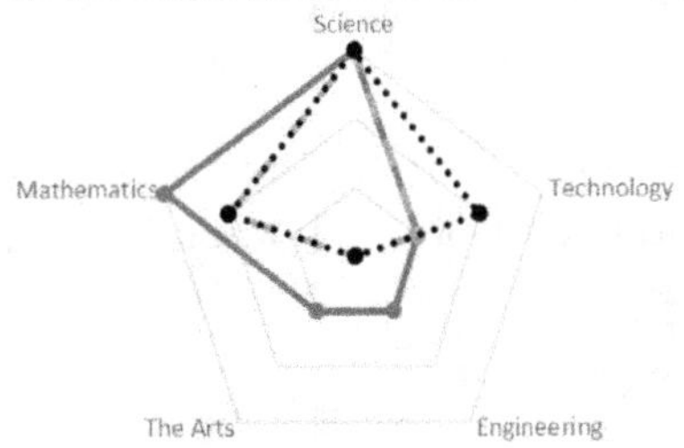
Crushing garbage	Trouble-shooting	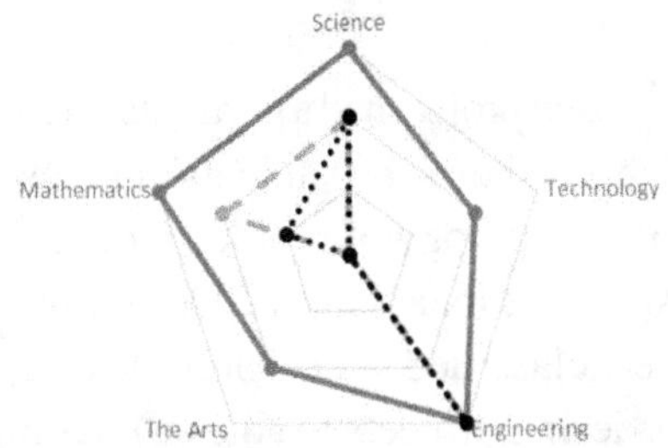
iSock WR	Design	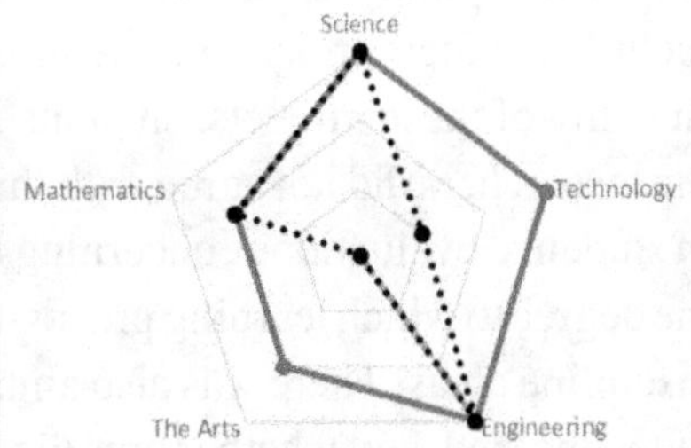
Potato battery	Decision-making	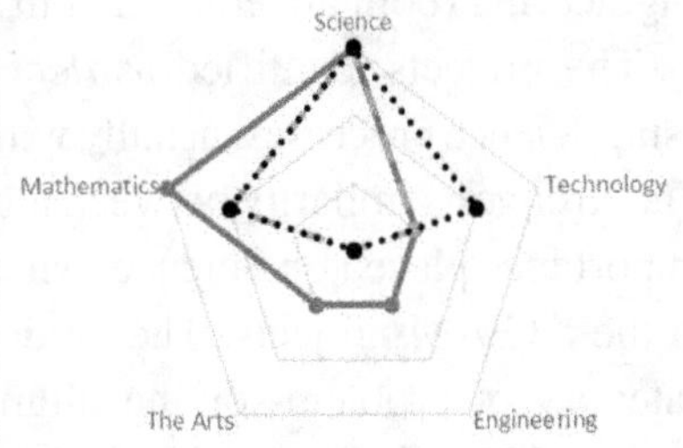

continues on following page

Table 6. Continued

Project name	Problem-solving category	STEAM classification
Water level alarm	Design	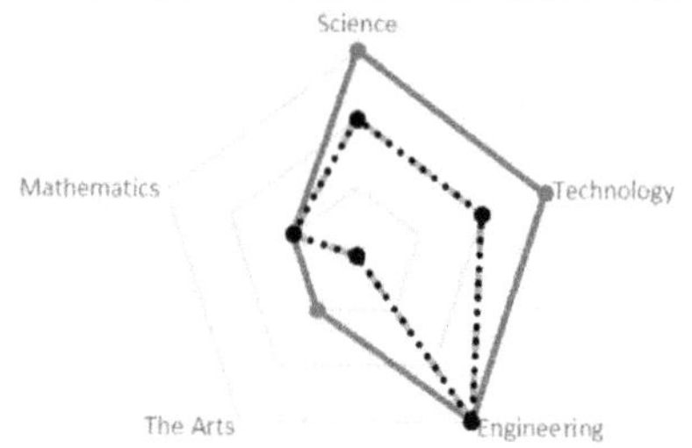

Each project had a particular area focus from which the investigation was generated. Table 6 shows the students' and researchers' classification of the projects according to the typology of problem solving as well as the focus of the STEAM disciplines. Applying Jonassen's (2000) typology of problem-solving, of the nine projects, six were classified as decision-making, one as trouble-shooting and two were classified as design projects (Tables 6). Examination of the classifications of the disciplines shows that students tended to over-estimate the number of STEAM disciplines included in their project (e.g. Atmospheric pressure or the Potato battery projects). In many of these projects, students identified arts tools as part of their project, but the researchers did not agree with this characterization for any. There was ambiguity in students' evaluations concerning the weighting of Mathematics and Science, and the degree to which learning processes (such as measurement) belonged to particular discipline areas. There was also ambiguity as to what constituted the Learning Area of technology, and whether non-digital tools were included in this discipline area or engineering (both are combined in the Australian national curriculum).

The projects identified as *decision-making* were identified by researchers as using science processes, usually with technology knowledge and mathematics tools. The strongest similarity between the groups that emerged from this analysis was the importance placed on Science, which was generally a heavily weighted component in the STEAM projects. The predominance of decision-making as the most used category of problem-solving aligned closely with the Science curriculum. The projects identified as *troubleshooting* and *design* all included engineering processes according to students and the researchers. In these projects, again, the students tended to mis-represent the nature of the knowledge or processes or tools used from other disciplines to solve the problem in question.

Given the nature of the projects, two different rubrics were provided for students, one for Technology and Engineering processes and one for Scientific processes. Each of these rubrics included three areas of assessment: Scientific or Technology/

Engineering processes (both with seven criteria statements), knowledge and conceptual understanding (both with one criteria statement), and evaluating and concluding (both with one criteria statement). Two groups did not include self-assessment through the rubric provided. Table 7 shows a comparison of self-assessed and teacher-assessed grades.

The teacher evaluation and the student self-evaluation used the same rubrics as a framework for assessment, and similar rubrics were used to evaluate Science and Technology projects. In all but one project (Potato battery), students self-assessed at a higher level than teachers. Examination of the elements of the rubric that teachers and students disagreed on in their assessment reveals that in the science projects, there was a misalignment concerned with the identification of equipment and description of the method amongst many of the groups. In the technology projects students and teachers had different expectations regarding the production of a prototype, and further misalignment was identified in the alignment with the brief and the evaluation of the project.

Table 7. A comparison of self-assessed and teacher-assessed grades

Project	Student	Teacher	Area of difference in the rubric
Bacteria resistance	A	B-	Identification of variables, hypothesis statement, identification of equipment, method description, presentation of data, poster construction, recommendations
Microwave radiation	N/A	B-	N/A
Bacteria growth	A	A-	Presentation of data
Atmospheric pressure	A-	B	Background research (student under-assessed), hypothesis statement, identification of equipment, method description, presentation of data, poster construction
Air pressure and distance	A	A-	Identification of equipment, method description, presentation of data, recommendations
Potato battery	A	A	hypothesis statement, method description, presentation of data,
Crushing garbage	A	C-	Research, design brief, identification of equipment, method description, prototype, poster, alignment between prototype and brief, evaluation
iSock WR	A	C+	design brief, prototype, alignment between prototype and brief, evaluation
Water level alarm	N/A	B	N/A

DISCUSSION

The research reported on in this paper aimed to identify student projects according to Jonassen's (2000) typology of problem solving and then the role of two computer-supported tools (a STEAM slider and an assessment rubric acting as boundary negotiating objects) were investigated to understand their role in mediating learning through consensus and self-assessment. Learning at the research sites aligned with Jonassen's (2003) guidelines for a constructivist approach, as students' self-regulated learning was active, problem-focused and project-based. The students were required to collaboratively set goals, determine strategies, collect data, engage in decision-making through discussion, draw conclusions and articulate their understandings. In the controlled environment at the research sites, researchers facilitated students in choosing their own constraints within the inquiry process. Students determined the equipment that was required, the way in which the equipment was used, and the strategies for addressing their chosen question/problem/topic. The STEAM projects provided clarity in learning, experience in developing stages of inquiry, the integration of skills from different disciplines, and an opportunity to synthesise past learning with emerging learning.

Each project relied on developing key understandings and skills and being able to access a specific category of problem-solving which was determined by the nature of the question/problem/topic. Three different project types were identified: decision-making, trouble shooting and design. Those that were classified as decision-making were closely aligned with science as the focus discipline area. Those classified as trouble-shooting or design were aligned with the Technology area. Students tended to over-estimate the number of discipline areas on which they would draw to answer the questions posed for their project, however, their focus area was, for the most part, aligned with that of the researcher. This may indicate that within their groups, students were inclusive of multiple discipline areas (or opinions) and able to reach consensus on the focus. The requirement to submit this as part of their portfolio meant that agreement had to be reached. Further requirements for students to describe how they reach decisions, perhaps as part of their reflections, may help to understand, and further scaffold, their decision-making process.

The criteria sheet enabled students and teachers to identify their understanding of demonstrated learning. In student projects that showed a close alignment between student and researcher identification of discipline areas (e.g. Bacterial growth), the assessment of learning was also similar. In projects in which there was a different (e.g. Bacteria resistance, Crushing garbage, iSock) the assessment was also different between teacher and student. This was mostly concerned with the design process and the framing of science problems. This misalignment in understanding of the requirements of the criteria sheet and will be used earlier in future iterations of the project to negotiate project scope or expectations with these students. In the next iteration of the project, the learning goals of the students will also be more closely aligned with the criteria sheet, negotiated with the teacher, and re-negotiated at key points during the Technology projects. By more closely scaffolding the different Science and Technology (design) processes, the criteria sheet could be used to facilitate learning associated with metacognition, motivation and self-assessment.

FUTURE RESEARCH DIRECTIONS

These findings show the need for further research to explore the relationship between tools for student learning and engagement in STEAM projects. There was substantial variability between student projects that could be related to the students' ability to scope and understand the nature of the disciplinary knowledge needed to be integrated in STEAM projects. Certainly there was a misperception among students about the way in which the arts could be incorporated. Greater clarification is required regarding the concept of STEAM in education, especially in refining a model suitable for implementing rigorously designed and assessed units of work. Investigation regarding the gap between student and teacher perceptions of STEAM units of work could consider the defining features of a tool which accurately reflects interdisciplinary learning through a constructivist approach. There is a need for further enquiry into the development of an approach to build teacher capacity in STEAM education as the basis for facilitating students' engagement with interdisciplinary learning. Additional research into strategies for preparing teachers to facilitate student learning in ways that promote and adequately support student-centered activity would also be beneficial. Further exploration into the metacognitive and disciplinary-specific understandings which underpin STEAM Learning Areas may assist teachers and students in developing problem-solving strategies and making assessment judgements which are more closely aligned. This process could be enhanced through exploring the relationship between tools for student learning and their engagement in STEAM projects. The reliability and effectiveness of such a tool could also be tested regarding the degree to which the outcomes may be generalized to new contexts. Clearly learning in inquiry-based STEM/STEAM projects is complex, and several other factors, including facilitation, need further examination.

CONCLUSION

Project based learning is a complex learning situation with great potential for impact on student learning. The design of this project gives participating students the opportunity to engage in shared tasks and goals that emphasize collaboration rather than competition. In complex learning environments such as these, design-based research approaches such as this allow researchers and teachers to make evidence-informed decisions about changes to the units of work. This examination of the role of two tools to support collaboration and self-assessment has revealed the potential of enabling further student agency in STEAM units of work, and in subsequent iterations additional processes will ensure that this potential is realized by all students. The effectiveness of project-based learning, and in STEAM units of work is dependent on the willingness of educators to allow students to regulate and control their own learning, and the willingness of students to accept this responsibility.

ACKNOWLEDGMENT

We acknowledge support and funding provided by the Step Up project, an Office of Learning and Teaching funded project, and the Griffith Institute of Educational Research.

REFERENCES

Akkerman, S. F., & Bakker, A. (2011). Learning at the boundary: An introduction. *International Journal of Educational Research*, *50*(1), 1–5. doi:10.1016/j.ijer.2011.04.002

Alhadad, S. S. J., & Thompson, K. (2017). Understanding the mediating role of teacher inquiry when connecting learning analytics with design for learning. *Interaction, Design, & Architecture (Washington, D.C.)*, *33*, 54–74. Retrieved from http://www.mifav.uniroma2.it/inevent/events/idea2010/doc/33_3.pdf

Andrade, H., Du, Y., & Mycek, K. (2010). Rubric-referenced self-assessment and middle school students' writing. *Assessment in Education: Principles, Policy & Practice*, *17*(2), 199–214. doi:10.1080/09695941003696172

Black, P., & Wiliam, D. (2009). Developing the theory of formative assessment. *Educational Assessment, Evaluation and Accountability*, *21*(1), 5–31. doi:10.100711092-008-9068-5

Blumenfeld, P., Soloway, E., Marx, R., Krajcik, J., Guzdial, M., & Palincsar, A. (1991). Motivating project-based learning: Sustaining the doing, supporting the learning. *Educational Psychologist*, *26*(3-4), 369–398. doi:10.1080/00461520.1991.9653139

Boud, D. (1989). The role of self-assessment in student grading. *Assessment in Higher Education*, *14*(1), 20–30. doi:10.1080/0260293890140103

Boud, D. (1995). *Enhancing learning through self-assessment*. London: Kogan Page.

Boud, D., & Falchikov, N. (1989). Quantitative studies of self-assessment in higher education: A critical analysis of findings. *Higher Education*, *18*(5), 529–549. doi:10.1007/BF00138746

Burton, R. R., Brown, J. S., & Fischer, G. (1984). Skiing as a model of instruction. In B. Rogoff & J. Lave (Eds.), *Everyday cognition: Its development in social context*. Cambridge, MA: Harvard University Press.

Carvalho, L., & Goodyear, P. (2014). *The architecture of productive learning networks*. New York, NY: Routledge.

Costin, D., Thompson, K., & Chapman, S. (2018). *Productive collaborative design of a STEAM unit of work in a multi-age class. Integrated education for the real world*. Presented at the 5th International STEM in Education Conference, Brisbane, Australia.

Dillenbourg, P. (1999). *What do you mean by collaborative learning?* Oxford, UK: Elsevier.

Flowers, N., Mertens, S. B., & Mulhall, P. F. (2000). What makes interdisciplinary teams effective? *Middle School Journal*, *31*(4), 53–56. doi:10.1080/00940771.2000.11494640

Goodyear, P., & Carvalho, L. (2013). The analysis of complex learning environments. *Re-thinking pedagogy for a digital age: Designing for 21st century learning*, 49-63.

Goodyear, P., Jones, C., & Thompson, K. (2014). Computer-supported collaborative learning: Instructional approaches, group processes and educational designs. In J. M. Spector, M. D. Merrill, J. Elen, & M. J. Bishop (Eds.), *Handbook of Research on Educational Communications and Technology* (pp. 439–452). New York, NY: Springer. doi:10.1007/978-1-4614-3185-5_35

Herro, D., & Quigley, C. (2017). Exploring teachers' perceptions of STEAM teaching through professional development: Implications for teacher educators. *Professional Development in Education*, *43*(3), 416–438. doi:10.1080/19415257.2016.1205507

Howland, J. L., Jonassen, D. H., & Marra, R. M. (2012). *Meaningful learning with technology* (4th ed.). Boston, MA: Pearson.

Jeong, H., & Chi, M. T. (2007). Knowledge convergence and collaborative learning. *Instructional Science, 35*(4), 287–315. doi:10.100711251-006-9008-z

Jonassen, D. H. (2000). Toward a design theory of problem solving. *Educational Technology Research and Development, 48*(4), 63–85. doi:10.1007/BF02300500

Jonassen, D. H. (2003). *Learning to solve problems with technology* (2nd ed.). Upper Saddle River, NJ: Merrill.

Jonassen, D. H., Peck, K. L., & Wilson, B. G. (2001). *Learning with technology: A constructivist perspective*. Upper Saddle River, NJ: Merrill.

Klenowski, V. (1995). Student self-evaluation processes: Empowering students in learner-centered contexts. *Assessment in Education: Principles, Policy & Practice, 2*, 145–163. doi:10.1080/0969594950020203

Lee, C. (2007). Boundary negotiating artifacts: Unbinding the routine of boundary objects and embracing chaos in collaborative work. *Computer Supported Cooperative Work, 16*(3), 307–339. doi:10.100710606-007-9044-5

Orsmond, P., Merry, S., & Reiling, K. (2000). The use of student derived marking criteria in peer and self-assessment. *Assessment & Evaluation in Higher Education, 25*(1), 23–38. doi:10.1080/02602930050025006

Panadero, E., Andrade, H., & Brookhart, S. M. (2018). Fusing self-regulated learning and formative assessment: A roadmap of where we are, how we got here, and where we are going. *Australian Educational Researcher, 45*(1), 13–31. doi:10.100713384-018-0258-y

Raca, M., & Dillenbourg, P. (2015). Camera-based estimation of students' attention in class. Lausanne. *EPFL, 2015*. doi:10.5075/epfl-thesis-6745

Reider, D., Knestis, K., & Malyn-Smith, J. (2016). Workforce Education Models for K-12 STEM Education Programs: Reflections on, and Implications for, the NSF ITEST Program. *Journal of Science Education and Technology, 25*(6), 847–858. doi:10.100710956-016-9632-6

Sandoval, W. (2014). Conjecture Mapping: An Approach to Systematic Educational Design Research. *Journal of the Learning Sciences, 23*(1), 18–36. doi:10.1080/10508406.2013.778204

Sandoval, W., & Bell, P. (2004). Design-based research methods for studying learning in context: Introduction. *Educational Psychologist, 39*(4), 199–201. doi:10.120715326985ep3904_1

Thompson, K., Gouvea, J., & Habron, G. (2016). *A design approach to understanding the activity of learners in interdisciplinary settings: Environment and Diversity.* Presented at the *International Conference of the Learning Sciences*, Singapore.

Thompson, K., Kanasa, H., & Chapman, S. (2019). Tools to Facilitate Teacher and Student Collaboration in Assessment. In Lund, K., Niccolai, G. P., Lavoué, E., Gweon, C. H., & Baker, M. (Eds.), A Wide Lens: Combining Embodied, Enactive, Extended, and Embedded Learning in Collaborative Settings, 13th International Conference on Computer Supported Collaborative Learning (CSCL) 2019, Volume 2 (pp. 967-968). Lyon, France: International Society of the Learning Sciences.

Timms, M. J., Moyle, K., Weldon, P. R., & Mitchell, P. (2018). *Challenges in STEM learning in Australian schools: Literature and policy review*. Academic Press.

Tousignant, M., & DesMarchais, J. E. (2002). Accuracy of student self-assessment ability compared to their own performance in a problem-based learning medical program: A correlation study. *Advances in Health Sciences Education: Theory and Practice, 7*(1), 19–27. doi:10.1023/A:1014516206120 PMID:11912331

Towler, L., & Broadfoot, P. (1992). Self-assessment in the primary school. *Educational Review, 44*(2), 137–151. doi:10.1080/0013191920440203

ADDITIONAL READING

Bazler, J., & Van Sickle, M. (Eds.). (2017). *Cases on STEAM Education in Practice.* IGI Global. doi:10.4018/978-1-5225-2334-5

Henriksen, D. (2014). Full STEAM ahead: Creativity in excellent STEM teaching practices. *The STEAM Journal, 1*(2), 15.

Liao, C. (2016). From interdisciplinary to transdisciplinary: An arts-integrated approach to STEAM education. *Art Education, 69*(6), 44–49. doi:10.1080/00043125.2016.1224873

KEY TERMS AND DEFINITIONS

Brainstorming: The process of generating as many responses as possible in relation to a prompt. This may be undertaken individually or in a group.

Constructivism: The family of educational theories that conceptualize learning as an active process undertaken by the learner.

Formative Assessment: Assessment that contributes to the learning, that takes places during the learning and is often in the form of immediate feedback.

Interdisciplinary Teams: Teams comprised of experts with overlapping and complementary knowledge and skills gathered to perform a particular function or achieve a particular goal.

Processes: When mentioned in relation to the STEAM sliders, refers to the dominant inquiry methods used in that field (e.g., the experimental method in the sciences).

Scaffolds: Any support that is provided to students to assist in the learning process. This may take the form of a worksheet or handout that breaks down a larger or more complex process. For example, a Venn diagram to assist students to compare and contrast.

STEAM Education: Pedagogical approaches that bring together science, technology, engineering, the arts, and mathematics in an integrated and interdisciplinary fashion.

Chapter 3
Artistic Biotechnology:
A Design Thinking Platform for STEAM Praxis

Krista Marie Stith
https://orcid.org/0000-0002-3274-2581
Ball State University, USA

Rachel Louise Geesa
https://orcid.org/0000-0002-8126-8118
Ball State University, USA

ABSTRACT

Biotechnology, the use of organisms or parts of organisms as tools to support and innovate human wants or needs, plays an extensive role in our daily lives. However, biotechnology is minimally addressed in K-12 educational environments. An issue with the lack of biotechnology in curricula is that we, as end-users, are becoming increasingly defined and dependent on biotechnological innovations. We should be able to think critically and form educated decisions about our medical care, the food we eat, and the biotechnological products we do or do not use. The end-user's acceptance of biotechnological solutions to solve a human want or need also may hinge on the ability of the problem-solver to be artistically creative. This chapter introduces artistic biotechnology and provides three exemplars of K-12 classroom lessons anchored in the Design Thinking Model (DTM) with purposeful art integration. The DTM guides instructors and students through solving ill-defined problems for end-users with artistic biotechnological solutions.

DOI: 10.4018/978-1-7998-2517-3.ch003

INTRODUCTION

Throughout history, humans have depended on the natural world to take care of our needs. The herbalist employed the medicinal properties of plants to care for the sick, the herdsman selectively bred animals to create new generations with desirable characteristics, and the baker and the brewer incorporated yeast that yielded food and drink. A significant similarity amongst the herbalist, the herdsman, the baker, and the brewer are their intentional uses of organisms (plants, animals, and microorganisms) as *tools* for fulfilling human wants or needs.

Fast forward to the 21st century, and we still use plants, animals, and microorganisms to improve the human experience. The research, development, and commercialization of these practices are situated in the STEM-focused field of biotechnology, which has exponentially become ingrained into our daily lives. How we take care of ourselves (i.e., penicillin-class antibiotics), what we eat (i.e., genetically modified foods), and how we power transportation (i.e., biofuels) have been revolutionized with biotechnological solutions. However, contemporary society has long been uninformed of the benefits and complications of continuously using and innovating the natural world to take care of our wants and needs (Ball, 2013; Hoban, 1998; McCaughey et. al., 2019).

Educators are tasked to prepare students for making informed decisions about their health, food, and lifestyle that are contingent on biotechnological capabilities, yet educators often do not address biotechnology curricula in the classroom (Gelamdin, Alias, & Attaran, 2013). We argue that through the implementation of design thinking, educators are better positioned to prepare students in what ways we can continue to make informed decisions through creative, innovative, and resourceful biotechnological approaches. The purpose of this chapter is to provide specific examples of design thinking in STEAM through the lens of biotechnology education. And art integration The authors address the following topics within the chapter: (a) provide evidence of biotechnology as an interdisciplinary field that integrates science, technology, engineering, art, and mathematics; (b) operationally define *artistic biotechnology* as a purposeful integration of art into biotechnology curricula; (c) anchor three artistic biotechnology lessons within the DTM; and (d) provide tools for educators to evaluate cognitive processes (i.e., creative thinking, critical thinking, design thinking). These lessons are based on the teaching experiences of one of the authors who implemented artistic biotechnology activities with K-12 students.

BACKGROUND

STEAM Education

Integrated science, technology, engineering, and mathematics (STEM) education has been a focus of teachers, school leaders, and policy makers in an effort to increase students' interests in STEM fields and better prepare students for success in STEM-based career paths through an integrative approach to teaching and learning. However, we are not solely a STEM society, but a STEAM society where learning should a metacognitive experience that incorporates all content areas. Through the addition of the arts in STEM education, "STEM" becomes "STEAM" as the "A" represents the arts in education. STEAM is still a very broad term though and lacks an operational definition. Colucci-Gray, Burnard, Cooke, Davies, Gray, and Trowsdale, (2017) highlights challenges for STEAM education's acceptance into curricula:

...STEAM retains this lack of educational clarity, indeed adds to it, firstly: by being itself a portmanteau term; secondly, by having varied modalities and associated purposes; and thirdly, because the terms 'art' and 'arts' are also used interchangeably, and often uncritically. Thus, it is unclear if 'art' or 'arts' refer to, for example, the arts, or specific art forms/ practices, or pedagogies used in teaching art subjects or arts and humanities. It is also unclear whether STEAM is intended to imply a reconfiguration of disciplinary relationships, interdisciplinary and/or transdisciplinary conceptions of the subject. Further, it is unclear whether an arts-infused or arts-integrated approach is implied. (p. 8)

Despite the unknowns of what STEAM is, how it should look and how it should be implemented in formal and informal educational settings, there is still usability and value. In our work, we consider a *STEAM-capable individual* as one who can think like a scientist, like a technologist, like an engineer, like an artist, and like a mathematician to solve a human want or need. The incorporation of the arts in traditionally STEM-focused lessons and curriculum provides students opportunities to explore and consider STEAM education as they "incorporate design, computer graphics, performing arts, art as expression, or even creative problem solving when exploring solutions" (Quigley, Herro, & Jamil, 2017, p. 3).

Design Thinking

Within industry and education, design thinking has many conceptualizations (Buchanan, 1992), however within the context of this audience's foci, design thinking is defined as a "systemic, intelligent process in which designers generate,

Table 1. Summary of Design Thinking Model (DTM)

Stage	Description
1.Empathy	Students learn to connect with the end-user. End-users are person(s) with a problem (i.e., a need or want that is known or unknown to the end-user). The purpose of this stage is to gather as much information as possible about the end-user through observations, interviews, or research.
2. Define the Problem	Students develop a specific problem statement from the empathy engendered in the prior stage. The purpose of this stage is to clearly identify a logical goal/problem.
3. Ideate	Students research, generate, modify, and co-construct new versions of ideas. The purpose of this stage is to brainstorm many ideas that may solve the specific problem.
4. Prototype	Students select an idea to create a prototype and justification is provided for the decision. The purpose of this stage is to design/engineer a model of the idea from the abstract to some sort of tangible or representative form
5. Test	Students test the prototypes to evaluate prototype functionality that addresses the problem of the end-user. The purpose of this stage is to return to previous stages to revise and redesign ideas and prototypes. Students will continue to re-test and innovate within the cyclical structure of DTM.

evaluate, and specify concepts for devices, systems, or processes whose form and function achieve clients' objectives or users' needs while satisfying a specific set of constraints" (Dym, Agogino, Eris, Frey, & Leifer, 2005, p. 104). The Design Thinking Model (DTM) provides a linear, yet recursive, five process-focused stage model: empathy, define the problem, ideate, prototype, and test (Plattner, 2010; Cook & Bush, 2017). Table 1 summarizes the five steps students undergo when solving an ill-defined problem for an end-user.

Design thinking is an impactful addition to the PreK-16 curricula to support students in the cognitive processes of problem-solving, critical thinking, and creative thinking (Shively, Stith, & Rubenstein, 2018), engagement and participant buy-in (Lee, 2005), and developments in analytical habits and growth mindsets (Long, 2012). Design thinking has also been identified in the literature within other areas of integrated STEAM learning. Cook and Bush (2017) provide exemplars of students engaged in design thinking with connecting elementary school anatomy and physics in designing a prosthetic limb for an end-user and connecting middle school physics and mathematics in designing a coat for giant.

Design thinking also provides a direction for curriculum design and identifying natural intersections (Henriksen, 2017). Students have opportunities to generate ideas, design, and create models in the DTM, aligning with the art-making process as well. Biotechnology education is a field of transdisciplinary problem solving and, coupled with design thinking, provides educators excellent opportunities to innovate meaningful STEAM experiences in the classroom.

Biotechnology as an interdisciplinary field. To conceptualize artistic biotechnology, we need to understand the underpinnings of biotechnology as an interdisciplinary field. The evolutionary process of biotechnology's incorporation into technology education began in the early 1990's with the creation of a Technological Method Model that categorized bio-related subjects into the field along with communication, production, and transportation (Savage & Sterry, 1990, p. 17). The subsequent reports stemming from Savage and Sterry's (1990) work, *A Conceptual Framework for Technology Education*, reflected biotechnology's progression of complexity through development of a taxonomical categorization. Instead of technology with biological concepts, the modification became an equal partnership of technology and biology within the field of technology education (Wells, 1994, p. 71).

Several definitions of biotechnology have been proposed over the last few decades as the field rapidly developed a revolutionary role in modern society. Savage and Sterry (1990) identified bio-related technologies as applying "biological organisms to make or modify products" that would result in a "higher quality of life for all members of our global society" (p. 17). Their definition includes bio-related technologies as a means "to apply a specific procedure to cure or correct a disease or pathological condition. When the human, plant, or animal output is improper, the necessary corrective measures must be taken" (Savage and Sterry, p. 17).

The U.S. Office of Technological Assessment (1984/1988/1991) and Federal Coordinating Council for Science, Engineering, and Technology [FCCSET] Committee on Life Sciences and Health (1993) define biotechnology as "including any technique that uses living organisms (or parts of living organisms) to make or modify products, to improve plants or animals, or to develop micro-organisms for specific purposes" (p. 60). The taxonometric structure for biotechnology education includes the following: foundations of biotechnology, environment, agriculture, bioprocessing, genetic engineering, biochemistry, medicine, and bioethics (Wells, 1994). See Table 2 for disciplines within biotechnology education.

MAIN FOCUS OF THE CHAPTER[2]

Artistic biotechnology. Biotechnology education serves as a powerful platform for STEAM experiences. Thus, the authors of this chapter operationally define *artistic biotechnology* as a pedagogical approach to purposefully integrate art through the natural intersections of design thinking and biotechnology education. Learning biotechnology with purposeful art integration can provide a platform for educators to motivate students to collaborate with one another, utilize science, technology,

Table 2. Taxonomy for Biotechnology Education (Wells, 1994)

Foundations of Biotechnology	Genetic Engineering
Defining Biotechnology Historical Background Relevant Terms Career Information Social Impacts of Biotechnology	Probing Techniques Genetic Engineering Applications Genetic Code Molecular Biology Techniques Analysis of DNA
Environment	**Biochemistry**
Bioremediation Biological Controls Biotreatment Systems Biorestoration Environment Safety	Enzymology Control and Regulation Proteins Methods of Analysis Carbohydrates
Agriculture	**Medicine**
Tissue Culturing Plants and Animal Applications Agrichemicals Aquaculture Food Science	Molecular Medicine Immunology Genetic Therapeutics Health Care Technologies Social Impact of Medicine
Bioprocessing	**Bioethics**
Fermentation Bioproducts Microbial Applications Separation and Purification Techniques Processing Design: Monitoring and Growth	Principles of Ethics Impacts of Using Biotechnology Potentials of Gene Therapy Patenting of Life Forensics

engineering, and mathematics to enhance their art-making processes, and purposefully strive for critical and creative problem-solving solutions.

The incorporation of art into the biotechnology curriculum can be specific or open-ended. Elements of art (i.e., color, form, line, shape, space, texture, and value) and principles of design are introduced to students. Principles of design, as described by the J. Paul Getty Trust (2011), include the following: balance, emphasis, movement, pattern, proportion, repetition, rhythm, unity, and variety. Both groups of concepts and components used and referred to in the art-making and analysis processes spur conversations and provide educators and students opportunities to consider the aesthetics, historical connections, and production process as it relates to the DTM.

Artistic biotechnology differs from art within biotechnology (referred to here as bioart). According to Mitchell (2012), bioart is a "mode of artistic practice that integrates bioengineered living organisms or tissues into artwork itself" (p. 90). An example would be using bacterial colonies of different colors and textures as the substrate for an illustration. Through bioart, art gallery visitors and viewers of art engage with works while they look, walk, and interact with them. These works of art promote the opportunity for viewers to be present while something is occurring that instant (Mitchell, 2012), and bioart provides viewers the opportunity to consider different forms of art. Artistic biotechnology is an integrative design thinking approach of using plants, animals, or microorganisms as tools to make products and then purposeful consideration of the artful components of the products.

So in what ways does art integrate into biotechnology in a STEAM society? As biotechnology is the use of living organisms, or parts of organisms, as tools to create or innovate a product, the arts are critically important in the creation, innovation, and entrepreneurism of that product. Color, texture, space, and value are examples of factors that one might consider when working in biotechnology to develop a product because the product is designed for the end-user. Consider the artistic expression of the baker as she decorates and a row of beautiful yeast-raised pastries. Consider the visual elements of genetically-modified fruit that may be scaled larger and more colorful next to its organic counterparts at the grocery store. Consider the creativity of pharmaceutical commercials which placates the soul with aesthetically pleasing imagery, calming music, and precise cinematography (while also listing potentially debilitating or fatal adverse side effects of the pharmaceutical product). Human-designed biotechnological solutions to real world problems have been used throughout human history and art has always been tied into those solutions- so in retrospect- STEAM has really been around for a very long time.

Issues, Controversies, Problems

Since Savage and Sterry first identified bio-related technology as one of the major technical content areas, little progress towards implementing biotechnology within technology education and other preK-12 curricula has taken place. In addressing biotechnology in classrooms, Brown, Kemp, & Hall (1998), share the following:

The traditional delivery mechanisms and disciplinary content structures do not appear to serve this emerging content area appropriately. Instructional materials designed to be delivered by teachers in isolation from teachers of other disciplines may detract from the richness and interdisciplinary nature of the content. Disciplinary separation also increases the probability that the content will not be taught well, since relatively few teachers have the background and expertise needed to teach biotechnology. Collaboration across disciplines and integration of content represent natural solutions to this problem. (para 33)

To increase biotechnology in classrooms, support for teachers is critical (Smalley, Hainline, & Sands, 2019) and STEAM-capable educators can implement biotechnology activities with collaborative knowledge, skills, and attitudes. We present three exemplars anchored in the DTM that integrate art into biotechnology education.

Table 3. List of biotechnology challenges

Grade level	Process and organism used	Challenge
3rd grade	Phytoremediation (Plants)	End-user need: A large construction project has led to an oil leak in a local park's soil. The town government has requested the soil to be cleaned up with a biotechnological solution that beautifies the park. Biotechnology: Specific plants are safely introduced to the contaminated site to break down the oil into carbon dioxide and water. Art integration: The town government requests that the biotechnological solution blends with the natural surroundings of the park.
7th grade	Enzymes (Animals)	End-user need: A company that creates household cleaners aims to decrease the toxicity of their cleaning products by releasing a "more natural" solution. Biotechnology: The use of animal-derived enzymes in cleaning products Art integration: The new enzymatic cleaning product packaging must be attractive and competitive in the market.
11th-12th grade	Microbial Fuel Cell (Microorganisms)	End-user need: Homeowner is interested in bioelectricity generation to lower electrical costs and carbon footprint in the home. Biotechnology: Design, construct, and test a microbial fuel cell using electricity generated by microorganisms. Art integration: Prototype must be aesthetically pleasing as it will be displayed in the end-user's home.

Each exemplar overviews the five steps of the DTM with students considering the technological, biological, and artful components to the biotechnological solution per the wants or needs of an end-user. See Table 3 for a list of artistic biotechnology challenges. From the instructional objectives of each lesson, standards were addressed from the Common Core (National Governors Association, 2010), Standards of Technological Literacy (International Technology Engineering Association, 2000/2002/2007), National Coalition for Core Arts Standards (2014), and Next Generation Science Standards (NGSS Lead States, 2013) (See Appendix 1).

Phytoremediation Lesson: 3rd Grade

Phytoremediation is an exciting example of how plants can be used as tools to solve a human want or need in the context of environmental conservation. The United Nations Environmental Program defines phytoremediation as "the efficient use of plants to remove, detoxify or immobilize environmental contaminants in a growth matrix (soil, water or sediments) through the natural biological, chemical or physical activities and processes of the plants" (Erakhrumen, 2007). For example, Brassicaceae species (i.e., Indian mustard) extracts heavy metals from the soil and Salix species (i.e., White willows) extract heavy metals and diesel from soil and water.

In this five day project, third grade students in their science period received instructions and constraints of a phytoremediation design challenge. For the design challenge, the scenario given was that a large construction project has led to an oil leak in a local park's soil. The town government has requested the soil be cleaned up with a biotechnological solution that beautifies the park. Upon completion of this activity, the students were expected to (a) explain the unique characteristics of plants, or parts of plants, that may be used to fulfill a human want or need, (b) design a phytoremediation system for the town government, and (c) evaluate the effectiveness of the prototype to blend into nature.

Empathy. There were three strategies (Interview, Engage, and Research) that supported students to develop empathy for the townsfolk who were impacted by the oil spill. Students conducted interviews with a local conservationist, observed pictures of oil spills, and video-conferenced with a government official to learn about government roles in environmental sustainability.

Define. After familiarizing with oil spills and developing empathy with the townsfolk who have experienced destruction of their park, students collaborated in small groups to define the problem and create a problem statement. The instructor assisted the students by asking what were the goals of the town government and asking them to think about the townsfolks' needs to clean the park. After sharing in small groups, as a class, the problem statement was conceptualized as "How might we develop a method for removing oil contaminates from the soil?"

Ideate. Independently, students sketched their initial brainstorming ideas on individual white boards. Students were encouraged by the instructor to develop as many ideas as possible. If the whiteboards were filled with sketches, students would take pictures of their whiteboards with their school-issued iPads. Students were then paired together to discuss each other's ideas and develop a prototype solution.

Prototype. Using a storyboard approach, each group sketched their ideas of how they would add plants in a park to phytoremediate the soil. On the last day of the design challenge, the class hosted a party and used food items to design the park with their phytoremediation plan represented by candy.

Test. An important consideration is that feedback is received by the end-user. The end-user may have suggestions or additional requests that will require the student to return to a previous stage in the DTM. An administrator of the school served as a representative of the town's government and provided feedback to the students' prototypes.

Project Conclusion. The 3rd grade students appeared to enjoy the action-based activity of creating a 3D model of their prototypes out of food, but the earlier steps were challenging for them. The groups were prone to selecting their first ideas or using the ideas of another group instead of their own solutions. For example, one student shared that his group watched another group build their prototype prior to their own construction because "we didn't want to mess up." The students, teacher, and researcher had rich discussion on what ways they could serve the town's needs by creating a tailored solution to their problem, but there was not necessarily a right or wrong answer. Students expressed enjoyment in learning that plants can do a variety of phytoremediation actions without humans being able to visualize the actions in progress. The students asked many questions on "what else" in nature can help the earth heal from pollution.

Enzymology Lesson: 7th grade

There are many household cleaners available on the market. While many current cleaners are known to be effective in cleaning various surfaces in the home, their ingredients can cause indoor air pollution and be potentially harmful when in direct or chronic contact with humans and pets. In this lesson, students used an artistic biotechnological solution of enzymology to develop a safer household cleaner. Enzymes are derived from plant, microbe, and animal systems that can be used as tools to solve a human want or need in many contexts. Enzymes were used in this design challenge to explore human health and entrepreneurism, however. In this lesson, the students use animal-derived enzymes.

In this one-week project, students in a 7th grade technology education class received instructions and constraints of an enzymology design challenge. In this scenario, a company that creates household cleaners aimed to decrease the potential toxicity of their cleaning products by releasing a "more natural" cleaning product. The company is interested in using enzymes for the cleaning product and also wants a product that has attractive packaging to remain competitive in the market. Upon completion of the design challenge, students were expected to (a) explain how the unique characteristics of animals or parts of animals (i.e., enzymes) may be used to fulfill a human want or need, (b) identify what type of enzymes work to clean various surfaces using examples of the use of enzymes in the industry, (c) test enzymes on various surfaces and then develop a hypothetical all-purpose solution, and (d) evaluate the effectiveness of the prototype to communicate to consumers of the "more natural" product modification.

Empathy. There were three strategies (Interview, Engage, and Research) to develop empathy for both the company that is innovating their cleaning product and consumers who are using chemicals that may be potentially toxic or can lead to health complications. Students conducted interviews with the school custodian and a collegiate researcher, collected data (using some of their general and special senses when safe to do so) to experience different household cleaners as a consumer,

Table 4. Word Dance example

Initial Problem Statement	
How might we [fulfill] a [want/need] for [an end user]?	
Original word	Possible new words
Fulfill ®Create, construct, design, imagine, frame, shape… A want/need ®Cleaner, grease-remover, all-purpose cleaner, spray, paste, glass-cleaner, hydrophobic solution… End user ® A company, a Fortune 500 company, a Mom and Pop shop, an LLC, a not-for-profit…	
New Problem Statement	
How might we [design] an [enzymatic cleaner] for a [cleaning product company]?	

reviewed websites of companies that produce household cleaners, and watched videos about enzymes.

Define. Students used a "Word Dance" strategy to consider what words arranged within their problem statement best represented the desired outcomes.

Language within the problem statement influenced the types of solutions they developed. Two groups requested a thesaurus to help them think about the words. The students generally provided a problem statement similar to, "How might we design an enzymatic cleaner for a cleaning product company for the home?".

Ideate. Students ideated many different ideas to solve the company's want/need. The instructor encouraged wild ideas, to build on the ideas of what others have done before, and go for quantity. Students were prompted to consider the art aspect, particularly the principles and elements of design as the company needed a competitive product on the market. Half of each small group collaborated on the enzyme formula and the other half developed a marketing plan and purposefully attractive packaging. Each group re-formed and had subsequent discussions on how to further innovate the enzymatic formula and the packaging.

Prototype. Students developed an enzymatic formula after learning the necessary content about enzymes (i.e., amylases, cellulases, proteases, lipases), creating a base powder and then adding enzymes. Students used empty containers (i.e., spray bottles, jelly jars) and created labels for the containers using various materials.

Test. Students tested their enzymatic formulas on stains. Table 5 provides examples of enzymatic agents used in the classroom.

Students tested their enzymatic products on surfaces with the stains and rated the "effectiveness" of cleaning on a five-point scale. Students graphed their data in their mathematics period.

Table 5. Enzymatic Testing Stage

Enzymes	Stains	Surface
Amylase	Chocolate, gravy	Cotton cloth, glass, tile
Cellulase	Bananas	Cotton cloth, glass, tile
Protease	Grass, eggs	Cotton cloth, glass, tile
Lipase	Lipstick	Cotton cloth, glass, tile

Project Conclusion. Students learned the content of enzymes in order to create a solution for two end-user groups- a cleaning product company and consumers. The students experienced some challenges in the first few days in understanding how they were to follow a process to design a biotechnological solution. However, the students collaborated well as they were used to small groups and the strengths of their group mates. According to one student, "I liked that we were allowed to work with others on this project. I also like that we all have the same challenge to meet, but we are able to come up with our own answers. I liked doing this project." The testing stage of the enzymes and the designing of packaging for the end-user were more exciting to many of the students and so they hastened through the Empathy, Define, and Ideate stages.

Bioelectricity Generation: 11th-12th Grades

A microbial fuel cell (MFC) is a device that converts chemical energy into electrical energy by the catalytic reaction of microorganisms. The anaerobic microbes required for MFCs are commonly found in wastewater and highly organic soil. When implemented in the classroom, students were requested to find local ponds and collect "mud" with a lot of organic material. Many factors must be taken into account for the fabrication of this microbial fuel cell. A combination of ideal substrate with a plethora of microbial colonies, sustainability of anaerobic conditions, and utilization of available materials must provide a significant electrical current. As this project worked with live organisms, it was essential that all of the biological requirements were met to maintain microbial viability and the technological requirements were met to accurately measure energy output.

In this biotechnology design challenge, a homeowner aspired to decrease his carbon footprint and lower electricity bills. The homeowner had heard of MFCs as an emerging technology and was interested in this approach, however the homeowner has requested that any MFC prototype be aesthetically pleasing for his home. The purpose of this bioprocessing system challenge was to design and test a MFC that would successfully generate enough electricity to power a light source. Students were to be cognizant of the aesthetic nature of the prototype as requested by the end-user. The project took five weeks in an Environmental Education class with 11th and 12th grade students.

Upon completion of this activity, the students were expected to (a) explain the unique characteristics of microorganisms or parts of microorganisms that may be used to fulfill a human want or need, (b) identify what type of organisms works to generate bioelectricity and why, (c) design a microbial fuel cell system that harnesses electricity generated from microorganisms, (d) apply engineering techniques to build a system, and (e) evaluate the effectiveness of the prototype to generate electricity and be aesthetically pleasing to the homeowner.

Empathy. The generation of empathy was encouraged in three strategies (Interview, Engage, and Research) to support students learning about the homeowner and about the need and demand for alternative energy sources. By gaining empathy, students were positioned to better serve the homeowner's needs by creating a tailored solution to his problem. In this class, students conducted interviews with a local alternative energy instructor at the local community college and learned about the innovative approaches to engage in more sustainable energy sources. Students watched online videos from government and conservation organizations. They also read their electric meters for the week at their homes and their school, compiling the data and estimating costs associated with energy. Students familiarized themselves with local, state, and federal policies regarding electricity use. Lastly, students read literature from websites, books, and scholarly journals about new developments in bioelectricity.

Define. The students were guided by the instructor to self-direct their learning about traditional and alternative energy sources and participate in group discussion to share increased interests in bioelectricity. At this point, the students captured findings and defined the problem to create a problem statement. To help the students narrow down information learned in the Empathy stage, they were asked to determine what are some of the goals/desirable outcomes of homeowner. The class agreed upon the problem statement as "How might we develop an aesthetically-pleasing bioelectrical system?"

Ideate. With the problem statement determined, student were encouraged to come up with many different ideas to solve the homeowner's need. Students were reminded to consider wild ideas, build on the ideas of what others have done before, and go for quantity. Sketches were initially drawn individually on paper and then groups of three to four students agreed upon a final solution to build as the prototype. Students provided justification for why they chose the final solution.

Prototype. Students shared with their instructor a list of needed materials to build their prototypes and spent time in and out of class developing microbial fuel cells with their decisions based on functionality and aesthetics. The microbes that generate electrons close to their cell membranes and are excellent sources of bioelectricity may be found in aquatic environments with a lot of organic material and so students who lived near water sources (i.e., ponds and lakes) brought soil substrate to school. Each group developed a unique solution to the problem.

Test. Students tested their prototypes in two approaches. The first approach was to use micrometers to evaluate the effectiveness of the prototype in developing electronic currents to power a load- an LED lightbulb. Data were collected over the course of a week to inform of sustainability. The prototypes were displayed at a science fair event that the school hosts for the community. Students shared their prototypes with members of the community and got feedback on the aesthetics from real home and condo owners.

Project Conclusion. In this biotechnological design challenge, students had to consider the biological, technological, and artistic components for an effective prototype. To provide a few examples of the many solutions they had to make decisions for, students needed to plan how to keep the bacteria alive for an extended period of time, they needed to solve for how chambers would retain their respective aqueous substrates, yet be connected to allow electrons to flow, and they needed to ideate how to get a microbial fuel cell, historically not an attractive technology, to be aesthetically pleasing and functional in a home. All of the groups were able to finish a working prototype and collect data within the time frame, and one very efficient group started creating a second microbial fuel cell in the time available. According to one student who addressed his design thinking during the ideate stage, "we looked at several different methods and decided which method was the most practical, then we based our own design off of that. We had to create our solution for [the end-user] and the original method we saw wouldn't have been good enough." The instructor of the class felt the activity was very useful for the students in encouraging creative thinking and critical thinking.

SOLUTIONS AND RECOMMENDATIONS

The assessment component is critical for educators to argue for continuation of innovative STEAM learning. Therefore, we provide several examples to illustrate how artistic biotechnology, design thinking, critical thinking, and creative thinking may be evaluated in the classroom. These approaches may continue to help build the argument for STEM to STEAM paradigm shifts amongst educational decision-makers.

Artistic Biotechnology Artifacts

In all three exemplars, students concluded with a biotechnological artifacts. While many educators assess the effectiveness of the end-product (Shively, Stith, &

Table 6. Alternative Assessment Strategies

Daily assessor activities	Culminating activities	
Peer assessment Performances Rubrics Journals Portfolios Observations Checklists	1. Put on a fair/festival/conference 2. Take part in debate 3. Give a workshop presentation or group presentation 4. Attend a local conference 5. Exhibit personal portfolios	6. Present project on a social platform 7. Display prototypes 8. Perform a student-written play 9. Develop an improvement plan for local organization 10. Design and market a new product

Source: (Drake & Burns, 2004)

Rubenstein, 2018), we encourage educators to assess students on the design thinking process. For example, with the high school classroom with the bioelectrical activity, one group was successful in generating electricity with their first prototype. A second group worked extremely hard for two weeks and was never able to design a prototype that generated electricity as well as the first group. Should the second group be penalized for their inability to match their peers in electrical output? With quality documentation of the design thinking process, we feel the second group should not be penalized as long as they showed mastery of the content and skills targeted in the activity. Documentation was emphasized in these lessons and students maintained portfolios throughout the experiences.

The portfolio, in particular, is an excellent approach to have students document the design thinking process. A portfolio can involve a description of the problem, information gathered through research about the problem, multiple brainstorm prototype ideas with technological descriptions, a summarization of the biological components (i.e., species of the organisms used), a description of prototype construction and testing, an evaluation of the prototype, and reflection of a redesign. Older students' production processes and materials, aesthetical perspectives, historical references, and criticisms of the work addressing the challenges in-progress and the finished products related to the challenges can be further discussed and described.

As there are many different ways to assess learning. Table 5 provides additional examples of how students can provide evidence to educators of STEAM-capableness.

Rubrics may also be beneficial for educators to measure creativity thinking, critical thinking, and the design thinking modes of STEAM learning. Table 7 provides a rubric to assist educators in assessing creative thinking. Table 8 provides a rubric to assist educators in assessing critical thinking. Table 9 provides a rubric to assist educators in assessing design thinking.

Our hope is that these rubrics may be helpful for educators who are tasked with assessing evidence of growth in modes of creative thinking, critical thinking, and

Table 7. Sample Rubric Anchoring Assessment Criteria on the Definition of Creativity

	Novice	Developing	Expert
Fluency	Students considered one idea.	Students considered several ideas.	Students considered many[a] ideas.
Flexibility	Students considered one type of idea.	Students considered several types of ideas.	Students considered many types of ideas.
Originality	Students developed a common idea that many other students would have suggested and/or replicated an existing idea.	Students developed an interesting idea that several other students would have suggested and/or minimally added onto an existing idea.	Students developed a unique idea that few other students suggested and/or substantially built upon an existing idea in a unique way.
Elaboration	Students added minimal details and improvements to their ideas.	Students added a few details and improvements to their ideas.	Students added many significant details and improvements to their ideas.
Usefulness	Students proposed ideas that may meet the end-user's need's in certain conditions.	Students proposed ideas that would meet the end-user's needs.	Students proposed ideas that would meet the end-user's needs and add significant value to their lives.
Specific creative strategy[b]	Students randomly selected and implemented a creative thinking strategy, and/or they were unable to leverage the strategy to improve their ideas.	Students selected and implemented a creative thinking strategy to develop their ideas. They explained how the strategy supported their creativity.	Students deliberately selected and implemented a creative thinking strategy to develop their ideas. They explained how the strategy supported their creativity.

Source: Shively, Stith, and Rubenstein, 2018

[a]The quantity of ideas may be provided for a given project, or teacher may use general quantities, like "many" or "several" depending on class/project needs.

[b]Teachers should supply the specific strategy they introduced in class. Creative strategies could include brainstorming techniques like reverse brainstorming, attribute listing, and analogical thinking, among others. Other strategies could help focus their ideas, like hits and hot sports or decision matrices.

design thinking. The rubrics are generalized so they may be used for STEAM lessons beyond biotechnology.

Limitations. It is important to note that the sample of students in these artistic biotechnology activities were within a single state, and though the classroom activities provide exemplars of the possibilities of interdisciplinary artistic biotechnology into different classroom content areas, the results cannot be generalized to all K-12 programs. The instructors of these courses routinely work within their content silos and so the collaboration with the researcher was a new experience. Additionally, the depth of exposure of artistic biotechnology for the students varied considerably. In design thinking, the students should also be the ones to identify the problem and create a problem statement, however in these activities, the biotechnological

Table 8. Using Components of Critical Thinking to Develop Assessment Criteria

	Novice	Developing	Expert
Summarizes topic or argument	Does not organize information, learning to inadequate understanding.	Inconsistently demonstrates ability to organize information, leading to inadequate understanding.	Consistently demonstrated ability to organize information, leading to adequate understanding.
Considers previous assumptions	Assumptions are defined, but not explained as having significance to the position.	Assumptions are defined and linked to topic ideas, but not clearly explained or elaborated upon.	Assumptions are defined and linked to topic ideas; students can elaborated on assumptions and discuss implications.
Communicates point of view	Does not identify own position on the issue.	Identifies own position on the issue, drawing support from experience.	Identifies own position on the issue, drawing support from experience, and information not available from assigned sources.
Provides evidence of research	No evidence provided to support argument	Accepts evidence at face-value, even if incorrect, inadequate, or misrepresented to support argument.	Information is gathered from appropriate and credible sources to support argument.
Analyzes data	No analysis of topic. Student only lists or defines concepts of topic.	Demonstrates ability to analyze and make interpretations of topic.	Demonstrates ability to analyze and elaborate on interpretations of topic.
Considers other perspectives and positions	No identification of other perspectives and positions.	Identifies other perspectives and positions.	Identifies and assesses other perspectives and positions.
Draws implications	Cannot explain or testify to the impact of new information.	Explains or testifies to the impact of new information.	Explains the impact of learning new information, making predictions, and generates new ideas.
Assesses conclusions	No reflection of idea evolution on argument development.	Limited reflection of idea evolution on argument development.	Extensive reflection of idea evolution on argument development.

Source: Shively, Stith, and Rubenstein, 2018

solutions as a phytoremediation system, an enzymology system, and bioelectrical system were pre-determined. Students still had opportunities for innovation within these constraints.

The artistc biotechnology lessons in the chapter are to help conceptualize artistic biotechnology in action. Referring to the bioelectricity design challenge as an example, instead of students learning about microbes in biology class, redox reactions in chemistry class, and electricity in physics class, students mastered

Table 9. Sample Rubric Using the DTM[a] as Assessment Criteria

	Novice	Developing	Expert
Empathy	Students did not adequately collect data to explore the end-user's needs. The data may not be clearly presented or not carefully analyzes or synthesized. The data may only represent the students' perspective.	Students explored the end-user's needs through interviews, observations, purposeful engagement, and/or research. Students presented their findings but did not fully synthesize across sources. Students were able to describe how the end-user's needs are unique and present an opportunity.	Student actively and deliberately considered the end-users' needs through in-depth interviews, observations, purposeful engagement, and/or research. Students synthesized across multiple types of data to represent various perspectives. Students were able to describe how the end-user's needs are unique and present an opportunity.
Define	Students presented limited methods to define the opportunity. Students seemed to randomly choose the problem without a provided defense. The problem statement may be too broad to help direct the subsequent stages. Furthermore, there is little evidence that the data collected from the empathy stage was considered.	Students considered several ways to define the opportunity. Students choose the problem statement and could defend their choice. The problem statement may be too broad to help direct the subsequence stages or it may not incorporate the data collected from the empathy stage.	Students considered multiple ways to define the opportunity. The designer integrated the knowledge and gained from the empathy stage into the problem statement. Students carefully choose the problem statement to anchor their work and could defend their choice. The problem statement is meaningful and actionable. The problem statement provides structure for the next stages.
Ideate	Students considered a few different ideas as potential solutions for the opportunity. Students were able to elaborate upon one of their ideas and defend which idea they would use to develop their prototype.	Students considered several different types of ideas as potential solutions for the opportunity. Students were able to elaborate upon one of their ideas and defend which idea they would use to develop their prototype.	Students considered many ways to develop solutions for the opportunity. Students presented unique and flexible ideas to address specific needs. Students were able to elaborate upon a few of their best ideas and defend which idea they would use to develop their prototype.

coninued on following page

Table 9 continued

	Novice	Developing	Expert
Prototype and Test	Students documented a few iterations, changes, and adjustments to the design, experiment, or model. Students were satisfied with their first idea.	Students documented specific iterations, changes, and adjustments to their design, experiment, or model. Students continued to elaborate or adjust their first idea throughout the process.	Students documented in-depth, specific iterations., changes, and adjustments to their design, experiment, or model. Students continued to elaborate or adjust their first idea throughout the process. Students used actual data collection from an authentic audience to make positive changes.
Reflection[b]	Students briefly reflected upon the design process by addressing a frustration or triumph. The reflection includes one idea for future exploration. Students neglect to support their reflection with specific evidence.	Students reflected upon the design process by addressing some of the project's frustrations or triumphs. The reflection includes future avenues for exploration. Students provide some evidence for these reflections.	Students thoughtfully reflected upon the design process by specifically addressing the project stages, frustrations, improvements in processes and products, and future avenues for exploration. Students provide valid evidence for their reflections.

Source: Shively, Stith, and Rubensteins, 2018

Note. DTM= Design Thinking Mode.

[a]For more information on each stage, see Stanford's d.school's webpage for free resources ("Tools for taking action," 2018).

[b]If students used the DTM in the past, they could also reflect upon how current/application demonstrated growth or change from previous projects. This will encourage metacognition and awareness of growth within students.

the content through the application of designing solutions in a real-world context. Studens also came to the realization of how much they simply do not know. "As students design their systems, they may discover different problems, and arrive at several solutions to the problems at hand" (Wells, White, & Dunham, 2000, p. 245). Students endeavored solutions through an interdisciplinary approach and gained greater appreciation for the complexities of biotechnological processes and the importance of art when designing for an end-user.

FUTURE RESEARCH DIRECTIONS

Future research should be explored of how artistic biotechnology emerges through professional learning, implementation in the classroom, and/or impacts on K-12 assessments. Institutions of higher education are beginning to explore opportunities for collaboration between and among disciplines in the sciences and the arts. In 2013, "EcoScience + Art" was created at George Mason University as an opportunity for students and faculty to interact with one another and engage in teaching and research initiatives that promote inter-, multi-, and trans-disciplinary approaches to learning, innovating, and solving complex problems (Ahn, 2015). Students and faculty studying the arts, engineering, humanities, and environmental sciences were brought together to develop sustainable solutions for the future (Ahn, 2015). The integration of the arts with the sciences in this higher education initiative promotes STEAM education and opportunities for new partnerships within the university and the community. Virginia Tech's Design-Based Biotechnological Literacy course (Wells, 2013) prepares pre-service and in-service educators across the world to implement biotechnology in the classroom through technology/engineering design. Institutions of higher education should continue to provide formal professional learning experiences and continue to investigate the natural connections of STEAM and innovative pedagogies to support pre-service and in-service educators.

Educational leaders and policymakers may identify examples of STEAM learning in classrooms, schools, and/or school districts that may be referred to and disseminated with others to enhance familiarities with interdisciplinary STEAM curriculum and pedagogy. The impacts of STEAM learning experiences on student engagement, development of STEAM knowledge, skills, attitudes, and impacts on assessment scores on standardized tests, are valuable to inform future decisions regarding STEAM. What approaches should we continue to take to build the STEAM-capable individual.

CONCLUSION

The goal of this chapter was to (a) provide evidence of biotechnology as an interdisciplinary field that integrates science, technology, engineering, art, and mathematics; (b) operationally define *artistic biotechnology* as a purposeful integration of art into biotechnology curricula; (c) anchor three artistic biotechnology lessons within the DTM; and (d) provide tools for educators to evaluate cognitive processes (i.e., creative thinking, critical thinking, design thinking). Integrating biotechnological constructs and art constructs through the DTM eases the complexity of integrating content, and the inexperience of guiding students in a design-based biotechnological setting, which is a formidable topic for teachers. Artistic biotechnology curricula anchored within the DTM provides a guide for supporting classroom learning as students will solve ill-defined problems using STEAM knowledge, skills, and attitudes.

ACKNOWLEDGMENT

This research received no specific grant from any funding agency in the public, commercial, or not-for-profit sectors.

REFERENCES

Ahn, C. (2015). Ecoscience + art initiative: Designing a new paradigm for college education, scholarship, and service. *The STEAM Journal*, *2*(1), 1–10. doi:10.5642team.20150201.11

Ball, K. M. (2013). Meet Your meat: The argument for increasing education and public outreach regarding the regulation and safety of animal biotechnology. *Food and Drug Law Journal*, *68*(3), 281–307. PMID:24640611

Brown, D. C., Kemp, M. C., & Hall, J. (1998). On teaching biotechnology in Kentucky. *Journal of Industrial Teacher Education*, *36*(4), 44–60.

Buchanan, R. (1992). Wicked problems in design thinking. *Design Issues*, *8*(2), 5–21. doi:10.2307/1511637

Colucci-Gray, L., Burnard, P., Cooke, C., Davies, R., Gray, D., & Trowsdale, J. (2017). *Reviewing the potential and challenges of developing STEAM education through creative pedagogies for 21st learning: How can school curricula be broadened towards a more responsive, dynamic, and inclusive form of education?* London: British Educational Research Association.

Cook, K. L., & Bush, S. B. (2017). Design thinking in integrated STEAM learning: Surveying the landscape and exploring exemplars in elementary grades. *School Science and Mathematics*, *118*(3-4), 93–103. doi:10.1111sm.12268

Drake, S. M., & Burns, R. C. (2004). *Meeting standards through integrated curriculum*. Alexandria, VA: Association for Supervision and Curriculum Development.

Dym, C. L., Agogino, A. M., Eris, O., Frey, D. D., & Leifer, L. J. (2005). Engineering design thinking, teaching, and learning. *Journal of Engineering Education*, *94*(1), 104–120. doi:10.1002/j.2168-9830.2005.tb00832.x

Erakhrumen, A. A. (2007). Phytoremediation: An environmentally sound technology for pollution prevention, control and remediation in developing countries. *Educational Research Review*, *2*(7), 151–156.

FCCSET Committee on Life Sciences and Health. (1993). *Biotechnology for the 21st century: Realizing the promise*. Washington, DC: Federal Coordinating Council for Science, Engineering, and Technology.

Gelamdin, R. B., Alias, N., & Attaran, M. (2013). Students' and teachers' perspectives on biotechnology education: A review on publications in selected journals. *Life Science Journal, 10*(1), 1210–1221.

Henriksen, D. (2017). Creating STEAM with design thinking: Beyond STEM and arts integration. *The STEAM Journal, 3*(1), 1-11. Retrieved from https://doin.org/10.5642/steam.20170301.11

Hoban, T. J. (1998). Trends in consumer attitudes about agricultural biotechnology. *AgBioForum, 1*(1), 3–7. Retrieved from http://www.agbioforum.org/v1n1/v1n1a02-hoban.htm

International Technology Engineering Association. (2007). *Standards of Technological Literacy*. Reston, VA: ITEEA. Retrieved from https://www.iteea.org/42511.aspx

J. Paul Getty Trust. (2011). *Elements of Art*. Los Angeles, CA: The J. Paul Getty Museum. Retrieved from https://www.getty.edu/education/teachers/building_lessons/elements_art.pdf

Lee, H.-J. (2005). Developing a professional development program model based on teachers' needs. *Professional Educator, 27*(1-2), 39–49.

Long, C. (2012). Teach your students to fail better with design thinking. *Learning and Leading with Technology, 39*(5), 16–20.

McCaughey, T., Budden, D. M., Sanfilippo, P. G., Gooden, G. E., Fan, L., Fenwick, E., ... Hewitt, A. W. (2019). A need for better understanding is the major determinant for public perceptions of human gene editing. *Human Gene Therapy, 30*(1), 36–43. doi:10.1089/hum.2018.033 PMID:29926763

Mitchell, R. (2012). Simondon, bioart and the milieu of biotechnology. *INFLeXions, 5,* 68-110. Retrieved from http://www.inflexions.org/n5_mitchellhtml.html

National Coalition for Core Arts Standards. (2014). *National core arts standards*. Dover, DE: State Education Agency Directors of Arts Education. Retrieved from https://www.nationalartsstandards.org/

National Governors Association. (2010). *Common core state standards*. Washington, DC: Council of Chief State School Officers & National Governors Association Center for Best Practices.

NGSS Lead States. (2013). *Next generation science standards: For states, by states.* Washington, DC: The National Academies Press.

Plattner, H. (Ed.). (2010). *An introduction to design thinking process guide.* Stanford, CA: Institute of Design at Stanford. Retrieved from https://dschool-old.stanford.edu/sandbox/groups/designresources/wiki/36873/attachments/74b3d/ModeGuideBOOTCAMP2010L.pdf

Quigley, C. F., Herro, D., & Jamil, F. M. (2017). Developing a conceptual model of STEAM teaching practices. *School Science and Mathematics, 117*(1-2), 1–12. doi:10.1111sm.12201

Savage, E., & Sterry, L. (Eds.). (1990). *A conceptual framework for technology education.* Reston, VA: International Technology Education Association.

Shively, K., Stith, K. M., & Rubenstein, L. D. (2018). Measuring What Matters: Assessing Creativity, Critical Thinking, and the Design Process. *Gifted Child Today, 41*(3), 149–158. doi:10.1177/1076217518768361

Smalley, S., Hainline, M. S., & Sands, K. (2019). School-based agricultural education teachers' perceived professional development needs associated with teaching, classroom management, and technical agriculture. *Journal of Agricultural Education, 60*(2), 85–98. doi:10.5032/jae.2019.02085

U.S. Congress, Office of Technological Assessment. (1984). *Commercial biotechnology: An international analysis.* Washington, DC: Author. Retrieved from http://ota.fas.org/reports/8407.pdf

U.S. Congress, Office of Technological Assessment. (1988). *New developments in biotechnology: U.S. investment in biotechnology—Special report.* Washington, DC: U.S. Government Printing Office. Retrieved from http://ota.fas.org/reports/8840.pdf

U.S. Congress, Office of Technological Assessment. *(1991). Biotechnology in a global economy.* Washington, DC: U.S. Government Printing Office. Retrieved from http://ota.fas.org/reports/9110.pdf

Wells, J., White, K., & Dunham, T. (2000). *Technology education biotechnology curriculum.* Morgantown, WV: Biosens.

Wells, J. G. (1994). Establishing a taxonometric structure for the study of biotechnology in secondary school technology education. *Journal of Technology Education, 6*(1), 58–75. doi:10.21061/jte.v6i1.a.5

ADDITIONAL READING

Davis, G. (2004). *Creativity is forever*. Dubuque, IA: Kendall Hunt.

Plattner, H. (2010). *Bootcamp bootleg*. Stanford, CA: Design School Stanford, Palo Alto.

Starko, A. J. (2010). *Creativity in the classroom: Schools of curious delight* (4th ed.). New York, NY: Routledge.

KEY TERMS AND DEFINITIONS

Alternative Assessment: Non-traditional approaches to understand student mastery of a topic.

Artistic Biotechnology: Pedagogical approach to integrate art through the natural connections of design thinking and biotechnology.

Biotechnological Solutions: The use of plants, animals, and/or microorganisms as an approach to solve a human want or need.

Biotechnology: Any technique that uses plants, animals, and/or microorganisms as tools to make or modify products.

Creative Thinking: Multidimensional construct of fluency (quantity of ideas), flexibility (different types of ideas), elaboration (building upon ideas), originality (uniqueness of ideas), and usefulness of ideas.

Critical Thinking: Multidimensional constructs of analysis, evaluation, and reflection.

Design Thinking Model (DTM): An iterative process to solve problems through empathy, define, ideate, prototype, and test.

Interdisciplinary: Development in the knowledge, skills, and attitudes in two or more areas of learning.

STEAM-Capable Individual: One who can think like a scientist, like a technologist, like an engineer, like an artist, and like a mathematician to solve real-world problems.

Chapter 4
Science Cartoons (SciToons):
A Model for Integrating Arts With STEM

Oludurotimi Adetunji
Brown University, USA

Roger Levine
Independent Researcher, USA

ABSTRACT

The 21st century offers many technological advancements resulting from cutting-edge research being conducted by scientists and engineers at higher education institutions, industries, and national labs. As exciting scientific research is being conducted, the need to address the challenges of inspiring and engaging new learners in science, technology, engineering, and mathematics (STEM) fields continues to grow. The authors present the Science Cartoons (SciToons) program as a model for making science more accessible through the integration of the arts with STEM (STEAM). The SciToons model is based on a Multimedia Learning Theoretical Framework (MLTF), which facilitates co-creation of knowledge and takes into account the contributions of STEM and non-STEM experts. The SciToons model combines this knowledge and expertise with art, animation, high-quality multimedia, and storytelling in the development of science videos that are engaging to a broad audience.

INTRODUCTION

One of the challenges of the 21st century for the science, technology, engineering and mathematics (STEM) communities in the private and public sectors, particularly in the United States, is persistence, retention and sustainability of *human capital*

DOI: 10.4018/978-1-7998-2517-3.ch004

engaged in STEM fields. This challenge is not new, but the critical economic and innovation roles provided by the STEM workforce, coupled with an increase in global engagement over the last two decades, have increased the need for more human capital to address these current and future issues facing STEM communities (National Science Board [NSB], 2010).

The most recent demographic data available from the United States show that women and students from Historically Underrepresented Groups (HUG) are poorly represented within the current pools of graduates with a bachelor's, master's or doctoral degree in the physical sciences (National Center for Science and Engineering Statistics [NCSES], 2017). The recognition of STEM's human capital problem by major education and research institutions in the United States has resulted in the creation of initiatives or programs, offices or centers of excellence charged with finding ways to engage and sustain women and students from Historically Underrepresented Groups (HUG) in STEM fields. For example, it is now common to find programs and centers focused on the mentoring and supporting first-generation college students or high-need students, who are disproportionately underrepresented in STEM fields, through degree completion. The approaches that have been developed to address this challenge are mainly centered on STEM-focused curricula or domain-specific interventions, such as the Meyerhoff program (Stolle-McAllister, Domingo, & Carrillo, 2010).

Arts and STEM are two powerful domains for understanding natural and social phenomena. STEM represents a focus on scientific concepts and applied research; STEAM represents STEM plus the arts (Maeda, 2013), yet more broadly can include music and dance, visual arts and design, language and literature, narrative and storytelling, and digital media, among other humanistic disciplines. STEAM uses the creative process as a vehicle to critically explore, challenge, contextualize, interpret, and communicate scientific concepts and research applications. Therefore, through the lens of STEAM, groups of STEM experts and new learners work collaboratively to produce new knowledge, in some cases resulting in improved ways to communicate scientific concepts and research.

The purpose of this chapter is to describe an innovative and inclusive approach for integrating the Arts with STEM through the development and dissemination of science animation videos called SciToons. The animations are designed to be engaging and accessible to audiences with diverse academic backgrounds.

The chapter begins with a brief description of the models (Cognitive Load Theory, the Cognitive Theory of Mulitmedia Learning (Mayer, 2014), and the Multimedia Learning Theoretical Framework (Adetunji and Levine, 2016)) upon which the production of SciToons are based. This is followed by a brief review of video animation research, including studies examining specific components of effective videos. As SciToons are produced by diverse, interdisciplinary groups, the composition of these teams and the approach used in developing scripts for the animation (storyboard generation, voice-overs, animatics, graphics and visual generation, and animation of the storyboard) follow, along with a discussion of how these animations are disseminated. Next, evidence of their impacts on both viewers and the students involved in the creation of the video are provided. The chapter concludes with a brief discussion of future research directions and a summary/ conclusion.

BACKGROUND

SciToons are distributed through YouTube™. A Google™ search for "STEM YouTube videos" will produce over 200 million results. Since anyone who signs in to YouTube™ can post a video, the quality (and accuracy) of videos found here varies tremendously. This creates problems for both the consumers of the videos (*How do I choose a good video?*) and the producers of these videos (*How do I produce a good video?*). SciToons' technical underpinnings include design principles based on Cognitive Load Theory, the Cognitive Theory of Multimedia Learning, and the Multimedia Learning Framework, as well as the limited research on YouTube™ videos. This research is presented following a discussion of these theories and framework and the interdisciplinary approach used to produce SciToons.

Cognitive Load and Cognitive Theory of Multimedia Learning

Cognitive Load Theory (Sweller, 1988; Sweller, 1989; Sweller, Van Merrienboer & Paas, 1998), and the Cognitive Theory of Multimedia Learning (Mayer & Moreno, 2003) provide explicit principles for instructional media creation based on a model of the cognitive processes involved with human memory and learning. Information processed through verbal and visual memory enters working memory, which has a limited storage capacity (Miller, 1956). For learning to occur, an individual must choose which information should be added and integrated with existing knowledge.

Visual and audio information are processed differently in each channel's working memory, both of which contribute to a mental model. So, videos must attend to the cognitive load they create and optimize the balance resulting from the different types of information processing (called essential processing, extraneous processing, and generative processing) critical for learning. Essential processing is concerned with providing the information essential for understanding the constructs being demonstrated. This can be done via the pre-training principle (providing information before the video, as with provision of a glossary of terms) and the modality principle (presenting words aurally rather than visually).

Given the limited capacity of working memory, extraneous processing should be minimized. This can be accomplished through adherence to a number of design principles: the coherence principle (presenting only relevant information, "weeding out" unnecessary or distracting information); the redundancy principle (avoiding narration that merely repeats prose presented visually); the signaling principle (providing an overview of the organizational structure of the material prior to its presentation); the temporal contiguity principle (matching narration with visual presentations); and the spatial contiguity principle (placing labels close to the reference objects). Generative processing can be facilitated through application of the interactivity principle (providing opportunities for viewers to control elements of the animation) (Yue, Kim, Ogawa, Stark, & Kim, 2013).

Multimedia Learning Theoretical Framework

The MLTF takes into account that most people will be considered to have no or limited science background. Therefore, the framework argues that it is important to take into consideration the approach by which "scientists" and "non-scientists" solve problems when developing an initiative or program for engaging a broad audience (technical or non-technical) in STEM fields. The MLTF also introduces the concept of "intermediate" audience. The intermediate audience represents an individual with science background other than the science domain currently of interest. For example, an advanced physics student with no or limited biology background working on a neuroscience problem would be a member of an "intermediate audience". This audience will bring their past knowledge and ideas to a different scientific domain. Materials that are developed should recognize these different problem-solving styles to maximize their utility to these diverse audiences. The framework also centers the use of narrative for developing science communication, teaching, and learning.

Video Research

Video animations are being used for a variety of educational purposes, with varying effectiveness. Health-related educational uses (of demonstrated effectiveness) have included preparing children for non-sedated Magnetic Resonance Imaging, enhancing the children's knowledge and reducing their anxiety (Szeszak et al., 2016) and in-patient fall prevention, reducing the incidence of falls, particularly for patients over 65 years of age (Nakagami-Yamaguchi et al., 2016).

They are more commonly used for general educational purposes. Animations incorporated into the curriculum can enhance learning as well as create positive attitudes about different STEM areas (Yisa & Ojiaku, 2016). Mohammed Jawad's doctoral dissertation (2018) developed a tool that combined art, animation, and code-sharing, enhancing students' ability and interest in computer science. They have also been shown to enhance 11th graders learning of chemistry (Al-Balushi & Al-Hajri, 2014). Similarly, in a study involving 15 teachers, animations and supplementary activities resulted in significant improvements in science thinking skills in fourth and fifth graders in contrast with the use of textbooks and pictures (Barak & Dori, 2011). Byers (2017) reviewed five different studies assessing the effectiveness of animation within education, concluding the effectiveness of animation was established in these studies, dealing with learning about shot put styles, general science education, electronics learning, geometry, and Energy and Environmental Design. Much of this research involved small numbers of classes or teachers, making it difficult to attribute effects solely to the animations in contrast to other factors associated with instruction.

In a well-designed study, entailing a complex, year-long design process including a website review, an expert working group of subject matter experts, an on-line experimental pilot study, think-alouds with patients, the involvement of community consultants, animation development by an outside contractor, and focus groups were used to develop a 10-minute video on whole genome sequencing for patients, research participants, and the general public. In comparison with written information, the animated video was rated easier to understand. However, knowledge gains were not significantly different from individuals who received comparable information in a written format, but were significantly greater than those receiving no information (Sanderson et al., 2016).

The quality of the animations will, of course, have a huge impact on their effectiveness. Morrison, Tversky, et al., (2002) noted some animations that were too complex and distracting to be effective. Since that time, a great deal of research has taken place to identify important considerations for the development of successful animations.

Based on work on Massive Open Online Courses' (MOOCs) videos, video length was shown to be important: videos should be short. As videos exceed six minutes in length, students are more likely to disengage (Guo, Kim, & Robin, 2014). Similar results were noted by Saurabh and Gautam (2019) in a study of their YouTube™ channel (which has thousands of subscribers and millions of views). They noted that lengthy videos can bore the audience while short videos cannot always effectively explain complex concepts. Although the popular videos they studied averaged between 10-15 minutes in length, they noted that most were watched for only 30-50% of their length.

Ten Hove and van der Meij (2015), using a popularity metric as an indicator of quality, analyzed 75 videos to determine how unpopular declarative knowledge development videos differed from popular ones. Factors associated with quality included higher production quality (resolution and less background noise), a greater frequency of static pictures, frequent presentation of both dynamic and static pictures, short text presented on screen, background music, and a greater speaking rate. However, the association of background music is inconsistent with the coherence principle. This may reflect a finding specific to the declarative knowledge videos or study design issues. Shoufan (2019) questioned the popularity rating they employed and created a different metric (Video Cognitive Value) based on the number of likes and dislikes and the total number of views.

Through a survey of 428 students, Shoufan (2019) showed the most powerful reason for students liking (or disliking) a video was whether or not they understood the material that was presented. Poor video quality was strongly associated with dislikes, as was speaker quality (incompetent, unfriendly, unenthusiastic narrators). Length of the video was also a factor, with "too long" videos a source of dislikes. An analysis of 105 digital logic design videos was performed to determine whether or not each of the ten different cognitive theory of multimedia learning principles were employed. The presence or absence of six production style characteristics as well as length, talking speed, gender, and whether or not the narrative spoke unaccented English was also assessed. Significant predictors of video cognitive value including pretraining (explaining new system components prior to explaining the whole system), modality (presenting words as narration rather than on-screen text), spatial contiguity, and embodiment (drawing a diagram instead of explaining an existing diagram). Features significantly associated with video cognitive value were the length and whether the narrator sounded like a native English speaker. The importance of narration style was also noted by Meyer (2008) who showed that a conversational style of narration (the personalization principle) promotes engagement.

Digital Video Production by Diverse Teams

The creation of a SciToon is an *interdisciplinary* process. The integration of knowledge and skills among the group members collaborating on a task can be labeled multidisciplinary, interdisciplinary or transdisciplinary (Begg & Vaughan, 2011). Multidisciplinary activities involve experts from different disciplines collaborating to perform a task. It is considered the weakest of these collaborative approaches because of the minimal amount of integration (Borrego & Newswander, 2010). In interdisciplinary activities, team members inform and shape each other's perspective, working together to adapt and apply a novel approach to solve a problem or explore scientific phenomena or produce a creative work. This integration of participants' perspectives and creative approaches is a key part of the SciToons model.

The use of project teams, comprised of individuals with different skills and interests, to collaboratively produce digital scholarship, such as videos and other research, teaching, and learning products, is clearly not unique to SciToons. The

Digital Youth Network (digitalyouthnetwork.org) works with young people in low-income areas, providing access to the digital tools and expertise of individuals that help them to develop the skills and knowledge to succeed in today's world. Similarly, notable projects are found elsewhere, including HIVE in New York City (hivenyc.org) and Bridge21 in Dublin (bridge21.ie). However, many of these projects' approaches tend to be multidisciplinary, with distribution of the resulting products not considered to be a major priority.

Interdisciplinary teams have successfully combined STEM and the arts and humanities to produce a variety of novel, creative works, including videos, music, and other digital products in efforts to deal and also visualize the 2014-16 Ebola outbreak in western Africa. Photos of healthcare workers were taken and attached to their protective suits to humanize them; songs were written and played on the radio to help teach people how to avoid infection (National Academies Press [NAP], 2018). Conversely, the failure of engineers to appreciate the cultural context of cook stoves has greatly hindered their adoption among the world's poor. The lack of understanding the cultural context of food preparation with the appropriate technology has resulted in minimal impact of technology on a problem (smoke from cooking and heating) with substantial negative health impacts on individuals and on the environment (Parthasarathy, 2016).

The SciToons creation model is an interdisciplinary model which engages both new learners (novice), experts, and artists in the development of content on emerging scientific research or concepts not well known among the public so that they can be presented in a way that aligns with principles comprising the Cognitive Theory of Multimedia Learning, and Cognitive Load Theories, which posit that pictures, sounds, and words communicate better than words alone and the Multimedia Learning Theory Framework, which recognizes the diverse audiences comprising the intended audiences for these videos. The basic principle of the SciToon model is that both new learners and experts, who are diverse with respect to gender and race/ethnicity comprise a SciToons Creation Group (SCG). The STEM experts are faculty or domain practitioners (investigators or researchers). The new learners (or novices) are students or individuals with limited knowledge in a particular domain. Generally, the experts in a STEM domain are new learners in animation and creative domains and vice versa.

A subset of the SCG members are selected as the core individuals (team) working on a particular project. The core teams include at least one STEM expert (usually a STEM faculty member and/or practitioner) and one illustration/animation expert (usually these are advanced students or practitioners), and two other SCG members. Through an iterative feedback process, experts and new learners work together to co-develop the script and gather feedback from the entire SCG along the way. They

Figure 1. A schematic image of infusing arts into STEM to produce visually appealing products, such as SciToons videos

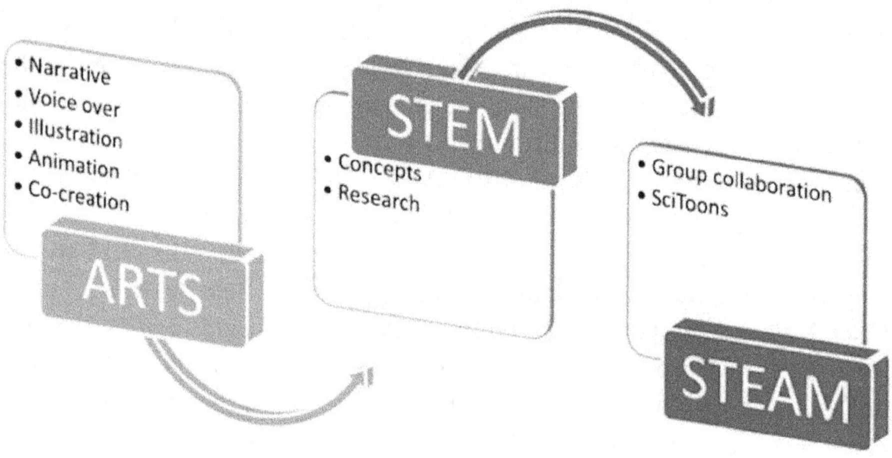

also co-create the storyboard and animations in conjunction with SCG members who are the experts in illustration and animation.

MAIN FOCUS OF THE CHAPTER

The SciToons model allows for the exchange and integration of ideas from both the STEM and arts domains as the SCG members work on co-authoring the narrative and visual representation of the scientific research or concept, resulting in an illustration of the narrative, and animation and production of an explainer video. The SciToons model epitomizes an innovative science, technology, engineering, arts and mathematics (STEAM) project. Figure 1 shows the SciToons' process of integrating arts with STEM.

THE SCITOONS CREATION PROCESS

Important elements comprising the SciToons process are listed below:

I. SciToons Creation Groups
II. Scaffolding
III. Iterative Review Process

Table 1. Gender and Racial/ethnic Composition of Student SCG members, by year

Year	Female	Male	Asian	Black	Hispanic	White
2014	14	4	3	0	2	13
2015	12	4	4	3	0	9
2016	13	3	5	5	0	6
2017	17	3	6	4	1	9
2018	30	6	7	7	1	21

IV. SciToons Production Pipeline
V. Distribution Platforms

SciToons Creation Groups (SCG)

The Science Creation Group (SCG) composition is a key component of the SciToons model. A small (3-5) member subset of the SCG members are assigned to individual SciToons projects in a particular scientific domain. In the initial stages, the project ideas are on broad scientific research or concepts such as *charge transport in organic materials, composition and characterization of graphene, self-assembly in macromolecules and advancement in alternative sources of energy.*

The larger SCG members then work together to develop an *outline* of the domain concepts, drawing from the collective background knowledge of the group. As the outline is developed, important concepts are highlighted and group members begin the process of asking basic questions as they try to decide what type of emerging scientific research or concept they would like to explore further. The outline then becomes a draft to which the group members provide feedback. Subsequent revisions are made until a majority of the SCG members are satisfied with the outline.

The students comprising the SCG teams have consistently represented a group of gender and racially/ethnic diverse individuals. (See Table 1.) We believe that this diversity can only have positive impacts on the final products. We encourage future research into the extent of diversity in other STEAM projects as well as the impact of diverse teams on the quality and utility of the products produced.

Feedback is collected from each group member during each script writing, voice over, storyboard, and animation phases. Group members are able to contribute their thoughts and ideas to these project phases during group meetings or through the SciToons' shared cloud drive where the projects' files are saved. Group meetings can be characterized as a very lively discussion among non-STEM students, STEM students, and faculty who are interested in contributing their expertise to the development of co-created SciToons videos.

SCG members have reported that those discussions are very inspiring for them, facilitating their understanding of the science behind a given topic and identifying any misunderstandings that they might have. The meetings also gave SCG members the opportunity to better communicate scientific concepts or research to other members, and learn from them. The SciToons' group composition allows groups of diverse experts and learners to co-create an effective way of communicating a STEM concept or research using creative process in the arts such as storytelling, music, and visualization. The co-creation process helps to introduce arts and design concepts in STEM settings. This also provides the opportunity for arts and design experts to contribute their knowledge and insights, enhancing their own STEM education as well as the audience's through STEAM projects.

Scaffolding: A Roadmap for Script Development

The outline is used as a guide for scaffolding the initial draft story or script. However, unlike "scaffolding", as used by psychologist and instructional designer Jerome Bruner (Wood, Bruner & Ross, 1976), referring to a process in which experts model how to solve a problem, without actively participating in solving the problem by being active participants throughout the problem solving steps, the authors use the term to describe the SciToons process for script development. The initial script is co-developed by a set of the SCG members and reviewed and revised by the entire SCG. Having scripts co-created by both experts and novices allows for the reduction of jargon while keeping the scientific concepts and ideas correct. Through the scaffolding process knowledge exchange dominates the lively discussions. Questions typically include:

- "What is the narrative hook?"
- "What are the potential applications of this research?"
- "Why would an average person want to watch this video?"
- "What would a better example for explaining these concepts look like?"

The discussion allows the SCG members to work on creating an outline that is robust and clear and serves as a roadmap for the development of the script. Table 2 shows the outline for the Gene Drive project. This outline was used to develop a script which was revised multiple times before the SCG members agreed on a final script. The title of the final video published was called *Population Engineering: Gene Drive by CRISPR-CAS9* and it can be accessed via the SciToons' YouTube™ channel (https://www.youtube.com/SciToons).

During the process of creating the script, some outlined topics were eliminated after they were discussed and considered by the SCG members. For example, the

Table 2. Initial SciToons' project outline for the Gene Drive project

I. Introduction A. What is the narrative? What can we use to maintain the audience's interest in the story? B. The term "gene drive" applies to any gene or technique that boosts the inheritance of a specific gene in order to increase its frequency in a given population. C. Gene drives can occur naturally. Some genes have developed mechanisms that increase their likelihood of being passed down to greater than 50%. 1. Consider explaining Mendelian genetics/traditional 50% inheritance pattern. D. Scientists are now making use of these naturally occurring gene drives and a recently developed molecular technique called CRISPR-Cas-9 to introduce targeted mutations into organisms that will spread throughout the population. 1. Link to CRISPR video [https://www.youtube.com/watch?v=9q8AGst7KiY]? 2. Link to mutations video [https://www.youtube.com/watch?v=Jl1Aa7Iq4tc]?
II. Tech background A. CRISPR (or a different natural mechanism) is used to insert a modified gene into one chromosome of an organism. Also inserted is a cas-9 enzyme and a guide RNA. B. Before being passed to offspring, the guide RNAs direct the cas-9 to cut the Wild-Type gene. Then, the cell repairs the cut wild type Wild-Type chromosome by copying both the modified gene along with the cas-9 enzyme and the guide RNA. C. Now the organism has two copies of the same gene, and will pass the modified chromosome onto future generations who will undergo the same process.
III. Ethical questions A. Should we be tinkering with genetics this way? 1. What other possibilities does this open up? What would it be used for? Eugenics, disabilities, etc. (though looking at humans may be too much) B. Testing on human populations where humans live — how to gain consent, etc.? C. How to prevent business/militaries from using as weapons or misusing the technology?
IV. Future use/challenges A. What if gene drive "jumped" to another species? B. How to contain to one area? Ex: mosquitoes can fly/travel to other countries and even continents C. Could it be used for other diseases with insect vectors? D. Adaptations within pathogen species/unintended consequences 1. Malarial parasites moving to another vector if living in mosquitoes stops being an option.

final script did not address the ethical concerns regarding the potential impact of the gene drive technology on humans and the future challenges section did not make the final cut. The process of grappling with what to include or not include was not unique to the development of the Gene Drive video. In fact, because the final script cannot be more than one and a quarter pages long, to keep the length of the video to about 5 minutes, the SCG members always have to debate what should be included or eliminated from the outline as the script is being finalized.

The narrative hook was the first sentence of the first paragraph of both the initial and final Gene Drive script. The narrative hook was very well conceived and remained unchanged throughout the iterative review process. The SCG members assigned to the project presented their progress to the overall SCG, who provided feedback on the different versions of the initial draft and there was universal agreement that the hook was excellent.

Table 3. Introductory paragraphs of the initial and final Gene Drive script

Initial Script: *Has anyone ever told you that you have your father's eyes and your mother's nose? Well, you inherit these traits from your parents through chromosomes. Human cells are diploid; that is, they have two copies of each chromosome, one from each parent. Now, let's imagine two people,* ***both with brown eyes****. Person A is a homozygote, which means that they have two chromosomes both containing the* ***same gene*** *— in this case, both chromosomes carry the* ***brown-eyed gene****. Person B is a heterozygote, which means that each chromosome carries a different version of the gene. Here, one chromosome carries* ***a brown eyed gene and one carries a green eyed gene****.*
Final Script: *Has anyone ever told you that you have your father's eyes and your mother's nose? Well, you inherit these traits from your parents through chromosomes. Human cells are diploid; that is, they have two copies of each chromosome, one from each parent. Now, let's imagine two people, who* ***both have chin dimples****. Person A is a homozygote, which means that they have two chromosomes both containing the* ***same version of the gene*** *— in this case, both chromosomes carry the* ***chin dimple gene****. Person B is a heterozygote, which means that each chromosome carries a different version of the gene. Here, one chromosome carries* ***a chin dimple gene and one carries a smooth chin gene****.*

Both the initial and finalized introductory paragraphs clearly communicated the scientific concepts envisioned in the outline correctly. However, as the SCG members discussed other factors such as illustration and aesthetics of the Gene Drive video being developed, it became clear that another example instead of ***brown eyes*** (which can be easily and prominently illustrated) might be better. That discussion led to the idea of ***chin dimples*** as a better alternative to eye color. This change made the storyboard easier to create.

The last paragraph of SciToons' script usually ties up the whole narrative so that the important application of a scientific concept or research is effectively communicated to the audience. The SCG members were able to emphasize three very important points in the final version that the initial draft did not communicate effectively. The first point was ethical concerns of using gene drive technology; the second was the

Table 4. Last paragraph of the initial and final Gene Drive script

Initial Script (third paragraph): While it is unlikely that Gene Drive would ever actually be used on humans, there have been proposals to use it on species like mosquitoes to eliminate malaria, yellow fever and more. Scientists will need to be very careful when using gene drives because, if released into the wild, they could alter entire species with a few generations, and reducing genetic diversity can decrease a species' ability to adapt to changing environments. These are the questions that scientists, lawmakers, and the public will need to begin making in the very near future.
Final Script (third paragraph): While it is unlikely that Gene Drive would ever actually be used on humans **for ethical reasons,** there have been proposals to use it on mosquitoes to eliminate the many diseases they carry, including malaria, yellow fever, and Zika. There are **several environmental concerns** that need to be addressed because if released into the wild, gene drives could alter entire species within a few generations. Additionally, there are ethical concerns. Should humans be altering genomes in such a dramatic and impactful way? What should the **legal barriers** surrounding gene drives be, and how will countries come to agreements on regulations? These are the questions that the public, scientists, and lawmakers will need to begin addressing in the near future.

environmental concern; and the third was the legal barriers surrounding gene drive technology. These are very crucial points that need to be debated and clarified in order for this technology to be widely accepted and used broadly. They are also points that the overall SCG agreed were effectively communicated in the finally produced video titled *Population Engineering: Gene Drive by CRISPR-CAS9*.

SciToons Production Pipeline: Steps

The production processes are guided by the iterative reviews, which allow SCG members to provide on-going feedback at each step of the production process. The first step is the development of the storyboard.

Storyboard

During this phase, simple sketches of the script are developed concurrently as the script writing is being finalized. The SCG members working on the storyboard (illustrators) provide their initial sketches to the scriptwriters. In some cases, an individual can be both an illustrator and scriptwriter. The reason for working concurrently at this stage is to help make the script more animate-able. As the SCG members go through this step, scripts are often minimally edited so that the illustration can be conveyed clearly and easily. This step ensures that the language used in the script is comprehensible to a diverse audience and precise. This also allows animators to give feedback on wording, structure and story. Storyboard concepts that can be communicated purely visually (and don't require explicit verbalization in the script) are sometimes discovered.

Storyboards are reviewed by the content experts (who are part of the SCG members assigned to a specific project) to make sure that the visual elements are accurately depicted or scientifically sound. Figures 2, and 3 show an example of a storyboard developed from the Gene Drive script showing character depicting Person A and Person B referenced in the second paragraph of the final Gene Drive script. Both of these individuals also have the ***chin dimples***.

Voice-Over

During the voice-over phase, a recording of the finalized script is made. Usually three takes are made and brought to a group meeting. The SCG members then provide feedback on pronunciation, intonation, and pitch to the voice-over artists. These are usually professionals, but students have also served as voice-over artists. The artist then incorporates the feedback in recording the final voice-over which is

Figure 2. This shows the storyboard of the characters representing Person A (homozygote) and Person B (heterozygote) with chin dimples

Figure 3. This shows the process of division of Person A (homozygote) and Person B (heterozygote) gametes

Figure 4. This shows the process of gamete replication for Person A (homozygote) and Person B (heterozygote)

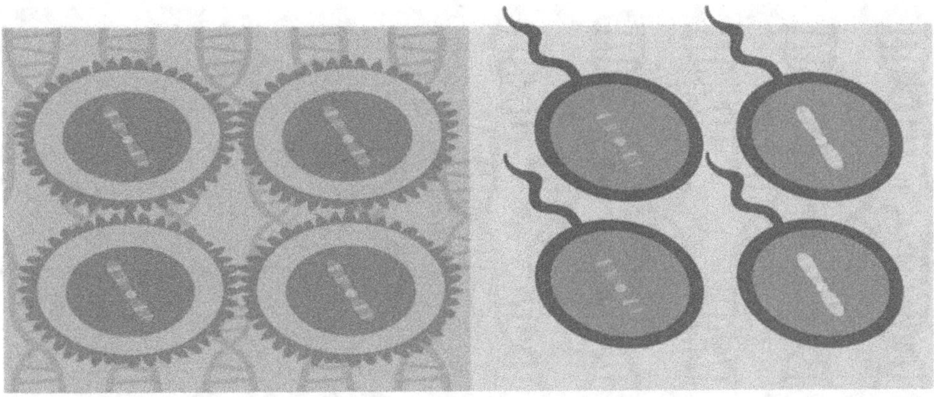

subsequently reviewed and approved by the SCG. The voice-overs give animators an idea of the timing and pacing of the script.

Animatics

As more characters are being illustrated, successive sections of a storyboard are made into a preliminary version of the SciToon. The preliminary animated version of the video is known as an animatic. The animatics can be set to match the voiceover and to demonstrate the general flow of the final video. Figures 3 and 4 show how the characters' gametes, either eggs or sperm, are produced by a process of division and then replicate, producing six gametes containing the chin dimple gene and two containing the smooth chin gene.

Graphics and Visuals

The development of the final graphics to be animated in the video are created during this phase. All aspects of the characters are developed by the SCG assigned to the project, and reviewed by the overall SCG with feedback provided as needed. At this stage everything that will ultimately appear in the final SciToons video: characters, props, backgrounds, settings, color palettes, and typography is developed. The SCG members also discuss and determine the "look and feel" of the final product. Discussions about digital and geometrical feel of the characters and their placement within a scene is considered. There is also a conversation about whether 2D or 3D illustrations would be better to effectively communicate the script ideas and to "bring the scene to life." Figure 5 shows an example of color background palettes (blue)

Figure 5. A blue background color palette showing seven characters featured in the SciToons video titled Weather and Climate

and different animal characters developed for the weather and climate SciToons. This SciToons video effectively communicated the difference between weather and climate making the concept easily accessible to a general audience. This video has been used by middle school teachers to introduce both concepts to their students before they listen to meteorologists visiting their schools. The video can also be accessed via the SciToons' YouTube™ channel.

Animation

The collections of storyboard visuals that have been created are animated to match the pacing of the voiceover. This step takes the most time, as animating a five-second scene can take anywhere from ten minutes to multiple hours. The graphics are usually created in Adobe Illustrator™ and then animated in Adobe After Effects™. As each scene (usually one or two paragraphs) is being finalized, the clip is reviewed by the

SCG members and iterative feedback is provided in order to have full participation in guiding the look and feel of the final video.

Dissemination Process

Public dissemination of science requires time and effort that could be devoted elsewhere. Energy that could be devoted to research and the pursuit of research funding is a scarce resource. Eagleman (undated) notes several excellent reasons for public dissemination. These include inspiring critical thinking, stemming the flow of bad information, informing public policy, clarifying what science is (and isn't) and sharing the beauty of the scientific pursuit with others. In order to facilitate wide distribution, the SciToons program has relied heavily on YouTube™. In addition, SciToons has coordinated with a number of organizations that have featured selected SciToons on their websites (such as the National Oceanic and Atmospheric Agency (NOAA) and the National Institutes of Health (NIH) as part of their National DNA Day website) and on their social media platforms.

Completed videos are published on the SciToons' YouTube™ channel. The channel currently has over 6,000 subscribers. After a video is published, a short announcement is posted on other SciToons' Instagram™ and Twitter™ accounts. When a new video announcement is made on the Twitter™ account, it usually generates retweets by social media influencers in the science communication such as the @ AmoebaSisters. The videos have also been tweeted by or added to the Facebook™ page of organizations such as TEDEd™ (TEDEd, 2018). Figure 6 shows a recent retweet by @AmoebaSisters of one of SciToons' post. In 2018 TEDEd featured a similar post about the SciToons' Echolocation video on their Facebook™ page. The TEDEd Facebook™ SciToons Echolocation post recorded 568 like and loves and generated 32 comments and 180 shares on Facebook. The ability to generate code for embedding YouTube™ content on webpages has allowed other science media sites to add SciToons to their site.

The creation and release of a SciToon seems to enhance the viewing of other SciToons. This "multiple viewing phenomenon," in which viewers of a SciToon look at other videos from the SciToon YouTube™ channel was first noticed by the authors in August, 2014 (Adetunji, 2015). This effect has also been noted since by others (Saurabh & Gautam, 2019). More importantly, evaluation data from Creation Group Member surveys and feedback from users, along with detailed data on dissemination, can be a major part of future research.

Figure 6. Retweet by STEM social media influencers

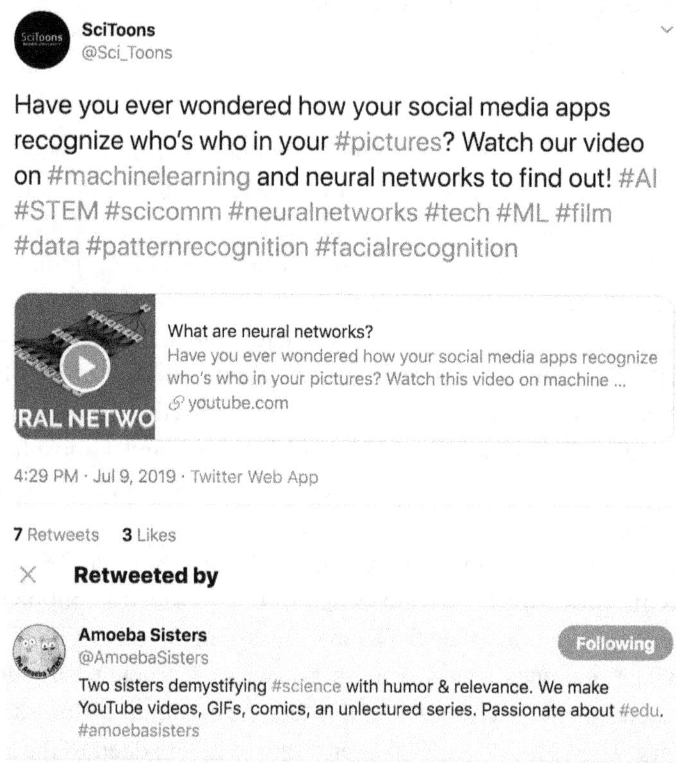

SCITOONS' IMPACT

As of November, 2019, there have been over one million (1,060,331) views of the 32 SciToons that have been published via YouTube™. The number of views per SciToon ranged from 400 - 257,000, with over 100,000 views for four of the SciToons (*How Do Solar Cells Work?*, *How Do We See Color?*, *Echolocation*, and *What is Solar Energy?*). Over the last two years, on average, a new SciToon reaches at least 1,000 views within 6 months of its publication date on YouTube™. A SciToon that has gained more than 1,000 views after 6 months typically reaches at least 10,000 views within one year of its publication date. A good example of a video poised to reach 10,000 views within one year of its publication date is the SciToon titled, *What are Neural Networks?*, published on March 27, 2019. As of July 2019, this video has been viewed over 3,000 times. The video introduces the science behind facial recognition, artificial intelligence, machine learning, and neural networks to non-technical audience. The recent social media coverage of facial recognition, such as the FaceApp™ that allows individuals to see a projection of how their face will look at an older age, or a reflection of how their face looked when they were younger, using photos uploaded from a smartphone has more than likely helped to increase the public interest in the video and learning about the science underlying the app. The authors project that this video will become one of the most viewed videos on neural networks on YouTube™, ranking among the top page neural network Google search results by the end of 2020. Google Analytics™ has provided data about the gender, ages, and location of viewers. Males were more likely than females to watch these videos (64.1% vs. 35.9%), but a substantial proportion of viewers are female. Viewers were most likely to range in age between 18 - 24 (accounting for 41.1% of the views) or between 25 - 34 (comprising 32.1% of the views). Only 4.1% of the views were by children aged 13 - 17, but this still represents nearly 40,000 views. About one-sixth of the views were by adults between 35 - 44. People 45 and over accounted for 6.3% of the views. These videos have been viewed in 108 different countries, with nearly half of the viewers in the United States (30.6%) and India (17.9%).

Assessing the impact of SciToons on learning of both the viewer and student involved with their production is a challenge. Traditional education program evaluation designs, such as random assignment to treatment and control conditions, for informal education experiences (such as video viewing) are rarely feasible. Accordingly, The National Academies encourages the use of evaluation incorporating multiple types of evidence in the assessment of programs integrating the humanities and arts with science, technology, engineering, math, and medicine (STEMM), particularly in situations when randomized controlled trials are neither feasible nor practicable. Qualitative evidence is likely to provide more information about impacts. Evidence of impacts is developed in stages; anecdote can be the start of a pathway toward knowledge (NAP, 2018).

Impacts on Viewers

Although future research is planned for assessing the learning impacts of SciToons, there have been assessment of other types of impacts. For example, the after viewing the *Conductive Polymers* SciToon, impacts were assessed through a survey of 34 high school students participating in the Brown Science Prep Program. After viewing this video, they completed a brief, voluntary survey which included items asking about their level of agreement with various statements about the SciToon. Nearly all felt the SciToon was interesting (94%) and easy to understand (88%). Most felt that they learned a lot (88%) and that watching it was a good use of their time (88%). Nearly all would recommend it to friends who like science (94%), but less than half (44%) would recommend it to friends who do not like science (Adetunji & Levine, 2015).

Table 5. Selected positive comments provided by viewers on the SciToons YouTube™ channel

Videos	Comments from viewers
How Do Solar Cells Work?	"Best explanation from the internet yet!" "Great video, thank you.:)" "This was so helpful I am using it on my science project for school."
Cosmic Microwave Background (CMB)	"Please continue making such superb videos" "Videos such as this renews my interest in science and learning in general. Thank you very much" "Great video. That was the best explanation of the CMB I have found yet." "Seriously awesome video. Well-paced and understandable explanation. Using it for class tomorrow!"
What is Solar Energy?	"Great stuff! Nice work. The public really needs to see more of this." "That was so helpful!! I am going to ace my science quiz." "Superb - simple yet effective explanation."
Conductive Polymers	"This helped thanks a bunch!!" "It is a good explanation with animations. Thank you." "Amazing animation with a beautiful explanation cleared it all Thank you"
Color by Size: Quantum Dot	"Best video ever" "Wooo one of the best videos." "Amazing video!!! keep up the awesome content!"
What Is Molecular Self Assembly?	"Thanks for presenting a cutting-edge topic in knowledge that is not trendy yet." "Congratulations on such a well-thought-out, easy-to-understand animation on molecular self-assembly! Excellent work!"
Echolocation	"Finally found out how echolocation is really goes!!!"

Table 5 shows a sample of comments voluntarily provided online by viewers on the SciToons YouTube™ channel for selected videos. The authors infer that these comments are a good reflection of the positive contribution that SciToons is making to the public's understanding of science.

Several SciToons videos are now ranked number one or two on Google™ or YouTube™ searches. These videos include *Conductive Polymers*, *Echolocation*, *How Do We See Color?* and *How Do Solar Cells Work?*. These is a strong indicator of the quality of the videos and their educational value; comprehension of the video being one of the major reasons that an individual likes a video (Shoufan, 2019).

Impacts on SCG Members

A brief electronic survey was emailed to students who were SCG members in 2014-2015. Ten of the 23 students responded. All of the respondents (half of whom were non-science majors) agreed that participation in SciToons improved their ability to understand contemporary science issues and to explain scientific principles to others; most felt participation either moderately or greatly enhanced these abilities. Additionally, all of the non-science major respondents reported their SciToons experience either slightly or moderately increased their likelihood of pursuing a STEM-related career (Adetunji & Levine, 2015).

In March 2019, students and alumni from several SCGs created a public event to share about the history of SciToons and their experiences being a part of the SciToons project (Adetunji, 2019). The event encouraged audience participation and engagement by having the members of the public use their cellular phones to access a URL in a web browser for online polls using the Kahoot™ platform. SCG members who had been involved in the various stages of script writing, storyboarding, and animations walked the audience through the process. In addition, they provided some online demonstrations of the graphics and illustration software involved. Figure 7 highlights attempts to get the word out to the public to attend and talk to students and faculty involved in SciToons. Students representing the audiences of STEM experts/novice learners, arts and creative design experts/novice learners also presented on their perspectives and impacts from their experiences participating in a SCG and creating a SciToon. These presentations were followed by a panel of STEM faculty who participated in the creation of a SciToon based on a scientific concept in their field or on emerging research in their lab. They presented on their perspectives and impacts from their experiences collaborating with the students and participating in a SCG and creating a SciToon. At the close of the event the students asked the audience to vote for types of scientific concepts and emerging research that they would be interested in learning more about in future SciToons.

Figure 7. Advertisement for SciToons Behind the Scenes event

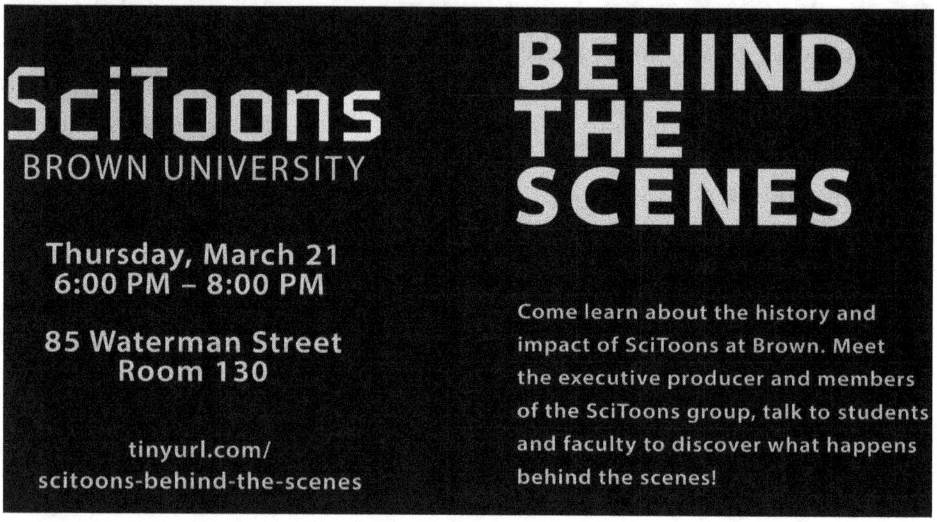

FUTURE RESEARCH DIRECTIONS

In addition to future assessments related to diversity and inclusion and learning impact, there is further intention to continue exploring effective communication of science to broad audiences. In order to better understand what words to avoid when communicating science to non-technical audiences, a qualitative analysis of the evolution of the different SciToons project scripts from the initial to final drafts will be conducted. In addition, the authors plan to explore tools such as Zooniverse or Amazon Mechanical Turk™ (MTurk) (https://www.mturk.com) to engage participants in a longitudinal impact study of SciToons videos. The Zooniverse (https://www.zooniverse.org/) has been used widely internationally by researchers in multiple fields to allow volunteers to participate, and contribute to research projects, and MTurk is now a well-respected platform for researchers to conduct online experimental research. One of the authors (RL) has an extensive background in conducting survey research, invaluable expertise and experience that will be consulted as the authors develop the impact study protocols.

CONCLUSION

Diverse learners are engaged in STEM through the SciToons production; diverse audiences learn about STEM through viewing these videos. The production process incorporates arts into STEM through the interdisciplinary efforts of diverse teams comprised of scientists and STEM students and non-STEM students who engage in creative processes including narrative design, illustration, music and animation to co-create videos. The use of iterative feedback has greatly increased the likelihood of content understanding by these learners. The SciToon creation process can be adopted into college and university courses to enable students to experience the co-creation and iterative processes as part of class projects. This could be especially beneficial for those seeking to integrate art into their STEM curricula, potentially fostering an enhanced appreciation of the Arts by STEM students and an enhanced appreciation of STEM by students involved in the Arts.

REFERENCES

Adetunji, O. O. (2015). UMRI Alumni: From Characterization of High Temperature Solution Growth Cr 2+:CdSe to Developing Videos to Enhance Metacognition and Diversity in Materials Science and STEM (Science, Technology, Engineering, and Mathematics). *MRS Online Proceedings Library Archive, 1762*. doi:10.1557/opl.2015.188

Adetunji, O. O. (2019). *SciToons Behind the Scenes*. Presentation, Brown University, Providence, RI. Retrieved July 19, 2019, from https://repository.library.brown.edu/studio/item/bdr:934244/

Adetunji, O. O., & Levine, R. (2015). Developing Effective STEM Animations: Application of a Multimedia Learning Theoretical Framework. *Journal of Research in STEM Education, 1*(20), 106–124.

Adetunji, O. O., & Levine, R. (2016). Toward a New Model of Science Learning, Teaching, and Communication. *MRS Advances, 1*(56), 3709–3714. doi:10.1557/adv.2016.105

Al-Balushi, S. M., & Al-Hajri, S. H. (2014). Associating animations with concrete models to enhance students' comprehension of different visual representations in organic chemistry. *Chemistry Education Research and Practice, 15*(1), 47–58. doi:10.1039/c3rp00074e

Barak, M., Ashkar, T., & Dori, Y. J. (2011). Learning science via animated movies: Its effect on students' thinking and motivation. *Computers & Education, 56*(3), 839–846. doi:10.1016/j.compedu.2010.10.02

Begg, M. D., & Vaughan, R. D. (2011). Are Biostatistics Students Prepared to Succeed in the Era of Interdisciplinary Science? (And How Will We Know?). *The American Statistician, 65*(2), 71–79. doi:10.1198/tast.2011.10222

Borrego, M., & Newswander, L. K. (2010). Definitions of Interdisciplinary Research: Toward Graduate-Level Interdisciplinary Learning Outcomes. *Review of Higher Education, 34*(1), 61–84. doi:10.1353/rhe.2010.0006

Byers, F. (2017). A Critical Evaluation of the Effectiveness of Animation within Education. *Selected Computing Research Papers, 6,* 15-19.

Eagleman, D. M. (2013). Why public dissemination of science matters: A manifesto. *The Journal of Neuroscience, 33*(30), 12147–12149. doi:10.1523/JNEUROSCI.2556-13.2013

Guo, P. J., Kim, J., & Rubin, R. (2014). How Video Production Affects Student Engagement: An Empirical Study of MOOC Videos. *Proceedings of the First ACM Conference on Learning @ Scale Conference*, 41–50. 10.1145/2556325.2566239

Maeda, J. (2013). Stem+ art= steam. *The STEAM Journal, 1*(1), 34. doi:10.5642team.201301.34

Mayer, R. (2008). Applying the science of learning: Evidence-based principles for the design of multimedia instruction. *Cognition and Instruction, 19*, 177–213. PMID:19014238

Mayer, R. (2014). Cognitive Theory of Multimedia Learning. In R. Mayer (Ed.), The Cambridge Handbook of Multimedia Learning (pp. 43-71). Cambridge, UK: Cambridge University Press. doi:10.1017/CBO9781139547369.005

Mayer, R. E., & Moreno, R. (2003). Nine ways to reduce cognitive load in multimedia learning. *Educational Psychologist, 38*(1), 43–52. doi:10.1207/S15326985EP3801_6

Miller, G. (1956). The magical number seven, plus or minus two: Some limits on our capacity for processing information. *Psychological Review, 63*(2), 81–97. doi:10.1037/h0043158 PMID:13310704

Mohammed Jawad, H. (2018). *Exploring how integrating art & animation in teaching text-based programming affects high school students' interest in computer science* (Doctoral dissertation). Retrieved from Eastern Michigan University Digital Commons@EMU: https://commons.emich.edu/theses/916

Monn, M. A., Weaver, J. C., Zhang, T., Aizenberg, J., & Kesari, H. (2015). New functional insights into the internal architecture of the laminated anchor spicules of Euplectella aspergillum. *Proceedings of the National Academy of Sciences of the United States of America, 112*(16), 4976–4981. doi:10.1073/pnas.1415502112 PMID:25848003

Nakagami-Yamaguchi, E., Fujinaga, K., & Batard, A. (2016). The effect of an animation movie for inpatient fall prevention: A pilot study in an acute hospital. *Safety in Health*, *2*(3). doi:10.118640886-016-0014-9

National Center for Science and Engineering Statistics. (n.d.). *Women, Minorities, and Persons with Disabilities in Science and Engineering*. Retrieved July 19, 2019, from https://ncses.nsf.gov/pubs/nsf19304/

National Science Board. (2010). *Preparing the Next Generation of STEM Innovators: Identifying and Developing the Nation's Human Capital*. Retrieved from https://www.nsf.gov/nsb/publications/2010/nsb1033.pdf

Parthasarathy, S. (2017). Grassroots innovation systems for the post-carbon world: Promoting economic democracy, environmental sustainability, and the public interest. *Brooklyn Law Review*, *82*, 761–787.

Sanderson, S. C., Suckiel, S. A., Zweig, M., Bottinger, E. P., Jabs, E. W., & Richardson, L. D. (n.d.). Development and preliminary evaluation of an online educational video about whole-genome sequencing for research participants, patients, and the general public. *Genetics in Medicine*, *18*(5), 501–512. doi:10.1038/gim.2015.118

Saurabh, S., & Gautam, S. (2019). Computers and mobile phones are the top two devices for viewing videos. *Computers & Education*, *128*, 145–158. doi:10.1016/j.compedu.2018.09.003

Shoufan, A. (2019). Estimating the cognitive value of YouTube's educational videos: A learning analytics approach. *Computers in Human Behavior*, *92*, 450–458. doi:10.1016/j.chb.2018.03.036

Sweller, J. (1988). Cognitive load during problem solving: Effects on learning. *Cognitive Science*, *12*(2), 257–285. doi:10.120715516709cog1202_4

Sweller, J. (1989). Cognitive technology: Some procedures for facilitating learning and problem-solving in mathematics and science. *Journal of Educational Psychology*, *81*(4), 457–466. doi:10.1037/0022-0663.81.4.457

Sweller, J., Van Merrienboer, J. J., & Paas, F. G. (1998). Cognitive architecture and instructional design. *Educational Psychology Review*, *10*(3), 251–296. doi:10.1023/A:1022193728205

Szeszak, S., Man, R., Love, A., Langmack, G., Wharrad, H., & Dineen, R. A. (2016). Animated educational video to prepare children for MRI without sedation: evaluation of the appeal and value. *Pediatric Radiology*, *46*(12), 1744. doi:10.100700247-016-3661-4

TEDEd. (n.d.). In *Facebook*. Retrieved July 22, 2019, from https://www.facebook.com/TEDEducation/posts/are-bats-really-blind-not-exactly-bats-use-a-special-process-called-echolocation/2218872224792638/

Ten Hove, P., & van der Meij, H. (2015). Like it or Not. What Characterizes YouTube's More Popular Instructional Videos? *Technical Communication (Washington), 62*(1), 48–62.

Tversky, B., Morrison, J. B. & Betancourt. (2002). Animation: can it facilitate? *International Journal of Human - Computer Studies, 57*(4), 247–262. doi:10.1006/ijhc.2002.1017

Women, Minorities, and Persons with Disabilities in Science and Engineering: 2019 NSF - National Science Foundation. (n.d.). Retrieved July 19, 2019, from https://ncses.nsf.gov/pubs/nsf19304/

Wood, D., Bruner, J. S., & Ross, G. (1976). The role of tutoring in problem solving. *Journal of Child Psychology and Psychiatry, and Allied Disciplines, 17*(2), 89–100. doi:10.1111/j.1469-7610.1976.tb00381.x PMID:932126

Yisa, N. C., & Ojiaku, F. C. (2016). Effectiveness of Computer Animation on a Progressive Learning Achievement of Secondary School Biology Students in Niger State, Nigeria. *International Journal of Education and Evaluation, 2*(4), 72-81.

Yue, C., Kim, J., Ogawa, R., Stark, E., & Kim, S. (2013). Applying the cognitive theory of multimedia learning: An analysis of medical animations. *Medical Education, 47*(4), 375–387. doi:10.1111/medu.12090 PMID:23488757

ADDITIONAL READING

Adetunji, O. O., & Renoe, S. D. (2017). Assessing Broader Impacts. *MRS Advances, 2*(31–32), 1681–1686. doi:10.1557/adv.2017.136

Brame, C. J. (2015). Effective educational videos. Retrieved July 17, 2019, from http://cft.vanderbilt.edu/guides-sub-pages/effective-educational-videos/

Jaeger, A. J., & Wiley, J. (2014). Do illustrations help or harm metacomprehension accuracy? *Learning and Instruction, 34*, 58–73. doi:10.1016/j.learninstruc.2014.08.002

Kim, S., Yoon, M., Whang, S.-M., Tversky, B., & Morrison, J. B. (2007). The effect of animation on comprehension and interest. *Journal of Computer Assisted Learning*, *23*(3), 260–270. doi:10.1111/j.1365-2729.2006.00219.x

Mayer, R. E. (Ed.). (2005). *Cambridge handbook of multimedia learning*. Cambridge University Press. doi:10.1017/CBO9780511816819

Mayer, R. E. (2008). Applying the science of learning: Evidence-based principles for the design of multimedia instruction. *The American Psychologist*, *63*(8), 760–769. doi:10.1037/0003-066X.63.8.760 PMID:19014238

National Academies of Sciences, Engineering, and Medicine. (2018). *The integration of the humanities and arts with sciences, engineering, and medicine in higher education: Branches from the same tree*. Washington, DC: The National Academies Press. doi:10.17226/24988

Orús, C., Barlés, M. J., Belanche, D., Casaló, L., Fraj, E., & Gurrea, R. (2016). The effects of learner-generated videos for YouTube on learning outcomes and satisfaction. *Computers & Education*, *95*, 254–269. doi:10.1016/j.compedu.2016.01.007

Shoufan, A. (2019). Estimating the cognitive value of YouTube's educational videos: A learning analytics approach. *Computers in Human Behavior*, *92*, 450–458. doi:10.1016/j.chb.2018.03.036

KEY TERMS AND DEFINITIONS

Co-Creation: This refers to the process by which expert and non-expert members of the SciToons Creation Group members collaborate on the co-development of script, storyboard, and animation.

Historically Underrepresented Groups (HUG): This refers to diverse persons whose demographic intersections (ability, color, ethnicity, gender, race, sexuality, sex) historically have not been well represented.

Integration: This term refers to the process of infusing the arts and humanistic domains into STEM fields.

Interdisciplinary: This term refers to collaborators from multiple academic fields informing each other's approach towards a novel method or product that integrates knowledge of collaborators.

Multidisciplinary: This term refers to collaboration of persons from multiple academic fields such as chemistry, humanities and engineering.

Multimedia Learning Theoretical Framework (MLTF): This model facilitates co-creation of knowledge and takes into account the contributions of STEM and non-STEM experts on an interdisciplinary multimedia project.

SciToons: This term refers to Science Cartoons.

SciToons Creation Group (SCG): Refers to the STEM and non-STEM students and faculty who are involved in the production of SciToons videos.

Section 2

STEAM Education in Schools and School District Initiatives

Chapter 5
Becoming STEAM:
Perspectives From School Leaders

Kelli Thomas
University of Kansas, USA

Douglas Huffman
University of Kansas, USA

ABSTRACT

This chapter shares a brief history of the STEM to STEAM movement, shares two case descriptions drawn from the perspectives of leaders in two school districts in which schools adopted a STEAM focus, describes challenges and opportunities associated with implementation of a STEAM initiative, and proposes five features to consider when implementing models to becoming a STEAM-focused school or school district. The five features drawn from analysis of the two cases are intentional efforts by school districts to gain buy-in; adequate time for teacher learning and planning through authentic and relevant professional development; community connections, real-world and problem-based or project-based; mutual decision-making and support between teachers and administrators; and budget planning and allocation.

INTRODUCTION

The world in which we live is constantly changing. In our increasingly dynamic global economy, it becomes critical for educational systems to focus on the goals of helping students learn to think for themselves, analyze relationships among concepts and disciplines, reason about broader challenges, and solve problems confronting an ever more global society. These goals have developed overtime in the United States

DOI: 10.4018/978-1-7998-2517-3.ch005

through a national focus on improving education in the sciences and mathematics that was prompted in the 1940s following World War II. The forming of the National Science Foundation (NSF) in 1950 marked a paradigm shift in support for education initiatives in science, technology, engineering, and mathematics. The educational focus on these disciplines has evolved overtime to recognize and acknowledge the interconnectedness of the four separate fields. The acronym STEM is now widely used and understood in education to represent the integration of science, technology, engineering and mathematics.

More recently, the arts have been added to STEM initiatives in an attempt to enhance the creative and innovative aspects of solving problems as well as a means for motivating a more diverse range of students to pursue careers in STEM fields. To date, there is not a consensus on a precise meaning of "the arts" in STEAM. Some view the arts as the fine arts and design, language arts and more recently as architecture. In 2019, the United States Congress passed the Carl D. Perkins Career and Technical Education (CTE) Act, formally recognizing architecture as a STEM subject. According to a statement by Robert Ivy, FAIA, EVP/Chief Executive Officer of the American Institute of Architects (AIA), "The CTE is intended to encourage a more diverse workforce and see that the promise of design as the synthesis of art and science are fulfilled through education," (AIA, 2018)

Because this new conceptualization of science, technology, engineering, arts and mathematics, known as STEAM, is still in its infancy there are many questions about how these different disciplines inter-relate, what it means to add the arts to STEM, what the A or the arts includes, and how schools can implement a STEAM-based education. This chapter shares a brief history of the STEM to STEAM movement, shares two case descriptions drawn from the perspectives of leaders in two school districts in which schools adopted a STEAM focus, describes challenges and opportunities associated with implementation of a STEAM initiative, and proposes five features to consider when implementing models to becoming a STEAM-focused school or school district. The five features drawn from analysis of the two cases are:

- Intentional efforts by school districts to gain buy-in from teachers and school administrators;
- Adequate time for teacher learning and planning through authentic and relevant professional development;
- Community connections, real-world and problem-based or project-based;
- Mutual decision-making and support between teachers and administrators; and
- Budget planning and allocation.

BACKGROUND

History of the STEM to STEAM Movement

The Rise of STEM

More than 50 years ago, astronaut Neil Armstrong stepped onto the moon's surface and uttered the now famous words, "that's one small step for a man, one giant leap for mankind." In the decades preceding the moon landing on July 20, 1969, international events including World War II and the launch of the satellite Sputnik into space by the Russian government, prompted the United States government to begin taking an active role in supporting science. Prior to the end of World War II, then President of the United States, Franklin D. Roosevelt requested recommendations from Vannevar Bush, the Director of the Office of Scientific Research and Development, for how the work of the office could continue during peacetime "for the improvement of the national health, the creation of new jobs, and the betterment of the national standard of living," (Roosevelt, 1944). Bush's 1945 report to President Roosevelt, *Science: The Endless Frontier*, was a catalyst for the formation of the National Science Foundation (NSF). On May 10, 1950 President Harry S. Truman signed the bill creating the NSF and among other directives, "the law directed the agency to encourage and develop a national policy for the promotion of basic research and education in the mathematical, physical, medical, biological, engineering, and other sciences; to initiate and support basic scientific research in the sciences; and to evaluate the scientific research programs undertaken by agencies of the federal government," (Mazuzan, 1994). The increased focus on and support for innovation in STEM fields was evidenced by continued advancement across various industries. The 1970s and 1980s ushered in many firsts in STEM fields including: NASA's Space Shuttle Program with the launch of the first orbital spaceflight, cellphones, personal gaming systems from Atari, personal computers from IBM and Apple, email communications, development of the network of networks that became the World Wide Web, and medical advances like the permanent artificial heart transplant. These innovations are among the pioneering advancements that have changed the way humans in developed countries function on a day-to-day basis.

Today, scientific and technological advances in areas like artificial intelligence (AI) and cloud computing are propelling global economies and pushing boundaries between humans and technology. There is wide spread agreement among educators, policy makers, and industry leaders about the importance of creating opportunities for students to engage in solving real world problems that require application of knowledge and processes from the STEM fields which has led to more integration across disciplines (Katehi, Pearson, & Feder, 2009). The landmark 1983 report, *A*

Nation at Risk, was a catalyst for changes as it criticized the American education system for failing to prepare students to enter a competitive workforce. The report called for educational reforms such as the adoption of academic standards, curricular changes in the basic subjects as well as art and computer science, and the implementation of assessments to measure achievement. The standards-based reforms that were fueled by *A Nation at Risk*, have been numerous and were initially grounded in academic standards for mathematics and science. These standards supported the notion of science, technology, engineering, and mathematics (STEM) education and a focus on how these multiple disciplines are integrated in education initiatives gained traction. Individual states across the United States are now engaged in adopting new standards, and re-thinking the goals of PreK-20 education. The Next Generation Science Standards (NGSS) and the Common Core State Standards for Mathematics (CCSSM) have increased emphasis on technology and engineering, and this new emphasis has the potential to change education programs, and what they mean from both a theoretical and practical point of view.

Since the publication of the National Council of Teachers of Mathematics (NCTM) standards for mathematics curriculum, assessment, and instruction in the late 1980's and early 1990's followed by the National Science Education Standards publication in science education, there have been significant reform efforts in STEM education in our country. We have learned much about what constitutes effective instruction in mathematics and science through research related to standards-based reform, such as curriculum development and implementation, as well as, professional development initiatives (particularly those funded by the NSF) (i.e. Huffman, Thomas, & Lawrenz, 2003; Tarr, Reyes, Chavez, Shih, Confrey, & Silver, 2006; Weiss, Pasley, Smith, Banilower, & Heck, 2003; Weiss & Pasley, 2004; Wenglinsky, 2002). We know that classrooms in which students are engaged in higher levels of thinking about important mathematics and science content and processes while solving meaningful problems support greater student achievement (Lesh & Zawojewski, 2007; Thomas, 2013; Weiss, et al., 2003; Wenglinsky, 2002). Furthermore, real problems in science and mathematics are complex, with effective solutions often crossing disciplinary boundaries making STEM integration an important element of classroom teaching and learning (Wang, Moore, Roehrig, & Park, 2011).

The Emergence of STEAM

During the same timeframe as the release of the NGSS and the CCSSM, STEAM emerged as a new paradigm intended to build student interest in STEM fields during the Americas for the Arts National Policy Roundtable in 2007 (Perignat & Katz-Buonincontro, 2019) and as the Rhode Island School of Design began using the term to promote the role of the arts and design in STEM K-20 education. The inclusion of arts within STEM education expands discipline specific conceptions of science, technology, engineering, and mathematics to include creativity, innovation, and design from the arts. The STEAM paradigm continues to develop in educational settings however, a common understanding of what the "Arts" represents has not materialized to date (Perignat & Katz-Buonincontro). A few of the more famous definitions include the creativity and originality of the arts. Frank Llloyd Wright wrote that, "Art is a discovery and development of elementary principles of nature into beautiful forms suitable for human use," (Popova, 2012.) Wright's view links art to science through reference to development of science knowledge from the "principles of nature." The great furniture designer, Charles Eames wrote, "Art resides in the quality of doing; process is not magic," (Popova, 2012). Eames view connects the arts to both scientific processes and the processes of doing mathematics. Both of these definitions fit well with the fields of Science, Technology, Engineering and Mathematics. In some ways, the STEM fields have always included an artistic component. Science, technology, engineering, and mathematics always included a creative and original component as scientists and mathematicians developed explanations and models for natural phenomena; and engineers take what they know about the natural world and apply that knowledge to their designs of technology to solve human problems. We are now finally recognizing that art is an integral aspect of STEM and deserves to be formally included in the STEAM movement. The formal inclusion of arts has the potential to appeal to a broader population of people than have been traditionally drawn to STEM fields and to result in higher quality solutions to real world problems that cut across STEM fields.

Use of the term STEAM has become more prevalent in American K-12 education and the shift from STEM to STEAM is changing what it means for students to learn within and across these disciplines. Lacking a clear definition and conceptualization of what "Arts" refers to in the STEAM acronym has not prevented American schools from adopting STEAM initiatives. It is an exciting time in schools where the Next Generation Science Standards focus on "three-dimensional learning" and the integration of science and engineering processes along with disciplinary knowledge. The integration of multiple perspectives and processes is a key part of the Next Generation Science Standards. While not directly mentioned in the Next Generation Science Standards, the Arts are part of the integration and the integration of various aspects of Arts is viewed as an integral part of having students engage in both science and engineering.

MAIN FOCUS OF THE CHAPTER

STEAM-Focused Schools Case Comparison

The following sections present two cases in which different school districts undertook initiatives to become STEAM-focused. School leaders from two school districts were interviewed about their perspectives for implementing STEAM initiatives. The name of the school, Truman Elementary, in the first case and the school district, Washington School District, in the second case are pseudonyms. For the Truman Elementary case, a semi-structured interview that lasted approximately 90-minutes was conducted with the principal of the school. For the Washington School District case, a semi-structured interview that lasted approximately one hour was conducted with each of three school leaders in the district including: the high school principal, a teacher leader in the middle school, and a teacher leader in the elementary school. The interview questions focused on the process for becoming STEAM as well as school leaders' perspectives and perceptions about the STEAM initiative in each school district. The interviews were recorded and transcribed verbatim to provide accurate data for analysis of the perspectives of the school leaders. The interview data were carefully examined individually to create a case description of each STEAM initiative. Then a cross-comparison of the interviews across cases was completed to identify important features for becoming STEAM-focused. The first case describes a school district approach where the district decided to begin with a STEAM-focused elementary school that will be referred to as Truman Elementary in this chapter. This endeavor has been successful and continues to grow throughout the school district. The second case describes a school district-wide endeavor that ultimately was not successful and does not continue as a formal initiative. The school district in the second case will be referred to as Washington in this chapter. A comparison of the two cases led to the identification of features for schools and districts to consider incorporating into models for becoming STEAM-focused.

Truman Elementary School

Features of the Initiative to Become STEAM in Truman Elementary

Truman Elementary School is located in a suburban area near a mid-sized city in the Midwest United States. Truman Elementary is designated by the school district as a STEAM-focused school that places an emphasis on the integration of science, technology, engineering, arts, and mathematics. Students at the school learn through a project based learning (PBL) approach that integrates the disciplines of STEAM where arts includes visual, performing, and language arts. The school has an extended academic year with students attending for 20 additional instructional days beyond other elementary schools in the district. This elementary school was opened as a STEAM-focused school. Enrollment capacity at Truman Elementary is 500 students K-6th grade. Attendance at the school is determined by open selection with applications for attendance accepted from across the district. Students are randomly selected for attendance according to their current school attendance area. Each elementary school attendance area is represented proportionally through the selection process. Enrollment in the school is based on a family household lottery system where families complete one application for all children in the household. When a child is selected for attendance through the lottery drawing, additional children in the household are eligible for attendance based on available open seats at the child's grade level.

Before Truman Elementary was opened as a STEAM-focused school, the school district spent two years investigating other STEAM initiatives and engaging with administrators, teachers, parents, and community representatives to lay the groundwork for launching a STEAM-focused school. During the first year of planning the district initially determined that the school would be STEM-focused. With input from parents and community representatives, including business leaders, STEM was expanded to STEAM because stakeholders believed that adding arts would lead to a more well-rounded and relevant education. From the beginning, the concept of arts included language arts as well as the more traditional disciplines of visual and performing arts.

A planning committee of stakeholders was formed that included administrators, teachers, parents, and community members to lead the efforts related to designing the school's model for STEAM. The committee organized visits to three STEAM elementary schools in the region for a broader representation of stakeholders. Several teams of teachers visited other STEAM schools to observe and ask questions about what it means to have a STEAM focus. After approximately a year and a half of investigation and planning, the school district determined that there was sufficient buy-in from all stakeholders to proceed with opening the new elementary school as a STEAM-focused school. A school mission statement was created and is posted on the school website. According to the Truman Elementary website, the primary goal of the mission is for the school to be a leader of STEAM innovation and to be a school community distinguished by

- A culture of independence and collaboration
- A global outlook based on a foundation of character
- A focus on problem solving and critical thinking
- And, real world experiences with integration throughout all subjects.

Shortly following the formal decision for Truman Elementary to be a STEAM-focused school, one of the planning committee leaders was chosen to be the school's principal. Elementary teachers already employed by the school district were invited to apply to be teachers in the STEAM-focused school. There were 21 regular classroom teacher positions available in the school and 30 teachers from across the district applied. Teachers were interviewed to evaluate their alignment with both the STEAM integration and PBL philosophies upon which the school was be based. Planning with the teachers selected to launch Truman Elementary began during the school year prior to its opening. The district paid the teachers and building administrators to engage in two weeks of professional development during the summer to begin writing curriculum. The first three days were spent learning about STEAM integration, the PBL approach and lesson planning with the 5E model from science education. The teachers selected the 5E model, originally developed in the late 1980s through the Biological Sciences Curriculum Study, because it promotes active, collaborative learning. The model supports students working to solve problems and develop understanding of new concepts by asking questions, investigating, developing explanations, and drawing conclusions for practical application of that new understanding. The model has been modified but the Truman teachers followed the Engage, Explore, Explain, Elaborate, and Evaluate phases for STEAM lesson planning. The remaining two weeks were dedicated to curriculum writing for the beginning of the school year. The school district had a Project Lead the Way engineering program that teachers found helpful for providing a model for the curriculum writing and development. The teachers also benefited from examples provided by one of the other area STEAM schools. All of the newly selected Truman Elementary teachers visited area STEAM-focused schools to observe, ask questions, and learn from other teachers, aspects of effective practices when implementing STEAM throughout the school. The teachers at Truman Elementary built the curriculum from the ground up and these supports were important to the launch of the STEAM-focused school.

The school principal explained that for the first year teachers primarily used science as the anchor content discipline and found ways to integrate the other STEAM disciplines into science projects, units, and lessons. She said, "the first year of implementation, I wouldn't say we had true STEAM integration. It took a few years and it's been a journey, but I would say now going into year five, the teachers have got it and projects bring in all STEAM areas with authentic integration across the disciplines." She expanded about this transformation and said the teachers are now "addressing the five areas in a project based collaborative learning environment and that was the biggest mind shift" because most of the teachers came from very traditional classrooms where each subject was in its own silo and addressed separately at designated times during the school day.

Five years into the STEAM initiative the school now devotes approximately half of the school day to project based learning that integrates all STEAM disciplines through a dedicated PBL daily block of learning time. The other half of the school days is used to focus on specific content and skills in areas like reading, language arts, mathematics, science, and social studies that are later applied during the STEAM-focused projects. However, the principal said that teachers are "integrating STEAM areas throughout the day in all subject areas." The principal found that one of the exciting things for her to watch was how teachers shifted from posting classroom daily schedules with blocks of time for separate subjects to integrating the subjects all day. She said, "if you walk into a classroom now you really won't be able to say, 'Oh, this is clearly designated as mathematics lesson time.'" Instead, the half day devoted to specific content and skills is labeled by the teachers as "learning time" and as the principal said, "it is a very flexible schedule depending on the students needs." The classroom environment in this STEAM-focused elementary school has become more of a collaborative learning environment guided by both the STEAM-focused project planned by the teacher and by the students based on their interests and needs. The teachers design the project allowing for flexibility to respond to students input. The principal added that, "what ends up happening is the teacher will develop the project overall, but then the students guide it. So, the projects end up looking different in all the classrooms." In addition to the continued STEAM-focus in Truman Elementary and the use of some of the STEAM-focused project based units by elementary teachers at different elementary schools in the district, the school district is in the process of transitioning a middle school to a STEAM-focused school. The district high school adopted an academies approach to reorganize how subject specific courses are grouped and offered. To build on the success from Truman Elementary, one of the academies in the high school is a STEAM academy.

Benefits of the Initiative to Become STEAM at Truman Elementary

Truman Elementary School's journey to becoming a STEAM-focused school was carefully planned with input from not only administrators and teachers in the district but other stakeholders like parents and community members as well. The broad support and buy-in contributed to the benefit of the school developing **close connections to the community**. Because stakeholders outside the school are included in its day-to-day operation, students and teachers at the school benefit from partnerships with local businesses and organizations that donate time and resources for the STEAM projects and curriculum implemented by the school. The students in the school also give back to the community through a community garden that is incorporated into the curriculum for all grades Kindergarten through 6th grade. The garden has 21 beds that are planned, planted, tended, and harvested by students. The produce is sent home with students for their families and also offered to the community on a first come first served basis. During the summer break the Parents and Teachers Association (PTA) helps organize a volunteer schedule. Families in the school sign up to come and take care of the garden when school is not in session. The garden is an outdoor classroom ripe with opportunities to integrate learning across the STEAM disciplines.

Another benefit of the STEAM initiative at Truman Elementary is the way that the true **STEAM integration transformed teaching and learning**. The principal described as a positive outcome, the shift from classrooms structured around teaching content in separate blocks of time by subject matter discipline to a focus on problems to be solved through a project based approach to learning. She believes that students are now "getting the most out of each instructional minute and are taught to think critically." Students are solving real world problems and gaining a deep understanding of content through application within cross-disciplinary STEAM projects. The principal sees evidence of high student engagement in learning and student ownership of the products of learning.

The STEAM-focus is well established in the elementary school and teachers have collaboratively developed the curriculum iteratively over a period of three to four years. The benefit for Truman Elementary is a curriculum that is flexible enough to be changed each year to **engage students in meaningful and relevant real world problems**. One example is a problem with energy use by the school. The 6th grade teachers used this problem as the basis for the STEAM Tank PBL unit that has been implemented in multiple years. The essential question of the unit was: How much energy is wasted at Truman Elementary School and what can you invent or access to improve this human impact on the environment? During the unit students researched energy waste at Truman Elementary School. They designed and created an invention and found funding sources to combat the problem. Findings were presented with charts, graphs, and technical writing; including a plan for marketing the invention with a website, the product, and marketing tools. The presentation was done in the style of the popular U.S. based television show, "Shark Tank." A panel of teachers, administrators and parents were the SHARKS who determined if the product would be funded. The unit focuses on students deepening their understanding of concepts related to energy; using engineering and design processes; collecting, analyzing, and presenting data; school and community service; and producing solutions that have far-reaching and long-term results. One example of a result was designing special fans that turn on at certain temperatures to assist the school's cooling system to keep classrooms a comfortable temperature. The students designed the mechanisms for the fans and built them for use in classrooms. In another STEAM unit, students noticed a problem with water drainage from the school property into an adjacent creek.

Students noticed erosion and how it was impacting the school playground and the creek. They studied the watershed, investigated solutions to improve the drainage, developed plans to improve the slope of the ground from the school property to the creek, presented the plans to the school district maintenance department, and finally worked with the maintenance department to see their plans implemented and the correct slope built. The principal noticed that by engaging in the STEAM-focused PBL units Truman Elementary students are developing skills to think critically and creatively. She indicated that one important benefit she observed is that "students can think and solve problems better than any students I have ever worked with in my life. They are phenomenal at it."

Challenges of the Initiative to Become STEAM at Truman Elementary

The STEAM initiative launched by opening a new STEAM-focused elementary school is viewed as a success by the school district and is being expanded to a district middle school. While successful, the initiative to become STEAM did come with challenges. The biggest challenge identified by the principal was the **time required to create PBL units that integrate all STEAM areas** in ways that are meaningful and relevant. The principal and teachers found that it took dedicated effort over a period of multiple years to realize true STEAM integration through PBL units designed around real problems. The principal acknowledged that during the first year the 5E lesson plan format and PBL approach was not utilized well and the units were not particularly successful. In the first year teachers also believed they should follow their lesson plans exactly as they had written them. The principal said, "Those projects kept failing." At the same time, the teachers and both principals were investing more time during the first year to get the school started and to write and implement the curriculum. The days in the summer gave the teachers a good start but were not enough to develop the curriculum for the entire school year. The principal admitted, "I'll be honest in that first year, it was a lot of hours, I mean it was non stop." The principal and the assistant principal helped teachers plan and also supported them by doing things like taking purchase orders and buying lesson supplies that teachers needed for the next day's lesson at night. Once teachers had more time in the second and third years to analyze and reflect on STEAM unit planning and implementation, combined with increased comfort with flexibility to change the plan, they started seeing better outcomes.

Another challenge for Truman Elementary was **changing how the budget was managed to accommodate increased costs for the STEAM initiative**. The principals and teachers found that implementing the STEAM curriculum they developed required different and additional supplies than in the past. The principal stated, "the biggest hurdle honestly is just the supplies and budgeting to make sure you have everything you need throughout the school year." The teachers were each allotted a classroom budget to purchase instructional supplies. A classroom budget for each teacher is common practice in the school district but building administrators discovered that changes were necessary to meet the needs of the STEAM initiative. In addition to the regular classroom budget that all teachers in the district have, the principals asked teachers to submit monthly project needs lists. The building administration used the lists for decisions about reallocation of funds to meet the teachers' classroom needs. The lists also helped the principals predict future budgets. During the first year knowing what they needed for each project a month ahead was a stretch as teachers were still building the curriculum. The day-to-day operation of Truman Elementary was described as a team effort with everyone in the building contributing. The school administrative assistant played a key role in the supply chain for teachers during the early years. She was conscientious about pricing, would help fulfill the purchase orders for the teachers, and synthesized the needs of the teachers to identify and purchase base items for the STEAM initiative. By year two and three the project needs lists could be predicted and submitted by teachers quarterly. In order for funds to last for the entire academic year, the building administrators learned how to manage the school budget differently and the teachers learned how to manage the supplies they needed differently. The principal explained that everyone needed to get past the mindset that they had to buy what was needed for a project. They all had to rethink how to create, recycle, or reuse things. The principal attributed their ability to confront this challenge to a team effort when she said:

There's not a person in this school that isn't on board and all in. So that it makes it phenomenal. Also, we started bringing in more recyclables for our projects because there's so much value there. And our families, they're excited to bring in recyclables. They bring them in by the bagful and we actually had to ask them to stop because we got too much stuff. That was a way that we were able to kind of change some of our mindset too. Is how can we actually use what's already out there? And it's also teaching the students a valuable lesson of, 'We can reuse these items as opposed to just buying new for everything.'

With the STEAM initiative came additional tools for using technology in the classroom. The school was opened with access to the latest devices for teachers and students. The teachers had not previously had many tools for technology integration and use in their classrooms. When they started the school year at the new STEAM-focused school the teachers had their own computers, interactive whiteboards, and one-to-one devices for every student in the classroom. In Kindergarten through fourth grade the students are given iPads and in fifth grade and sixth grade the students are given laptops. This reality presented teachers with the challenge of **authentic use of technology** in their classrooms. The teachers did not have much background in how to integrate the use of technology to enhance learning rather than to merely substitute for business as usual practices like using the interactive whiteboard in static ways or electronic worksheets versus a paper and pencil version. The teachers also struggled with how often to have students use their devices. Initially teachers felt like students should be using the device most of the day. The principal explained, "when the students got the devices, the teachers' perception was that students needed to be using the device all the time and that was a big mind shift. For technology use to be purposeful and to enhance what students are doing one day you might use the device 10% of the time and another day 50% of the time." It took time for teachers to learn how to incorporate technology that allowed students to create new tasks and products rather than using technology just for the sake of using it. The school utilized support from a district technology specialist who came to the school regularly to work with teachers on how to incorporate technology into the STEAM units. She provided practice embedded professional development for teachers, assistance with finding applications to enhance particular STEAM projects, and co-taught with teachers. The librarian was a technology support resource for teachers as well. She assisted teachers in finding appropriate resources for their STEAM units and co-taught with them. The library space was transformed from a more traditional quiet space where students read and check out books to a learning commons space where they work on projects and have access to additional resources. The learning commons space includes a Maker Space and additional computers and other technical tools. The principal described the change when she shared, "changing their mindset about technology was a huge challenge and hurdle at the beginning and now we have actually redefined how we use technology in the classroom."

Washington School District

Features of the Initiative to Become STEAM in Washington School District

Washington School District is located in a rural area in the Midwest and is situated near a U.S. military installation. While the local community is rural, the school district has a diverse population of students due to the proximity of the military installation. Nearly three fourths of the students in the school district are military affiliated. With this military affiliation come additional funding opportunities that a rural school district in the United States might not be able to access. The school district superintendent made an administrative decision to apply for and was awarded a one million dollar federal grant to implement a STEAM initiative across the school district, Kindergarten through 12th grade. The grant was awarded during the summer when school was not in session with a start date of the coming academic year. The district curriculum coordinator was assigned to lead the efforts to implement STEAM within every school in the district. A lead art teacher in the district said, "our district took it pretty literally as integrated science, technology, engineering, and mathematics, meaning every subject matter, every teacher, every grade level, every student, whole school, K-12 was involved."

To launch the grant, the district hired an external consultant who was known for her work in STEAM school implementation to provide professional development for teachers and administrators. Initially, a few lead teachers from the two elementary schools attended a conference in the summer. The week before school started, the entire district attended a three-day in-service professional development provided by the external consultant that took place the week before school started in August. Teachers and principals across the district all learned about the initiative to be a STEAM-focused district at the same time. Rather than taking a gradual implementation approach with early adopters--every teacher, in every school in the district was expected to start implementing a STEAM focus in their classroom as the school year began. Describing the launch of the STEAM initiative, the high school principal explained that teachers could understand the value of engaging students in the STEAM disciplines in an interconnected way. However, the principal said "there was no chance to get buy in and there was no opportunity to plan for it; we all sat and heard the spiel all at the same time and it was like, 'we don't care if you want to or not, everyone is doing this.' So, that was our introduction to it and it was difficult." A lead teacher in the middle school said about the professional development provided, "It started out okay and progressively got worse and worse."

Following the initial back to school professional development, each school in the district divided teachers into STEAM teams. At the elementary schools teams were formed by grade level, at the middle school the teams were mostly cross disciplinary teams by grade level, and at the high school the teams were cross disciplinary based on teacher interests. The teams were instructed to develop a STEAM unit according to the guidelines provided during the initial professional development session. The unit was then submitted to the external consultant for scoring and an evaluation made to either approve the unit for implementation or to require revisions before the unit was implemented. Once approved for implementation, teams were required to follow the lesson plan format provided by the consultant and to document the implementation by completing forms provided by the consultant. The documentation also included formal lesson observations by administrators. All of the STEAM unit documentation was submitted to the consultant. Teams planned and implemented two STEAM units during the first year of the grant. Teams typically had about one hour of common team planning time a week.

The model the district adopted the first year was not teacher lead but rather framed and structured according to an external program developed by the consultant the district hired. This approach left teachers and administrators feeling like they had no ownership or professional license to develop and create a meaningful STEAM focus in their schools. Further, the requirements placed on the teams about unit and lesson development during the first year of implementation forced integration of each discipline in ways that was not authentic for each area. According to the high school principal this resulted in "lessons that were contrived" and diminished the potential for teachers in different discipline areas to truly collaborate and design learning experiences that utilized and fostered connections across content areas. There was no time for teachers on the teams to learn from each other about curriculum standards in each of the disciplines and to consider together common themes or concepts across disciplines. The content area that often suffered the most was the Arts. This may be in part because the initial interpretation of Arts was narrow as the external consultant told teachers that drawing a picture of some aspect of the unit or doing a painting could integrate art. A middle school art teacher's description of one unit developed by the team she was placed on during the first year illustrates this challenge. The art teacher said:

There were misunderstandings about subject matter and assessment methods. For example, the first team I was on, for their STEAM unit they decided they were going to concentrate on roller coasters. Science was going to be force and motion, teaching about that. For technology, students were going to use software to design the roller coaster. For engineering, students would actually build the coaster in the shop class, or industrial technology. And then for mathematics, students were going to figure the velocity and angles, and that would incorporate mathematic. And for art students were going to paint this roller coaster. Students were going to paint it! I was thinking, NO! You know, what am I going to teach that is in my curriculum when we paint the roller coaster? Maybe I can teach color theory, but that's really forcing and that's really reaching. And I had to point out to my group that, hey, I have state standards and curriculum too, and this doesn't fit anywhere in that. So I'm giving up two weeks of instructional time to paint roller coasters. So it was a forced fit and it just didn't work.

The rapid and forced integration of STEAM content from each of the disciplines into units may also have contributed to what was described as inauthentic integration of Arts into the units. Identifying authentic connections among STEAM disciplines by analyzing curriculum standards across content areas to synthesize concepts and skills is difficult and takes dedicated time. This is confounded by an educational system, like the one in the United States, characterized by separate content specific classes in middle schools and high schools and dedicated time for separate content lessons in elementary schools. As the high school principal said, "So, you have an art class, and this art class wants to do something really cool with the algebra class. Well, some of those students in algebra aren't in art. Some of those students in art aren't in algebra. You know, how do you even make that work? The logistics are very difficult."

Teachers and administrators felt like they were being forced to implement STEAM units that were not authentic and they were frustrated because adequate preparation and plan time had not been built into the implementation model. By the second year, central administration relaxed some of the strict requirements related to planning and implementing the STEAM units in schools. Teachers were told that they did not have to force fit every aspect of STEAM into every lesson like they were required to do the first year. And, rather than submit the STEAM units to the external consultant for evaluation, teachers could submit the STEAM unit as part of the district teacher evaluation system. Teachers in the middle school were allowed to self-select their STEAM team and that was viewed as a positive change. The district administration also recognized that the whole group professional development provided by the external consultant was not effective for all teachers in the district and provided teachers with the opportunity to self-select and request professional development more specific to their needs. A lead teacher said that in the second year, "the district started to listen to the teachers about what training they needed, the release time needed for preparation and then giving teachers the time to work." Despite some of these changes, the high school principal said about the STEAM initiative, "it didn't have much steam even after the first year." She went on to say that in the high school, "not one teacher continued with the structured lessons from the first year, that's how contrived we felt they were." When the superintendent who was awarded the grant left the district the following year, the formal STEAM initiative was dropped.

Benefits of the Initiative to Become STEAM in Washington School District

The formal STEAM initiative was short lived and is no longer a primary focus in the district. However, there are some lasting residual benefits for the teachers, the students, and the schools. Five years after the initiative began there are several changes that relate to the STEAM focus in the schools. A notable change was **increased attention to the connections across STEAM disciplines and an increase in programs that integrate STEAM disciplines**. The high school principal said that the teachers "do look for those connections, more so than they did five years ago and the connections are so much more natural now." Middle school and elementary teachers also shared that they noticed an increased attention to integration across the STEAM disciplines. A middle school art teacher said, "I found it helpful that teachers started considering subject matter outside their own." The art teacher explained that a science teacher she was working with was planning a unit on light and came to her to ask about a project students could make. The art teacher said to the science teacher, "Why don't you talk about what the impressionists did and how they studied light? Because some of them were very scientific about it." The art teacher shared various impressionist artists names and works with the science teacher to incorporate into the light unit. The art teacher said, "it was kind of eye-opening for the science teacher to realize that art is not just about crayons and tempera paint."

Programs like Project Lead the Way, Robotics, and Lego Labs have thrived in the district starting in the elementary schools and continuing through high school. The high school Robotics team recently had the opportunity to engage with engineers on the military base to help solve an issue the base was having with an important piece of military equipment. The general from the military base called the district and asked if the Robotics team could help solve the problem the base was having. The district agreed and the Robotics team was able to engage in finding solutions to a real world problem that was relevant in their community. As it turned out, the Robotics team designed an add-on switch plate cover for the military equipment because switches were accidentally getting knocked and causing automatic mechanical reactions that were not correct. The high school students produced the switch plate covers using a 3-D printer. The piece was tested and found to be an effective solution that is now being used by the military on the equipment. The high school principal talked about how this was a great application of integration across STEAM disciplines that students and teachers were excited about because it was a real world problem and relevant.

The high school Robotics team's collaboration with the military base was an example of **increased use of problem-based learning** that was another benefit of the STEAM initiative. Teachers began planning and implementing more units that developed around a theme with a problem or project to anchor the instruction and learning. This increase in problem-based learning was confirmed by the middle school art teacher's comment that because, when integrated, STEAM disciplines can generate interesting and relevant problems, the teachers adjusted some of their teaching to promote student engagement in problem-based or project-based units. She said, "Making in Maker Spaces, project-based learning, generating solutions to problems in all the subject matters is still emphasized [even after the STEAM initiative was no longer formally in place]."

Another benefit of the STEAM initiative was a more **authentic and expanded view of Arts in STEAM** to focus on concepts and skills of visual and performing arts and inclusion of English and language arts that was in turn, extended to business and digital media classes. The middle school art teacher described the change by sharing how she felt during the first year of the STEAM initiative compared to her collaborations with other teachers after the first year. She said, "I was pretty excited at first, because I thought, 'yay, finally we're incorporating art. Finally art is going to get what it needs.' And then I realized the science and mathematics teachers didn't really know much about art. And they didn't know much about the art standards that I taught, or my curriculum." In the second year of the STEAM initiative, the art teacher worked with a sixth grade team and the unit focused on the state and state history. To integrate arts, the team planned lessons for the students to study pottery and the Native American people who lived in the region and created pottery. In contrast to the beginning of the STEAM initiative, the art teacher said about this later unit "we were able to incorporate making pottery, studying the pottery, how these people lived. So that was a much better fit, because I was able to fit three-dimensional ceramics curriculum standards into that unit pretty easily." In the high school, teachers in English language arts, business, and digital media supported authentic integration of Arts into STEAM focused learning by facilitating students' learning to communicate with others about problem solutions, products, and results. Students produced video and print materials to share their work and the products of their work. These student communications about their work incorporated aspects of language arts, visual arts, and media arts.

Challenges of the Initiative to Become STEAM in Washington School District

Overall, the district STEAM initiative was not considered a successful endeavor and several challenges may have contributed to this outcome. A primary challenge the district faced was garnering **support and buy-in from building administrators and teachers**. An elementary teacher noted this when he said, "the biggest challenge was getting everyone on board and for central administration to understand that doing this is not a light switch kind of activity that can be easily turned on." The centralized approach to seeking funding to support a district-wide STEAM initiative and ultimately the launch of the initiative did not include building administrators and teachers in the decision-making process about any of the initial elements. Even if teachers could understand the value and intention behind a STEAM focus in their school and classrooms, a majority of the teachers were not excited or positive about the initiative. The high school principal critiqued this approach when she said, "All of us have so much on our plates, and in my experience, any successful change has come with looking at your early adopters, getting those people on board that are excited and say yes, we want to do that. Then, really digging in with them. And then, their enthusiasm creates the growth naturally. That wasn't allowed to happen at all."

The district adopted a top-down approach to implementation during the first year of the STEAM grant and required teachers to strictly follow a structure for unit and lesson design developed by an external consultant. The professional development provided by the external consultant was not well received. This created another challenge for the district because teachers did not have the **professional license to take ownership of planning and implementing STEAM units**. Teachers and building administrators perceived that the external consultant did not have a thorough understanding of curriculum standards in all content areas of STEAM and that the integration of the STEAM disciplines was narrow and contrived. A challenge this perception revealed was **misunderstandings about subject matter and assessment methods**. The elementary teacher sharing his thoughts about the authentic integration of the STEAM disciplines said, "I thought it was very arbitrary, to be honest with you, the way we were teaching things." The explanation of the roller coaster unit shared by the art teacher provides an example of both of these challenges.

Another challenge was providing teachers with the necessary **time to collaborate, plan, and implement** STEAM units using a different model of instruction than was the norm. The STEAM initiative called for integration of the separate disciplines within thematic units. With little time for building administrators to restructure teacher collaborative plan time the first year, each school used a one-hour block of collaborative time that was already designated for teachers to meet together. Teachers were grouped into their STEAM teams for that weekly collaborative time. However, teachers and administrators found that limited time was not adequate for meaningful teacher learning, collaboration, and change. Teachers at all levels found it difficult to shift from an individual discipline and individual classroom focus in their planning and lesson implementation. The art teacher said, "because what they were asking us to do, there was no way you could do it just on the existing collaborative plan time and release time they gave you." She continued to explain saying, "we were forcing fits. If you were unfamiliar with the concept and were teaching something new to you or using technology that was new to you, and hadn't had enough time where you felt prepared, that was bad. We did have some people that were really not happy with having to do that." The high school principal talked about this and other challenges from a broader lens saying about the district approach, "It was backward, there should have been time for a year or two for the teachers to begin collaborating together, generating ideas, and doing it on their own."

The decision to offer teachers and building administrators a three-day professional development session delivered by the external consultant the week before school characterized the challenge of a **lack of authentic and relevant professional development**. The art teacher identified this as a challenge when she said, "one of the problems I think we had was finding good authentic professional development. The whole group professional development did not seem to work well." Adopting a

STEAM focus and integrating the disciplines of STEAM was a change for teachers in the district. During the initial year of the STEAM initiative, teachers were not provided professional development that they believed was meaningful. "It felt like we were kind of pushed into it before we were ready to teach it," the art teacher said about the STEAM units. The elementary teacher attributed some of the challenges to inadequate professional development and said, "you have to give teachers the opportunity to learn how to build these lessons." After the initial semester of the STEAM initiative the district placed less reliance on the external consultant to provide professional development and responded to feedback from teachers and building administrators. This shift in approach to professional development was an improvement and helped address this challenge. As the art teacher said, "after the first semester that was pretty rough, when the district started to listen to the teachers about what training they needed, when we were provided with plan time and release time to develop the STEAM units, that definitely made things better." The shift allowed teachers to pursue professional development in areas that helped them develop STEAM units and lessons and time to test the lessons with each other before implementing them with students. The high school principal said that because of the lack of adequate time for learning and planning through authentic and relevant professional development, "the STEAM implementation lacked greatly," teachers and administrators in the district generally do not hold positive views about the notion of being a STEAM focused district.

FINDINGS

Features of Becoming STEAM

Schools and school districts can learn from both cases in this chapter important features to consider when becoming STEAM-focused. One important distinction between the two cases that affected the long-term success of the STEAM initiatives was **intentional efforts to gain buy-in**. In the Washington School District there was little opportunity for building administrators and teachers to learn about what becoming a STEAM-focused district would entail for their own professional practices and they were given no choice about whether to participate or even how to participate. Teachers and building administrators could understand the value of STEAM integration but felt their professional expertise was disregarded in the implementation. STEAM integration in the classrooms was forced and artificial leaving the STEAM initiative nearly non-existent after the first year. In contrast, Truman Elementary was opened as a STEAM-focused school only after the school district had spent two years investigating and preparing for the launch and implementation. A wide range of stakeholders were involved in the decision to create a STEAM-focused school and building administrators and teachers made the choice to work in the school. The start up time dedicated to building support and buy-in from stakeholders led to a level of motivation to participate in the STEAM initiative and a commitment to engage in the challenging work to change teaching and learning practices that was very different from those in the Washington School District initiative. Attention or, in the case of Washington, lack of attention to gaining support from key stakeholders for a STEAM initiative seemed to have an effect on the teachers motivation to continue the work to implement the change.

Another feature that warrants careful attention for schools that are becoming STEAM-focused is **adequate time for learning and planning through authentic and relevant professional development**. In addition to the limited opportunity that personnel in Washington School District were given to consider the implications of implementing a STEAM initiative, the professional development provided by the external consultant did not meet their needs. Further, the principal and teachers believed that because of the strict structure for STEAM units and lessons teachers were required to develop directly following the three-day professional development, the STEAM cross-disciplinary integration was both forced and contrived. They shared examples of how STEAM integration within curriculum units improved in the second year when professional development was better connected to areas that teachers identified and when they were given more sustained time to develop units. At Truman Elementary the teachers were paid to attend two weeks of professional development in which beginning to write their STEAM curriculum constituted more than half the time. Early on, teachers and other stakeholder went to established STEAM-focused schools outside their district to observe classrooms, review curriculum, ask questions, and learn from other educators. The school administration made adjustments to the teachers planning time to ensure that teams had more time to collaborate then is typical in schools. Schedules were restructured for lunch, recess, learning commons, and classes like PE, music, and art to support a more fluid and flexible school day for teachers and students. The school administration also knew that changing teaching and learning practices would take time and they understood that it would take multiple iterations, over a period of years, to implement STEAM projects, units, and lessons with fidelity. The principal explained these this shifts in how learning, both teacher and student learning, was viewed had a positive effect on the school's efforts to become STEAM. In many ways, the approach to professional development at Truman included elements identified in Elmore's (2004) seminal book on school reform where he noted, the best professional development for teachers is rooted in instruction, focuses on student learning over the long-term, is anchored in a specific domain (in Truman's case it was anchored in the interdisciplinary nature of the STEAM domains), and builds capacity for the organization to embrace a model of continuous improvement.

In both school districts, three additional features for realizing a robust STEAM-focus were consistent throughout the interviews. When asked about benefits and successes of the STEAM initiatives all of the interviewees described the value of **community connections to infuse real-world and problem-based or project-based** learning into their schools and classrooms. These connections to the community seemed more pervasive and stronger at Truman Elementary with their community garden, parents and community members sending in supplies and volunteering at the school, and collaborations with business and industry in the area to address real problems to name a few. However, many of the examples of benefits shared from Washington School District involved the community outside the school such as the military installation and professionals who contributed to their Project Lead the Way, Robotics, and Lego Leads programs. The importance of **mutual decision-making and support between teachers and administrators** was emphasized across the interviews. In Washington School District, the absence of mutual decision-making between schools and the district central administration during the first year of the STEAM initiative was viewed as a barrier that may have contributed to the demise of the formal STEAM focus in schools. Yet, STEAM focused activities do still continue in Washington schools in pockets, albeit without the STEAM label. When the principal and teachers talked about those continuing efforts, they always included a reference to school-level teachers and administrators collaborating and supporting each other. The Truman Elementary initiative to become STEAM began with mutual decision-making and support that continues to the present. In addition to the broader decisions and supports, the principal elaborated on examples of how all staff in the building function as a team to support the STEAM efforts. This included examples of how she and the assistant principal are in and out of classrooms most of everyday; not to observe the teachers but to join them in implementing the STEAM PBL learning activities. Finally, the necessity for changes in **budget planning and allocation** were highlighted in the interviews. The costs involved in the STEAM initiatives in both school districts required each to rethink a number of financial allocations through revised budget planning processes. Some examples of these included: how supplies were acquired and managed; when and how the budget for substitute teachers was used; and consideration of what resources were needed to fund projects like the community garden and producing prototypes of the new part for the military equipment designed by students.

CONCLUSION, SUGGESTIONS, AND FUTURE DIRECTIONS

Both STEM education and STEAM education initiatives are often prompted by policies or by-products of policies (e.g. discipline specific standards, funded grant projects) related to education and workforce development. The seminal Change Agent study conducted by the Rand Corporation in the late 1970s and revisited by Milbrey McLaughlin in 1990 discussed findings about implementation of change initiatives in education that hold true today. One such finding is the false notion of a one-to-one relationship between policy and change in educational practices. Additionally, according to McLaughlin (1990), policy does not dictate what is important in distinct local contexts. Whether or not an innovative effort is sustainable and successful is dependent on local capacity and will (McLaughlin, 1990). Interestingly, the degree of success and longevity of the two cases described in this chapter aligned with the Rand Change Agent study's findings about implementation strategies. Effective implementation strategies were those that supported mutual adaptation of the project and the educational setting to each other (McLaughlin, 1990). The Truman Elementary initiative to become STEAM-focused reflected this notion of mutual adaptation and utilized all of the effective implementation strategies from the Rand student and summarized by McLaughlin (1990, p. 12):

- "Concrete, teacher-specific and extended training
- Classroom assistance from local staff
- Teacher observation of similar projects in other classrooms, schools, or districts
- Regular project meetings that focused on practical issues
- Teacher participation in project decisions
- Local development of project materials
- Principals' participation in training."

Ineffective implementation strategies were those that ultimately were incompatible with teachers and principals needs, motivations, and interests (McLaughlin, 1990). The Washington School District initiative to become STEAM was characterized by the ineffective implementation strategies that included:

- "Reliance on outside consultants
- Packaged management approaches
- One-shot, pre-implementation training
- Pay for training,
- Formal, summative evaluation
- Comprehensive, system-wide projects," (McLaughlin, 1990, p. 12).

Local contexts and other mitigating factors such as financial and personnel resources, other priorities for initiatives (new teacher evaluation systems, focus on social and emotional well-being of students, etc.), and changes in district or building leadership also effect the implementation and sustainability of any STEAM initiative. Future longitudinal research is warranted to better understand and evaluate the elements and nuances of efforts to become STEAM in schools. The cases in this chapter provide a profile of both a successful and an unsuccessful initiative to become STEAM. School districts and schools can use the features of the implementation of each case in this chapter, of the findings from the previous change agent studies (McLaughlin, 1990), of the benefits and challenges of each STEAM initiative in the chapter, and of other features as each STEAM initiative was developed to guide and frame their own pursuits to become STEAM.

REFERENCES

American Institute of Architects. (2018, July 26). *Career and Technical Education bill to recognize Architecture as STEM*. Retrieved from https://www.aia.org/articles/206266-career-and-technical-education-bill-to-reco

Elmore, R. F. (2004). *School reform from the inside out: Policy, practice, and performance*. Cambridge, MA: Harvard Education Press.

Huffman, D., Thomas, K., & Lawrenz, F. (2003). Relationship between professional development, teachers' instructional practices, and the achievement of students in science and mathematics. *School Science and Mathematics*, *103*(8), 378–387. doi:10.1111/j.1949-8594.2003.tb18123.x

Katehi, L., Pearson, G., & Feder, M. (2009). *Engineering in K-12 education: Understanding the status and improving the prospects*. Washington, DC: National Academy Press.

Lesh, R. A., & Zawojewski, J. (2007). Problem solving and modeling. In F. Lester (Ed.), *Second Handbook of Research on Mathematics Teaching and Learning* (pp. 763–804). Charlotte, NC: Information Age Publishing, Inc.

Mazuzan, G. T. (1994). *The National Science Foundation: A brief history*. Office of Administrative and Public Affairs, the National Science Foundation. Retrieved from https://www.nsf.gov/about/history/nsf50/nsf8816.jsp

Perignat, E., & Katz-Buonincontro, J. (2019). STEAM in practice and research: An integrative literature review. *Thinking Skills and Creativity*, *31*, 31–43. doi:10.1016/j.tsc.2018.10.002

Popova, M. (2012). What is art? A few famous definitions, from antiquity to today. *The Atlantic*. Retrieved from https://www.theatlantic.com/entertainment/archive/2012/06/what-is-art-a-few-famous-definitions-from-antiquity-to-today/258871/

Roosevelt, F. D. (1944). *Letter to Dr. V. Bush from the White House November 17, 1944*. Retrieved from https://www.nsf.gov/od/lpa/nsf50/vbush1945.htm#letter

Tarr, J. E., Reys, R. E., Chavez, O., Shih, J., Confrey, J., & Silver, E. A. (2006). *A longitudinal study of mathematics curriculum and student achievement*. Paper presented at the NCTM research presession, St. Louis, MO.

Thomas, K. (2013). Changing mathematics teaching practices and improving student outcomes through collaborative evaluation. *Teacher Education and Practice*, *26*(4), 779–796.

Wang, H. H., Moore, T. J., Roehrig, G. H., & Park, M. S. (2011). STEM integration: Teacher perceptions and practice. *Journal of Pre-College Engineering Education Research, 1*(2), 1-13.

Weiss, I. R., & Pasley, J. D. (2004). What is high-quality instruction? *Educational Leadership*, *61*(5), 24–28.

Weiss, I. R., Pasley, J. D., Smith, P. S., Banilower, E. R., & Heck, D. J. (2003). *Looking inside the classroom: A study of K-12 mathematics and science education in the United States*. Chapel Hill, NC: Horizon Research. Available www.horizon-research.com/insidetheclassroom/reports/highlights/highlights.pdf

Wenglinsky, H. (2002). How schools matter: The link between teacher classroom practices and student academic performance. *Educational Policy Analysis Archives, 10*(12). Retrieved [2/27/2006] from http://eppa.asu.edu/epaa/v10n12/

ADDITIONAL READING

Bequette, J. W., & Bequette, M. B. (2012). A place for art and design education in the STEM conversation. *Art Education*, *65*(2), 40–47. doi:10.1080/00043125.2012.11519167

Guyotte, K. W., Sochacka, N. W., Costantino, T. E., Walther, J., & Kellam, N. N. (2014). STEAM as social practice: Cultivating creativity in transdisciplinary spaces. *Art Education*, *67*(6), 12–19. doi:10.1080/00043125.2014.11519293

Herro, D., & Quigley, C. (2016a). Exploring teacher perceptions of STEAM: Implications for practice. *Journal of Professional Development in Education, 43*(3), 416–438. doi:10.1080/19415257.2016.1205507

Herro, D., Quigley, C., & Cian, H. (2019). The Challenges of STEAM Instruction: Lessons from the Field. *Action in Teacher Education, 41*(2), 172–190. doi:10.1080/01626620.2018.1551159

Honey, M., Pearson, G., & Schweingruber, A. (2014). *STEM integration in K-12 education: status, prospects, and an agenda for research.* Washington: National Academies Press.

National Research Council. (2011). *Successful K-12 STEM Education: Identifying Effective Approaches in Science, Technology, Engineering, and Mathematics.* Washington, DC: The National Academies Press.

Vasquez, J., Sneider, C., & Comer, M. (2013). STEM lesson essentials, grades 3–8: integrating science, technology, engineering, and mathematics. Portsmouth, NH: Heinemann.

KEY TERMS AND DEFINITIONS

Problem-Based: The use of problems or tasks as a vehicle for student learning.

Problem-Solving: The process of analyzing a situation or problem to reach a solution.

Professional Development: The activities of teacher learning during a professional career.

STEAM: STEAM is an acronym for science, technology, engineering, arts, and mathematics. In education, the acronym signals the integration of the separate content discipline areas for teaching and learning and expands the STEM to include arts.

STEAM-Focused Initiative: The aspects of system-wide efforts that encompass the integration of science, technology, engineering, arts, and mathematics for teaching and learning.

STEM: STEM is an acronym for science, technology, engineering, and mathematics. In education, the acronym signals the integration of the separate content discipline areas for teaching and learning.

Chapter 6
Making Success:
Researching a School District's Integration of the Maker Movement Into Its Middle and High School

Keith W. Trahan
University of Pittsburgh, USA

Renata de Almeida Ramos
https://orcid.org/0000-0002-5649-9061
University of Pittsburgh, USA

Jeffrey Zollars
University of Pittsburgh, USA

Wei Tang
https://orcid.org/0000-0002-9981-4594
University of Pittsburgh, USA

Stephanie Maietta Romero
University of Pittsburgh, USA

Cynthia A. Tananis
University of Pittsburgh, USA

ABSTRACT

Increasingly, the maker movement has been pointed to as a means of bringing more innovation and creativity into education. As an educational program, making has pressed educators to question entrenched beliefs and assumptions about the structure of activities, lessons, and classes, pushing them to embrace a more student and experience driven learning environment. "Making Success" was a two-year research project to investigate and describe the integration of making into one school district's middle and high school. The starting point of the research was to learn and

DOI: 10.4018/978-1-7998-2517-3.ch006

describe the critical characteristics and capacities that allowed TRSD to integrate making so deeply into its secondary schools. A key lesson of the project was that many interconnected ideas and people played important roles in the initiative to bring about success.

INTRODUCTION

Making is one of the more recent initiatives espoused as a potential remedy for the persisting challenges of the United States of America (U.S) schools. Educators, researchers, and policy makers have initiated countless efforts to improve U.S. education and workforce development in Science Technology Engineering and Mathematics (STEM) (Cuban, 2018; Elmore, 2000). Increasingly, the maker movement has been implemented in U.S. schools to provide learning opportunities that spark student interest in STEM. As a teaching practice, *making* is a set of activities that uses high and low technologies to design, construct, test, and revise objects while integrating a range of disciplines (Bevan, 2017; Halverson & Sheridan, 2014; Peppler & Bender, 2013). Making has pressed educators to question entrenched beliefs and assumptions about the structure of activities, lessons, and classes, pushing them to embrace a more student and experience driven learning environment.

Found in both formal and informal education settings, makerspaces are the material representation of the Maker Movement. Makerspaces house the resources used to foster and support making. The Fabrication Laboratory (Fab Lab), from the Massachusetts Institute of Technology (MIT), is perhaps the most often cited example of a makespace, particularly regarding the promotion of high end technology (e.g., 3D printers, laser cutters, etc.). Nonetheless, a makespace can be any space where making is promoted.

Although making has become well accepted in U.S. education, the maker movement is not without critics. In the U.S, making is often depicted in its most narrow and branded version as mostly a pursuit of white, middle-class, male hobbyists with an abundance of technical knowhow, experience, and resources to create an array of artifacts (Barton, Tan, & Greenberg, 2017; Vossoughi, Hooper, & Escude, 2016). Critics have also cited a number of difficulties and challenges related to making: the considerable time it takes for educators to learn how to use the technology in the makerspace; how to connect making to the curriculum in a meaningful way; how to address constant battles with technology (e.g. troubleshooting) and students' misuses of technology; the transdisciplinary nature of making (e.g. some teachers may feel comfortable with some overlapping of crafting and computing, but others don't); and questions about how to assess making, especially for projects that are not finished within timeframes that work for the school or that fail altogether (Stevenson et al., 2019; Fields et al, 2017; Resnick & Rosenbaum, 2013).

On the one hand, many of these challenges are illustrations of how schools, steeped in traditional structures, procedures, and policies are often the antithesis of the energy and engagement of the maker movement. Making, with an uncommon popular appeal to different educational actors, such as students, parents, school teachers and leadership, policy makers, and researchers, may represent a unique opportunity to implement progressives changes at schools: more experiential, student-centered, real world connected approaches that were historically promoted by Rousseau (1961), Dewey (1902), Freinet (1993), Montessori (1912), Piaget (1964), and Vygotsky (1978) (Martinez & Stager, 2013). Thus, educators can learn much from research of a school district's extensive integration of making in its schools.

This chapter is organized as follows. First, the authors present the background on their theoretical framework for the study of the maker movement and how that relates to practice. Second, the school district and context of the study are described. Third, researchers describe the data collected, how it was analyzed and interpreted, and the limitations of the study. Fourth, the researchers present findings organized by the primary actors of the study: leadership, teachers, and students. Lastly, a discussion and conclusion are presented.

BACKGROUND

Making appears to always be in the process of redefining itself through trial and error. This process results in a synthesis of new formats as it continues to remake itself for new learning contexts. To develop a more comprehensive understanding of making, the authors researched the literature starting from a global epistemological perspective, in this case constructionism (Papert, 1991), and increasing the focus of the search in a series of steps from theory to practice. First, researchers narrowed the theoretical lens by delineating the perspective in terms of the Science, Technology, Engineering, the Arts and Mathematics (STEAM) and the most recent understanding of the maker movement in the formal education environment. Narrowing the focus further, the researchers saw value in acquiring a means to assess and evaluate teachers' experience with making as a practice and its relationship to student engagement. The focus on student engagement led to Flow Theory (Csikszentmihalyi, 1990) and the observation of the components of flow as occurring in classroom participants as one means to gain insight into the practical application of a student-centered, creative learning environment. Therefore, narrowing from the epistemology of constructionism to the application of flow components, the resulting insights and findings can provide a better-defined overall understanding of the phenomenon of making when implemented in schools. Thus, the researchers framework for supporting the use of making in schools draws from theories of learning, school reform, and engagement.

Constructionism (Papert, 1986) forms the theoretical foundation for making as pedagogy and connects to two other often cited learning theories in U.S. education, Dewey's (1902) experiential learning and Piaget's (1964) constructivism. More recently, a prevalent thread of school reform has been the STEAM movement, with which making is often tied. STEAM is an extension of the long standing U.S. focus on improving STEM education by going beyond the limitations of science, technology, engineering, and mathematics, through the addition of arts and its correspondent potential towards wonder, critique, inquiry, and innovation (The Institute for Arts Integration and STEAM, 2019). Collectively, making's connection to these theoretical foundations provides promise that it can help schools provide opportunities for deep STEM engagement and learning for students.

Constructionism (Papert, 1986, 1991) is commonly recognized as the epistemological learning theory at the foundation of making (Blikstein, 2019; Bevan, 2017; Martinez & Stager, 2013; Kafai & Resnik, 1996). Constructionism informs making in two important ways: knowledge is constructed uniquely by the student (e.g., constructivism) and this construction is better done while building a product and sharing the processes and results (e.g., constructionism). As such, constructionism is a "synthesis" of Piaget's constructivism epistemology that explored the "opportunities offered by technology" to engage students socially and emotionally on the "construction of an intelligent entity rather than on the acquisition of knowledge and facts without context" (Papert, 1986, p.8).

Constructivism is defined as a "genetic epistemology" (Bringuier, 1980, p.7) theory of learning developed by Jean Piaget. Constructivism holds that knowledge is constructed in continuous and incremental ways through processes of assimilation and accommodation. Therefore, knowledge cannot be transmitted. Educators who use this theory as a foundation for their practice are called constructivists. Constructivists aim to design learning environments that accommodate more complex and meaningful ways of learning; where students construct knowledge in an integrated and relevant way rather than fragmented or isolated manner. Students are at the center of the learning process and considered active participants in the construction of their knowledge, thereby making the learning process more personally significant. Since constructionism is the theoretical foundation of making and the theory underlying it is constructivism, making may be an effective way to promote a more constructivist learning environment in our schools.

STEAM (Science, Technology, Engineering, Art, and Math) is a similar and often synonymous pedagogy to making. There are many similarities between the two. For example, in terms of training or training methods, both are based on projects or problems; in the purpose of curriculum setting, they emphasize student-centered and hope to be interdisciplinary. Both focus on hands-on practice and strong connections with reality. Makers use the same skills that scientists and engineers use, which is to overcome challenges by exploring, innovating, and problem solving.

Through STEAM and making, students gain direct experience with STEM, which can help peak interest in college and career paths. Lent et al. (1994) developed the social cognitive career theory (SCCT) to predict interest and intent to pursue careers. The factors include background learning experiences, self-efficacy, and personal, environmental, and behavioral interaction. As a project-based learning, making prepares students with knowledge of STEAM in school, gives access to technology and guidance from teachers, which increases their awareness, intrinsic motivation, and interest in STEM careers.

To further unpack student engagement with making, the researchers adopted the lens of Flow Theory. Mikhail Csikszentmihalyi (1975) first introduced Flow Theory, or 'optimal experience' theory. The common, simplified understanding of this theory is expressed by the sports analogy of when an athlete is "in the zone," experiencing peak performance. Previous research identified nine components of flow, which while working in combination to varying degrees within the individual produces an 'optimal experience' (Moneta, 2012). The nine components contributing to a flow experience are a balance of the challenges to skills, the merging of action and awareness, goals, environmental feedback, focus, the paradox of control, loss of self-consciousness, perceived distortion of the perception of time, and an autotelic experience (enjoyment) (Moneta, 2012). The nine components of Flow Theory suggest that flow and engagement are closely related (Shernoff, 2013). Through the course of an activity, the higher occurrence in number and frequency of observed flow components by a learner suggests the likelihood that the learner will approach an optimal learning experience (flow). Movement in a positive trajectory would indicate meaningful engagement that may lead to optimal learning. On the contrary, the less frequent occurrence of flow component evidence observed suggests a move away from an optimal learning experience and toward disengagement.

DESCRIPTION OF THE CASE: TRSD'S INTEGRATION OF MAKING

This case study was a two-year (September 2016 - August, 2018) National Science Foundation funded research project to investigate and describe the broad integration of making into the middle and high school of Three Rivers School District (TRSD)[1]. TRSD is a school district in the mid-Atlantic, rust belt region of the United States. The region has been noted as an especially rich and diverse context for the maker movement. The region has a history in the innovation, discovery, technology, and creativity of making, going from the heat and grit of steel making and related heavy industries, to high-tech worlds of entertainment, healthcare, and transportation technology.

Since 2013, TRSD has fully embraced making in its middle and high school, coupling key making tenants with additional related innovations, such as gamification, embodied learning, and educational technology integration, while also embracing personal technology among teachers and students resulting in 1:1 iPad access to create a new culture and learning environment (see Trahan et al., 2019). This initiative was brought to the school district by its superintendent and vice superintendent. These leaders were inspired and supported by local leaders in technology, creative arts, education, and philanthropy to integrate making into the district. Yet, the implementation of the maker movement at TRSD was broad involving all secondary school actors (leadership: superintendents and principals, teachers, and students) and the community.

From there, TRSD's high levels of consistency and capacity in personnel allowed the initiative to grow, developing multiple maker classes, such as game design and multi-media productions, and renovating existing spaces into makerspaces. The process for the initiative involved a combination of professional development (PD), improving facilities, investing in technology, and adapting school curricula and schedules. Substantial time and effort were given for a wide range of PD so that teachers could lead the 39 making related classes and utilize the 15 spaces (six at middle school and nine at high school) that came with the integration of making.

To add so many new and innovative elements to school, TRSD developed a culture of learning, collaboration, and risk taking. One of the biggest efforts to integrate making into the learning culture was the creation of the Imagination Station. The Imagination Station is a complex of makerspaces at the middle school that provides original curriculum for students to learn about robotics, engineering, design, and programming with state-of-the-art tools and technology. Imagination Station is a required class for students during middle school, with limited exceptions related to student's schedule conflicts. In this high school the use of the makerspace is not compulsory for students, but teachers from all disciplines are encouraged to use those spaces.

The researchers undertook this study to answer the question: what were the characteristics and capacities of TRSD's integration of making into its middle and high school and which had been critical for success? From previous work in STEM reform in the region, the researchers came to know the basics of TRSD's integration efforts described above and many of the interconnected people who played important roles in the initiative. Thus, the development of the study was more collaborative than most. Researchers had known the leaders of TRSD for some time and all were interested in chronicling, describing, and conducting inquiry on their use of space, innovation, and making in their secondary schools. Based on prior work in the field, the authors felt that TRSD was integrating making to a high degree and were interested in describing that integration and studying the experience of administrators, teachers, and students; and conducting some exploratory research regarding outcomes.

METHODS

The goal of this study was to describe the story and identify critical building blocks of people, process, and culture for TRSD's integration of making. To understand TRSD's integration of making, the researchers conducted a mixed-methods case study. More specifically, to describe what the integration entailed and how it appeared, observation and documentation methods were used. To gather evidence of personal experience with the integration, interviews, focus groups, and surveys were conducted.

In Year One (Y1), the study began with a complete inventory of TRSD's makerspaces and equipment. Researchers conducted site visits at both the middle and high school, photographing and inventorying equipment in each of the classrooms and spaces related to making integration: six in the middle school and nine in the high school. In these spaces, researchers then conducted 40 observations of making classes: 18 at the middle school and 22 at the high school. Observation data were collected through field notes and a combined and adapted observation instrument drawn from Tushnet, et al. (2000) observation instrument for technology rich learning environments and Silk's (2015) STEAM observation tool for dynamic learning environments.

The researchers then followed up with semi-structured interviews and focus groups with TRSD administrators, teachers, students, and community partners. Interviews were conducted with the four key administrators involved with the development, leadership, and implementation of the initiative: the superintendent, assistant superintendent, high school principal, and middle school principal. Interviews were also conducted with the 13 teachers who had taught the making classes that were observed and subsequently focus groups were conducted with students of those teachers. The student focus groups were designed to collect data on students' experiences in making classes, their interest in taking additional making classes, their interest in making activities outside of school, and their interest in pursuing careers and post-secondary studies related to their experience in those classes. Four non-making teachers, who had been in the district before making integration, were also interviewed. Five community partners who were identified by TRSD administrators as instrumental in bringing making to the district were also interviewed. These interviews were designed to gather data on the partners' contributions to TRSD's efforts and their perceptions of how TRSD has contributed to broader regional making initiatives.

In Year Two (Y2), the researchers conducted surveys of TRSD teachers and students, and conducted secondary analysis of student data. The survey administered to the teachers was adapted from Becker & Anderson's (1998) *Teaching, Learning, and Computing: 1998. A National Survey of Schools and Teachers - Describing Their Best Practices, Teaching Philosophies, and Uses of Technology* (TLC). The survey was administered by email to all TRSD secondary teachers. It was completed by 90 teachers, yielding a response rate of 86.4%. The survey included forty-two scaled items and one open-ended question[2]. Seventy-four teachers were at TRSD prior to the integration of making. For only these teachers the survey included the open ended question (n=68) and a set of retrospective self-assessment items (n=72) on "traditional transmission instruction" and "constructivist-compatible instruction" (Ravitz, Becker, & Wong, 2000) examining the frequency teachers reported using these types of instruction before and since the integration of making and the perceived role that the integration had on those reported changes. Sixteen of the retrospective items were selected to recreate the Pedagogical Change Index (Becker & Riel, 2000), which was used to report teachers' perceptions of their pedagogical change. Descriptive and correlation analyses, and two one-way ANOVAs were run on the IBM SPSS Statistics software version 25; assumptions of homogeneity and normality were not violated. ANOVAs were used to explore possible differences between Maker teachers (e.g. Makers and non-Maker teachers) and school level (e.g. middle and high schools' teachers). For this study, maker teachers were those who taught at list one making class at the time of the survey administration. One teacher who taught at both schools was not included in the analysis by school level. Answers to the open-ended survey question were coded as follows: (1) teacher was affected by the initiative, yes or no; (2) teacher reported change because of the initiative, yes (positive and negative), no, little; (3) technology (activities and influences); (4) student affect or change.

To gather evidence of student experience and perception of the integration of making at TRSD, students in their first year of high school (9th grade) and last year of high school(12th grade) were surveyed and secondary data collected. The survey included questions pertaining to students' experience of making classes and school environment, as well as their perceptions of making subject areas. The survey also included sections from the STEM Career Interest Survey (STEM-CIS) (Kier, et al.,

2013) and the Constructivist Learning Environment Survey (CLES). Two hundred twenty-six students completed the survey. For the secondary analysis of TRSD student course taking and performance data were conducted. For this, the researchers worked with TRSD district staff to collect de-identified demographic, course taking, and performance data for the 12th grade of the 2016-2017 and 2017-2018 school years. The data were used to investigate rates and patterns of participation in making classes; relationships between making course participation and other student variables; and relationships of making course participation with student performance.

Lastly, the researcher conducted observation focusing on student experience with making. A sample making class of 14 eighth-grade students in the Imagination Station was selected. The teacher divided the class into four student groups, each with its own workstation, that functioned as an imaginary business designing and marketing speaker boxes. Recording occurrences of each of nine flow components observed in student behavior provided insight into student engagement that leads to learning (Zollars, 2018). The following is an example of how a flow component (e.g. perceived distortion of time) could be observed: if during a classroom observation, a small group of students intensely working on a project were surprised by the class change bell, then the action was tallied as evidence for the flow component perceived distortion of perception of time. Likewise, if during the student focus group interview a student indicated that even though the assignments are a challenge s/he is able to meet that challenge without too much frustration, then that action was tallied as evidence for the flow component balance of skills to challenges. In addition, evidence leading away from flow was recorded. Students reporting aspects of the class project as being too difficult evidenced an imbalance of skills to challenges. Likewise, students disengaged and talking about unrelated topics evidenced a lack of focus. Similar evidence was gathered for the remainder of each of the nine flow components over the period of the school term. Evidence gathered during the course of the term were analyzed for emergent patterns as occurring in individual students, in each of the four student groups, and as a whole class.

LIMITATIONS

A number of limitations should be considered with regard to this research study. This study was singular and not intended to be generalizable to all school settings, content areas, grade levels, or demographics. Authors acknowledged that the TRSD was an outlier in many forms. Furthermore, not all making classes were the same. Even making classes within the same building could vary in many ways, such as student expectations and environmental complexity.

The Teacher Survey had two main limitations, which are intrinsic to the TLC survey design. First, the survey measured change by asking teachers to compare their practice at the time of survey administration (January 2018) to what their practice looked like from three (2015) to five years (2013) prior. This type of retrospective self-reporting may have a response bias in what could be measuring socially desirable behavior, characterizing a possible validity threat to that part of the study. On the other hand, the instrument's high validation served to reduce this type of threat (Lam and Bengo, 2003). Second, the classification of items as "constructivist-compatible instruction" and "traditional transmission instruction," which also forms the Pedagogical Change Index, followed the definition and structure of the Becker (1999) study. By maintaining consistency with the original work, the researchers inherited the validity of the instrument and Index, but also thought that some items could have been classified differently. For example, the item "have students write a page or more on a single subject," said to be constructivist, could also be claimed to be an example of a traditional practice.

Limitations also existed in terms of observing for flow. Previous research did not yield strong results in capturing flow on a consistent basis for a majority of students in traditional classrooms (Shernoff, 2013). Classrooms with a traditional instructional framework and structure may not lend themselves to recognizing or fostering components of flow in students in that type of learning setting. Traditional schools modeled after early twentieth century industry conditioned students to a style of learning that reflects the measurement of learning as the subtraction from a perceived perfection. Assessment of skills and knowledge are often based on the subtraction from a 100% score. In semi-structured making learning environments based in constructionism, students and teachers may not possess an underlying framework for understanding this approach to learning, let alone the lexicon to explain it.

Lastly, the researchers combined experience as educators and researchers extends from three different countries (Brazil, China, and United States) and multiple positions in formal and informal education, both in K12 and higher/tertiary education. Given their diverse experiences, while working in this research, the authors converged in the belief that making provided a unique opportunity to rethink and change the design of learning environments, promoting a more engaging and transformative education to all students. However, the researchers did not subscribe to the idea that making was the only way to improve schools or that it was the best initiative for all schools. The decision on how to promote change and improve schools should be made based on the needs and hopes of the community. Nonetheless, if a school chooses to implement making as a driver of change, or one of them, the researchers firmly believe that it should be inclusive of all students, regardless of race, gender, immigration status, age, physical abilities, grades, or beliefs on natural inclinations for STEM.

FINDINGS

Findings from this research pointed to a complex and comprehensive initiative. TRSD's integration of making started with visionary leadership, supported by a regional network of education and technology organizations. The initiative took root because of the buy-in from principals, teachers, and students. The combination of these capacities and characteristics allowed the initiative to grow and flourish. The researchers found that a recurring mantra was to know and start where you are, with the people and resources you have, to take chances, and to learn together. Thus, findings are presented for the three primary participants with the making initiative: first, the leaders, followed by the teachers, and then the students. The researchers hope the findings may inform and provide hope to other schools and communities that change, although hard and labor intensive, is possible, especially when undertaken collectively.

Visionary Leadership

Findings suggest that the visionary leadership of TRSD administrators was the origin of the initiative. Interview insights indicated that district leaders were inspired by community partners to motivate teachers and students to engage with school in new ways. Administrators first had a vision of how school could be different, and second, provided teachers and students with the necessary time, space, and support to innovate. Characteristics that contributed to this reform initiative included a network of community partners, individualized teacher professional development, space renovation and equipment purchases, and curricular changes. Combined, TRSD fostered a culture where teachers and students could innovate and take risks.

Network/Local Milieu: Productive Nexus of Informal and Formal Education

TRSD worked with many leaders from informal and higher education, and industry to develop its making activities, classes, and spaces. The original spark for TRSD leaders came from the founder of the Entertainment Technology Center (ETC) at Carnegie Mellon University. ETC was founded as a studio, project-based program at the intersection of gaming, design, and interactive storytelling; a truly applied STEM learning environment. TRSD's goal was to transform how students learned in TRSD and to motivate students to come to school by creating innovative, colorful spaces; painting the walls, structuring the spaces so that there was collaboration. TRSD set forth to provide a more engaging culture in its schools; modeled after the open studio environments found in ETC, informal education spaces, and corporations, such as Google and Schell Games.

These community partners inspired TRSD leadership to focus more on preparing students for an emergent workforce focused on human-centered design[3]. A unique regional context provided TRSD with a milieu of initial inspiration, the continuing economic and demographic shift toward cutting edge technology-driven innovation and a younger and more nimble workforce, and focused external support, through the significant legacy of philanthropic investment of manufacturing families and companies in the region's past, to carry out the initiative. TRSD visits to places such as ETC and a regional children's museum were part of the initial inspiration and ongoing support in bringing making into its schools. Maturation of collaborative local organizations such as Remake Learning (2019), MAYA Design, the League of Innovative Schools, Schell Games, and the Tech Shop (just to name a few) provided many resources. Early and steady financial commitment from regional foundations helped the district to gain expertise and moral support provided seed funds that were leveraged to gain additional grant funding. Thus, mentors, thought partners, and cheerleaders were available through these and other organizations and played an important role in TRSD's efforts.

Capacity: Personnel

Leadership's decision and ability to first look inward for staffing played an important role in the success of the initiative. TRSD benefited greatly from having an existing staff that was, in many cases, qualified and excited to integrate making into their schools and classes. This was developed and fostered through the district's internal recruitment via invitation for current teachers to apply for new making positions. In addition to the highly qualified and motivated staff, the district generally has very low turnover in its entire system, including its school board, superintendent and central office staff, school administration, and teachers. This provided continuity for the district across longer-term changes and contributed substantially to the initiatives success.

Space and Equipment

On a more practical note, TRSD's leadership initiated substantial space renovations and equipment purchases that were planned and progressive over time. Although TRSD committed considerable internal funds, the initiative also benefited from external funding from local and regional foundations. To garner such support, TRSD leaders had to develop and communicate a vision, entice and engage external funders, convince and continue to inform their community and board, and maintain and extend more traditional measures of success (attendance, decreased behavioral issues, academic achievement, standards-based assessment proficiency, etc.). Through our interviews, we found that TRSD leadership sought out various examples of making across the country and were able to procure local collaborations and grants to support their purchases of equipment, space renovations, and staff professional development. Thus, TRSD was able to create in its schools' learning environments that are more often found in informal education; thus, affording students opportunities for open, self-driven learning that they otherwise might not have had. This is particularly relevant to semi-rural communities, like TRSD, with fewer informal learning opportunities than large urban centers. Our findings showed that TRSD students were not likely to travel forty minutes into the city to participate in similar informal learning opportunities, but were likely to participate in opportunities offered by the district or school.

Classes and Curriculum

TRSD leadership set up its curriculum so that nearly all students experience making in both middle and high school. TRSD approached making in their middle school with a curriculum that would first prepare the students with foundational skills in the sixth and seventh grades. The students in eighth grade complete a 6-week rotation in each of the three areas of art, computer coding, and technology education to create guided projects. The eighth-grade courses take students through the innovation process from idea generation, through design and prototyping, and finally, to product completion. At the high school level, Three Rivers Academy is the class that every student takes. This is a basic introduction to creativity, technology, and computational thinking. After the Three Rivers Academy class, students can choose to take additional making classes or not. Thus, in both middle and high school, nearly all students are able to engage in STEM-related making projects that are based on their own experience and interest.

Through this commitment, TRSD leaders position making and engagement with STEM as a regular curricular component, as opposed to a self-selected program engaged in by some or few students with pre-existing interest. This model holds promise to broaden opportunities for students to experience and build an interest in STEM as it is applied in a relatively individualized context. In terms of the courses that were added to the offerings at TRSD, some were pre-packaged curricula, for example, Zulama and Project Lead the Way, and others were developed organically at the schools, like Imagination Station and Multi-Media Productions. Others were developed to meet the needs of particular student groups. The Girls Maker Lab class was created when no girls were attending the general Maker Lab class; thus, addressing the gender disparity.

Teachers

Findings showed that the making integration promoted a more constructivist approach to teaching, especially among Maker teachers. For example, teachers reported increases in their use of group work and listening to students, along with a decrease in the use of textbooks. However, in this process of change, teachers maintained many and increased the use of two traditional practices. Furthermore, most of the teachers (n=58, 80%) indicated that their changes were promoted by the making initiatives; however, the level of influence varied greatly.

Characteristics

Most of the Teacher Survey respondents (n=90) were female (65%), white (80%), held a master's degree (62.5%), were an average of 41 years old and had 16 years tenure at the school. Teachers taught various subjects including languages (n=22, 26.8%), mathematics (n=15, 18.3%), science (n=16, 19.5%), social science (n=8, 9.8%), art and music (n=8, 9.8%), and other (n=13, 15.9%). Most teachers were identified as being non-maker teachers (n=70, 84.3%) while 13 teachers (15.7%) were identified as Maker teachers. Eleven maker teachers, three from middle school and eight from high school, and 63 non-maker teachers, 35 from middle school and 31 from high school, were included in the retrospective and Pedagogical Change Index analysis. Sixty-eight teachers answered the open-ended question, including 10 maker teachers, and 34 teachers from the middle school and 33 from the high school.

Work Environment

Teachers working in the TRSD during the implementation of the making integration reported high levels of engagement at work with their leadership and each other, and appreciation for the new learning environment. Almost all teachers (n=72, 93.5%) agreed with the item "I have the freedom to try new things in class" and that "the district is an exciting and innovative place to work" (n=69, 89.7%).

A majority of the teachers agreed that "my administrators support me in pursuing training(s) that fit my needs and interests" (n=67, 87.1%) and they "feel supported by my administrators" (n=66, 85.8%). Most of the teacher felt "encouraged by other teachers to try out new ideas" (n=66, 85.8%), that "most teachers here share my beliefs about what the central goals of the school should be" (n=63, 81.9%), and "my principal's values and philosophy of education are similar to my own" (n= 61, 81.3%). However, teachers were not so confident regarding some items related to their professional development, indicating more disagreement (from strong to slightly) than agreement, such as in the items above. Most of the teachers disagreed

Figure 1. Work environment in TRSD (n=77)

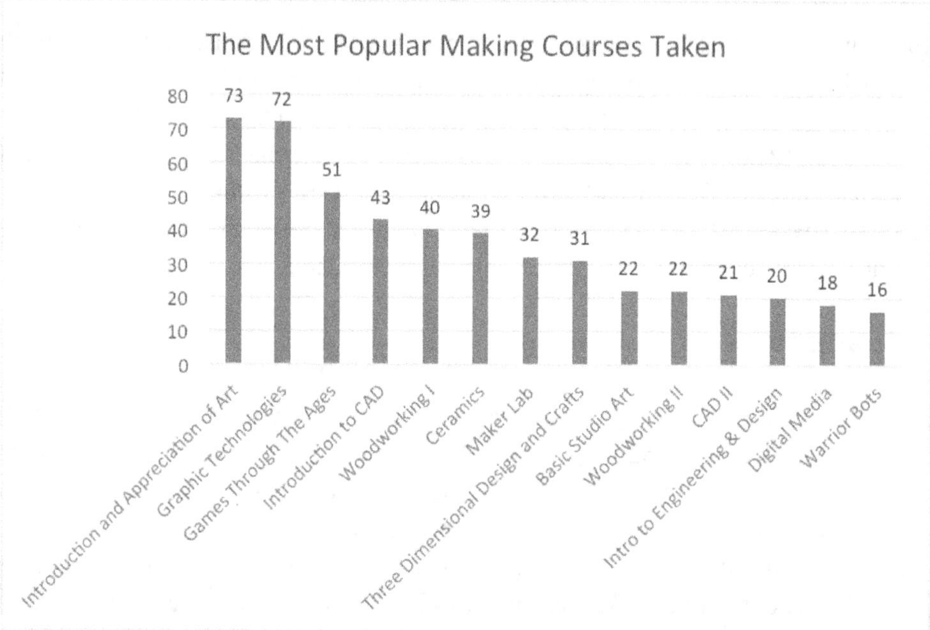

that "teachers play an important role in defining staff development activities" (n=47, 61.1%) and that "major staff development activities are followed by support to help teachers implement new practice" (n=39, 50.7%). Nonetheless, 89.7% (n=69) of the teachers agreed that they are "continually learning and seeking new ideas" (Figure 1).

Classroom Practices

Generally, utilization varied among constructivist teaching practices across the various teachers, including maker teachers. In support of the constructivist learning, most teachers indicated that they frequently had students verbally justify and explain their reasoning (n=77, 97.5%) and elicited students' ideas and opinions about the material presented (n=71, 89.9%), encouraged students to make conjectures about what they will learn or discover in the class (n=68, 86.0%), utilized small group discussions (n=67, 84.9%), had students relate what they are working on to their own experiences (n=63, 80.8%) or had students do hands-on/laboratory activities (n=54, 68.5%) (Figure 2). In addition, teachers reported to "rarely say" how a subject or activity will help them to get a good grade (n=63, 79.7%), check for homework completion (n=53, 67.1%), or used rewards for students doing well

Figure 2. Most frequently used constructivist practices (n=79)

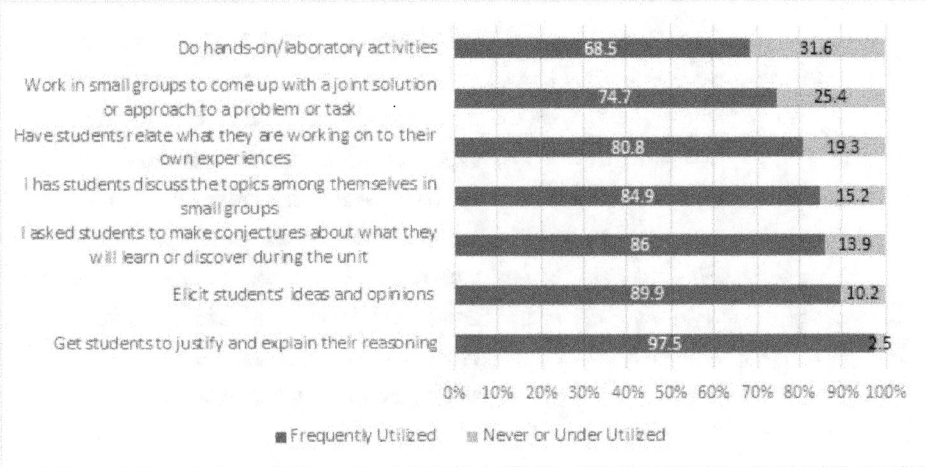

in their classrooms (n=50, 63.3%). A middle school, non-Maker, math teacher's experience illustrated such practice:

I think – especially I know math has become more collaborative. Not just [individual]... you know, "I do my math and keep to myself" kind of thing. It's more collaborative, and a lot of the activities that we're doing are more innovative. Like we give a problem at the onset so they do it themselves, they work in groups, we talk it out as a large group. So we are doing more of those kinds of things than just one-on-one. We still do the one-on – you know, by myself kind of work, but we do a lot more group work. A lot more, "how did you solve this problem? Could we solve it differently," that kind of thing. So – and make inferences about what could happen when we do some [unintelligible] math kinds of things. So it's different than it was before.

However, many constructivist practices were reported as rarely utilized or under-utilized by teachers. Most teachers indicated that they "never" or "sometimes/seldom" had students debate or argue over a particular point of view (n=70, 89.7%), demonstrate their work to an audience other than family and school peers (n=65, 82.3%), make a product that would be used by someone else (n=62, 78.5%), write an essay or hold a serious discussion assessing their own work (n=57, 72.1%), design their own problems to solve (n= 55, 69.6%), work on problems that had a non-obvious method or solution (n=55, 69.6%), suggest new classroom activities or topics (n=55, 69.6%), explain their thinking in writing (n=54, 68.3%), decide on their own procedures for solving complex problems (n=53, 67.1%), write in a

Figure 3. Reported least frequently used constructivist practices by teachers (n=79)

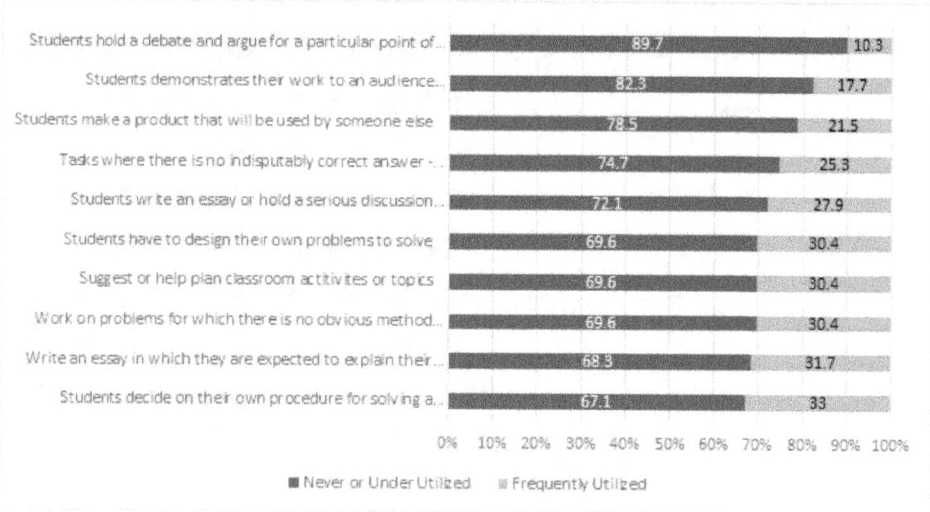

journal (n=50, 63.3%), or assigned tasks where there is no indisputable or correct answer (n= 59, 74.7%) (Figure 3).

Change in Teaching Practices and Assignments

Whenever leadership brings a new initiative to a school, some change is to be expected. Through surveys and interviews conducted at multiple time points, the researchers were able to compare teacher's attitudes and practices over time. Teachers indicated whether they used each teaching practice or assignment (traditional or constructivist) "less now than 3-5 years ago," "same as before," "more now than 3-5 years ago," "much more now than 3-5 years ago," or "never used the practice." Constructivist practices included, assignments (e.g., have students work in long projects, have students explore a topic on their own, have students teach or help other students) and practices (e.g., allow myself to be taught by students, evaluate students through their products instead of tests). An open-ended question also asked how the making initiative had affected them personally, as well as their teaching curriculum or classroom.

Specifically teachers indicated that they were utilizing the following constructivist practices or assignments "more" or "much more than before:" having students work in groups (n=46, 63.9%), let students partially influence the topics in the lesson (n=43, 59.8%), have students teach or help other students (n=43, 59.7%), have multiple activities going on in the room at the same time (n=42, 58.3%), allowed themselves to be taught by the students (n=39, 54.9%), and evaluate students through

Figure 4. Reported change percentages on teaching practices items: constructivist-compatible instruction and traditional-transmission instruction (n=72)

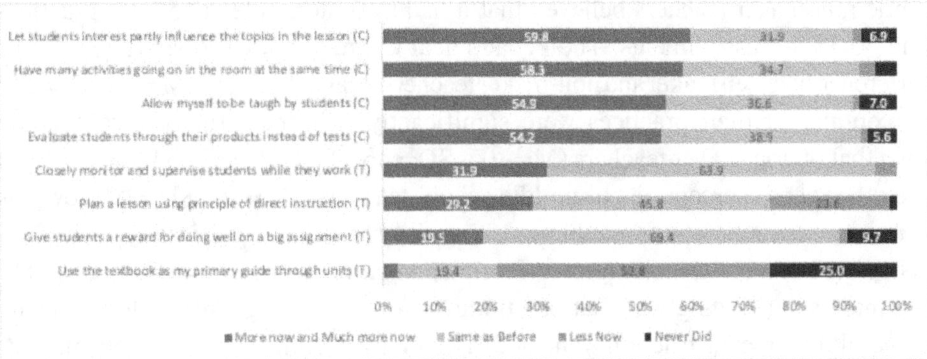

deliverables or products instead of tests (n=39, 54.2%). However, teachers did not necessarily decrease their traditional practices to make room for increases in the constructivist ones. The only traditional practice they did less than before was related to the use of textbooks, as a "primary guide through units" (n=38, 52.8%) or as in answering "questions in their textbooks" (n=32, 44.4%) (Figure 4) (Figure 5). Thus, teachers tended to increase or sustain their constructivist practices without reducing their traditional ones.

In addition, the researchers found that being a maker teacher was significantly associated with a positive change in constructivist assignments ($r = 0.31$; $p < 0.01$) and a positive change in constructivist practices over time ($r = 0.34$; $p < 0.01$), indicating that they were using more constructivist pedagogies in their teaching than

Figure 5. Reported change percentages on teaching assignment items: constructivist-compatible instruction and traditional-transmission instruction (n=72)

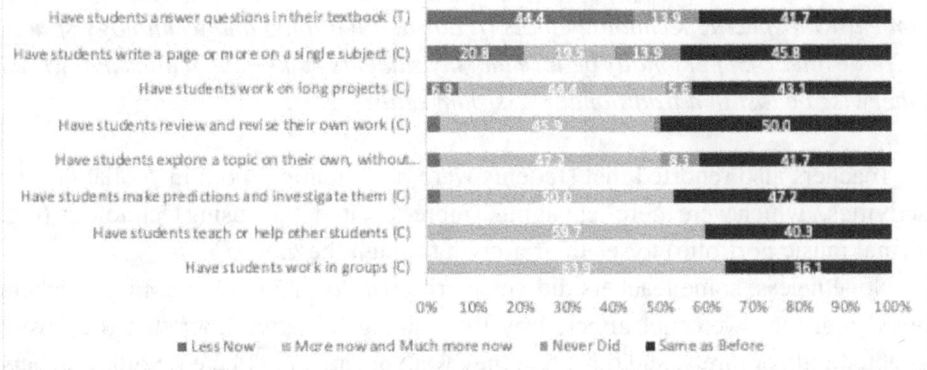

before. Maker teachers also indicated that the role of the making initiative affected them in a "major" or "substantial" way, 80% for both assignments and practices, while non-maker teachers believed that it had a "minor" role, 43.5% for practices and 40.3% for assignments. The Pedagogical Change Index further validates this difference between maker and non-maker teachers. Maker teacher's (M=.38, SD=.50) perceptions of their practices were significantly more constructivist-compatible than that of non-maker teachers (M=-.05, SD=.46) (F (1,71) = 7.134, p=.009). On the other hand, perceptions of middle school teachers' (M=.03, SD=.42) towards a more constructivist-compatible teaching did not differ significantly from high school (M=-.04, SD=.54) after the implementation (F (2,73) = .221, p=.802). An interesting finding was that one traditional practice (e.g. "plan a lesson using principle of direct instruction") and one traditional assignment (e.g. "closely monitor and supervise students while they work") increased for maker teachers, 30% and 60%, respectively. This may indicate that items recognized as traditional might be present in constructivist classrooms and take on a new meaning. For example, maker teachers working closing together with their students could be assessing their students more frequently and deeply to support their on going project than those who monitor students to locate right and wrong answers.

The related open-ended survey question revealed more detail about the impact of the maker initiative on the teachers. Most of the teachers described a positive impact (n=43) in themselves or in their practice. Teachers reported "realiz[ing] the importance of students being able to create," being "open to new ideas," and "not to be[ing] afraid of trying new projects with my students." Teachers described their teaching as becoming more "flexible," "fun," "meaningful," and "innovative," by doing things like, "cross curricular projects" and "turning the curriculum [into] more project-based engineering." As a middle school non-maker teacher wrote:

One of my goals for teaching [social science class] at [middle school] grade is to make the topics/lessons relevant. One way that I am able to accomplish this is by the integration of technology that is promoted within [TRSD]. By combining history content with specific technology tools (iPads applications) and technology spaces (iCreate Studio), I personally believe that my students make connections that would otherwise be lost in a traditional classroom setting.

Teachers also reported that students were participating more in "collaborative activities," which were more "engaging," more "excited," and using technology (e.g. virtual music portfolio) to see their growth through the grades.

Nonetheless, some teachers did voice criticism for the initiative. Five teachers noted that they were not affected by the making initiative, four that they were affected little or rarely, and three that they were negatively affected. Some teachers thought the "district's singular focus on technology literacy" was deviating students and faculty "from the overall importance of learning as a practice." Some teachers reported a "battle" with mobile phones and iPads and others cited "very little room for creative project" because their subject involved a state keystone exam and thus needed to be "very structured." Some teachers also described that the makerspaces became a "reward" for students, instead of a place of learning. Some teachers were also concerned at the time-consuming nature of implementing constructivist/maker practices. A high school non-Maker teacher noted that investing time to create new exciting lessons that integrated technology, to actually learn the technology themselves, and then planning an effective way to teach concepts using the new approach, actually "takes time away from true actual teaching and learning." Almost 67% (n=48) of the teachers reported that they spend much more time now preparing lessons. One of the maker teachers interviewed stated it is "a lot to juggle, [and is] daunting" to figure out how to squeeze more meaningful content in the classroom within the time they have. Interestingly, two teachers that reported making didn't affect them, expressed that their students were more "creative in their projects and seem to develop more innovative ideas in my classroom" and that the initiatives were good for the school district.

Resonating with the data presented above regarding teachers increased their constructivist practices, teachers in their interviews reported an ability to "grow" and do "what I wanted to do." They also noted that student attendance increased and that students were being more "innovative" and "creative," while "figure[ing] out" solutions in a "collaborative" manner and at their "own pace." In the voice of a veteran high school, English, non-maker teacher:

I had one of my students, when we did a project on Hamlet last year, we were comparing The Avengers in the story. So, he made a poster sort of and it had clocks on it. And so he had one clock moving very quickly and one clock moving very, very slowly and one that was going at normal time to compare The Avengers. And then yeah, he had all the other components, but he had free rein to design it how he wanted to. So he had working clocks on his poster. So I've seen them pull in little elements of things from the different areas, too, which I think has been really interesting. He did something else earlier that had lights in the year, too. So different kids, I think because it taps into that bank in their mind that they're like, "oh, I can do this." I'm like, "you can do whatever you want," so that's what I ended up with, which was truly awesome.

As this statement illustrates, the making initiatives promoted positive changes in the teachers' pedagogical practices. Since teaching is just one part of the learning process, students were also affected by the initiative. In the next section we present the insights and findings from the students.

Students: Course Taking and Engagement

The following section shows how TRSD making courses correlate with the constructivist orientation of the school environment. It outlines the patterns that emerged regarding student participation and perception of making as it is included in the school curriculum. In addition, making courses in TRSD's curriculum are disaggregated by student interest. A connection between TRSD making courses and subsequent STEM career interest is also explored. Subsequently, this section reviews the value of making courses in a constructivist learning environment as it occurs within a traditional school district. Finally, insights will be shared regarding student engagement as it occurred in a sample TRSD makerspace viewed through the lens of Flow Theory.

Student Course Taking and STEM Interest

The majority of students took at least one making course. Graphic Technologies, Evolution of Games, and Intro & Appreciation of Art were the three most picked making courses. Students chose making courses because its topic was interesting or prepared them for future careers. The same trend was found from students' perspective on making courses and makerspaces. Most students believed makerspaces made their school more enjoyable and making courses were helpful to their future career. The students who take at least one making course have higher career interest in science, technology, and engineering. The model of TRSD making courses exhibited that making courses could be delivered to every student and had further impact on the STEM career interest. It was also a good example that making courses at school were student interest driven.

Inclusive for All Students

Almost every student in TR high school took making courses. From course taking data, 154 students (95.7%) took at least one making course during their high school experience. Sixty-four percent of the students (n=103) took more than three making courses in the same period. Among students who took at least one making course, 45.5% (n=70) were female and 54.5% (n=84) were male, 145 (94.2%) students were white, and 29.9% (n=46) had free or reduced lunch. The average GPA for these students was 3.37 points and 70.1% of students (n=108) had a GPA above 3.0.

Figure 6. Top ten making courses in TRSD (N=161)

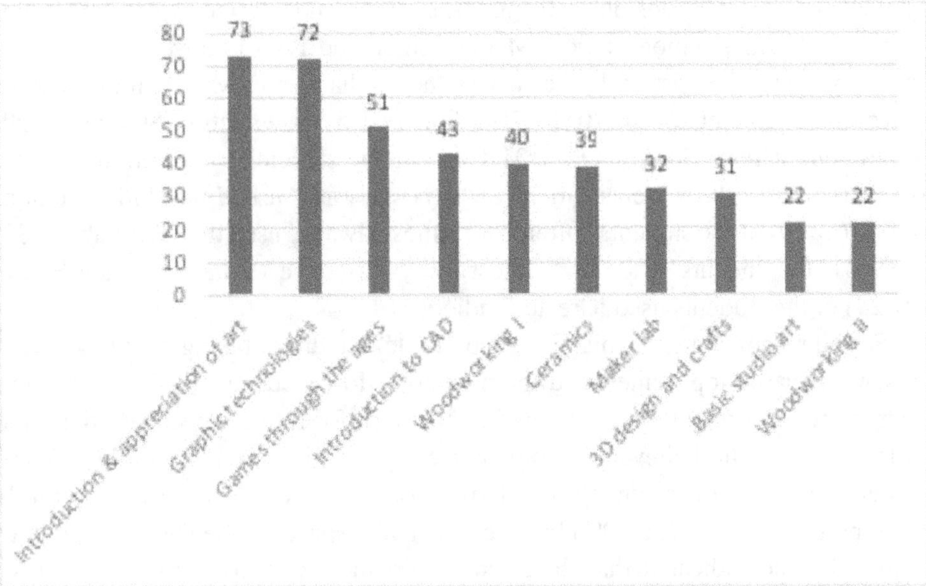

Although gender, ethnicity, economic status, absentee rates, and GPA were not significantly related with the number of making courses, the proportion of students taking more than three making courses during their high school years was high (n=103, 64%). Also, there was no difference of gender, ethnic, economic status, and academic performance between students who had taken more than three making courses and students who had taken less than three making courses. Thus, the access to the making courses was equally distributed (e.g., inclusive) between the students that showed interest in them.

Making Courses at TRSD

Making courses were among many other elective courses offered at TRSD, their number changed over time, and they were across different subjects. Making courses were categorized as such based on whether or not creating a new product was a key component of the course. In total 59 courses were identified as making courses, covering arts and design, computer programming and games, and manufacturing sectors. As shown in Figure 6, the most popular making courses taken were: "Introduction and Appreciation of Art" (n=73, 45.3%), "Graphic Technologies" (n=72, 44.7%), "Games Through the Ages" (n=51, 31.7%), "Introduction to CAD" (n=43, 26.7%), "Woodworking 1" (n=40, 24.8%), and "Maker Lab" (n=32, 19.9%).

In addition to making courses, TRSD created makerspaces where some making courses and activities took place. In our observations and interviews, iCreate Studio, Energy Lab, Imagination Station, Media Center, and FAB Lab were some of the most popular makerspaces. In the survey, the Media Center was the most popular place among students (n=170, 86.7%), followed by Imagination Station (n=90, 45.9%) and iCreate Studio (n=61, 31.1%). Among students who used the spaces, 62.3% (n=56) of the students "often" or "almost always" used Imagination Station, 44.2% (n=19) of the students "often" or "almost always" used the FAB Lab, 39.8% (n=90) of the students "often" or "almost always" used the Media Center, and 39.3% (n=24) of the students used iCreate Studio.

For all respondents, the biggest reason for taking or not taking making courses was whether the topic interested them or not. Most students also chose making courses because they thought it was fun or useful for future careers. At the same time, students who believed making courses would not help them in their future career or found them boring, opted not to choose making courses. Nearly two-thirds of respondents (n=137, 64.9%) believed the makerspaces made their school more enjoyable, and indicated that their school provided many opportunities to take interesting classes. Similarly, a majority of students (n=124, 58.7%) reported that their school was preparing them well for future careers. Over a third of the students (n=71, 33.7%) indicated an increased interest in attending school because of the makerspaces and courses that were available to them. Nearly half (n=94, 45.0%) of the students reported an increase in wanting to come to school because of having so many opportunities to explore a variety of topics. Generally, students had a very positive attitude toward making courses and makerspaces at TRSD.

STEM Career Interest

Findings showed that students who took at least one making course had higher career interest in science, technology, and engineering. The means for all students' career interest in science, math, technology, and engineering were 3.51 (SD=0.06), 3.59 (SD=0.06), 3.54 (SD=0.06), and 3.12 (SD=0.08), respectively. For 9th graders, the career interest in engineering, technology, math, and science was 3.06 (SD=1.08), 3.26 (SD=0.91), 3.43 (SD=0.80), 3.37 (SD=0.83), respectively. For 12th graders, the career interest in engineering, technology, math, and science was 3.20 (SD=1.16), 3.87 (SD=0.80), 3.77 (SD=0.90), and 3.69 (SD=0.97), respectively. The means for career interest in science, math, technology, and engineering of students who took at least one making course were 3.58 (SD=0.91), 3.64 (SD=0.90), 3.66 (SD=0.88), and 3.29 (SD=1.08), respectively. There was no statistically significant difference for career interest in math between students who took at least one making course and those who did not take any making course (p=0.081).

In the specific items, some differences between students who took at least one making course and students who did not take any making course were significant. Students who took making courses reported higher interest in and plans to use science in their future career than those who did not take any making courses. Students who took making courses rated that their parents would like it if they chose a science career and that they worked harder on math more highly. In a word, students who making courses reported higher interest in choosing science, technology, and engineering in their future career.

Student Engagement

To further the understanding of student engagement in making courses in TRSD in action, the researchers observed a sample TRSD makerspace for evidence of flow (Zollars, 2018). The Imagination Station, technology education makerspace included fourteen eighth-grade students divided into four subgroups, each assigned a group project of creating and marketing a wooden speaker box.The following is a summary analysis of the observations and student focus groups, focusing on insights gained from evidence of the more prevalently observable flow components.

A sampling of many essential questions listed below are based on components of flow and provided guidance during the classroom observations and focus group interviews to elicit the insights that follow: Did the students feel overly challenged, bored, or feel the challenge was adequately in balance with their skills? Did the students experience hyper-focus while engaged? Did they have specific goals in mind? Did they surprisingly run out of time? Did they enjoy the learning experience and why?

Each of the four student subgroups produced a body of evidence indicating to varying degrees the presence or absence of flow components. Small group dynamics created variations in intensity and frequency of evidence of flow components as manifested in each subgroup. The expressed learning experience and the quality of the project's final product paralleled the evidence gathered. Three of the student subgroups exhibited ample evidence of components leading to flow and found strong personal success in both meeting the objectives of the course and in making the learning experience meaningful.

In terms of flow, two components came to the forefront with each of three subgroups: 1. There existed a consistent balance between the challenges presented to them and the skill level of each of the three subgroups (a balance of the challenges to skills), and 2. The students of each of the three subgroups expressed satisfaction at being able to have control over the direction of their learning even though the project was structured by requirements (the paradox of control). For example, when students were asked regarding the difficulty level of the class if 'messing up'… making mistakes bothered them, one student (Alexandria) responded, "It is just part of the whole thing of making it. You have to mess up sometimes… Now, it [the end product] is really nice." Another student (Michael), when asked about outside-of-school experiences as related to his making experience, "I do sports. I play hockey, soccer, and track… [Why?] "… 'cause it's harder. It's more rewarding." When asked about whether they believe they have more or less control over what goes on in the maker class as compared to other classes, Michael replied, " More control—because we got to pick what we wanted to do…When, in a normal class, they would say 'ok, now we are going to do this', and then you just do it…[in Imagination Station] we just have more flexibility."

Of the three previously mentioned subgroups, in addition to evidence of many flow components, more than one group evidenced high instances of the components of focus and the perceived distortion of the perception of time than the other nine flow components: One student (Haylee) explained, " Um, like when we cut materials and stuff, we don't have time to look at the clock, we are really focused on our project so we don't know when the bell's going to ring." During one observation, Mr. C (the teacher) approached one subgroup that was highly focused on their project and hovering over their workstation fifteen minutes into the next period after the late bell and said, "I need you guys to get some passes and for you to go to your next classes so that your teachers don't get upset with me."

Another group exhibited a high frequency of enjoyment (autotelic experience) along with a balance of the challenges to skills and other components of flow. One student from this group (Nate) shared, "It is fun here because you are interacting with others. You are using some skills that some people may have, but some may not. Like, we sit in other classes. Like, we don't have chairs in this class because we are always working and making things. That is why I like this class." Another student (Malia) added, "The reason I like to take Imagination Station is because it is fun and you can work with others and make things that you maybe have not made before." Nate, explained his role in his group as similar to his feelings about playing baseball, "I like the energy when we are all together… You feel like you are really good at doing something good for the team—for your friends"

A fourth subgroup of students showed that under the same conditions and completing the same project assignment, an outcome distinct from the other three groups occurred. The tipping point for this last group occurred when it became obvious that the flow component of a balance of the challenges to skills was not present. The challenge of the project exceeded this group's collective skill level. Subsequently, this group cascaded into boredom, disengagement, and finally resistance.

DISCUSSION

The foundations of TRSD's initiative were a culture of innovation that started with constructivist, visionary leadership that allowed building principals and teachers to take the lead on implementation and explore their own areas of interest and expertise in making. This was embodied in the individualized and flexible PD that was offered and promoted among teachers. This constructivist approach with teachers allowed them the freedom to fail, which they then passed on to students. This culture of innovation transformed pedagogy, relinquishing control and allowing more freedom, creativity, and engagement throughout the schools.

Through our interviews, researchers found that TRSD leadership sought out various examples of making from across the country, and then looked for local collaborations and grants to help develop and support their own integration of making, which included extensive equipment purchases, space renovation, and staff training. They had a vision and set forth to create a "school that learns;" encouraging individualized professional development opportunities for teachers and permitting them to try new and innovative practices without fear of repercussions for failure. The atmosphere of being "free to fail" that was created for teachers was also passed down to students. This was observed in classes where the students had more autonomy in their project choices and skill building. As one middle school student explained, "We get to pick what we want to do. We have more control. We just have more flexibility." With these capacities present, leadership and teachers worked to develop and offer new courses, content, and pedagogy.

As these new making related courses, content, and pedagogy were implemented, teaching practices started to change. Most teachers, especially the maker teachers, reported becoming more constructivist in their practice and assignments, while maintaining or reducing traditional ones. Two arguments can be made to understand this scenario. First, as teachers used the makerspaces - a flexible and colorful space, full of gadgets and machines, where seats are commonly communal - they may need to rethink how to approach their teaching style, or "think outside the box [while] not...be[ing] afraid of trying new projects" (middle school non-maker teacher). In this process, just as students construct their knowledge by questioning and revising previous ones, teachers start to construct their own new knowledge of teaching (e.g. pedagogy) . Second, while living in this process of change, teachers don't simply let go of what they know. They sustain their old practices (what they know best) while trying new things. This may suggest that teachers begin to change by adding on some elements of the new approach (constructivism via constructionism) to gradually start to replace old with new and transform their practice. In this sense, these arguments may also help us understand why differences between maker and non-maker are so drastic. Maker teachers have the liberty to recreate themselves as teachers, and rethink their practice, on a daily basis. On the other hand, non-maker teachers are in a more complicated area, where risks are higher, for example students still need to take high-stakes exams, and more limitations are present, such as limited access to the makerspaces and time to develop long and demanding projects. These changes were supported by a reported work environment where teachers believe they have freedom to try new things while being encouraged by their peers and supported by their leadership.

Another element that may have supported changes in teacher practices was the environment promoted by making, the makerspaces. Makerspaces were reported to be utilized to support students' meta-awareness and collaboration processes. However, implementation and the use of those spaces differed from the theoretical framework (constructivism/constructionism) or initiative intent. Technology, for instance, was reported to support innovative lessons; yet, some of the described uses of the technology was for students to demonstrate their knowledge (recall), to allow teachers to go paperless, or no longer glue posters on the walls. These are not illustrations of active learning ("think with,"Papert, 1986) or solving problems ("program our reality", Rushkoff, 2011) envisoned by educators that promote making. Despite some differences between intent and implementation, students perceive these spaces positively. Students attribute the availability of the makerspaces to their increased enthusiasm to come to school and for providing a more enjoyable environment. This impact was exactly what the leadership of TRSD had hoped for when they were planning this initiative.

Furthermore, through TRSD's integration of making, students experienced a constructionist learning environment that promoted creativity, initiative, and exploration. Students had opportunities to build knowledge in deep and meaningful ways. The inclusive school environment and supportive faculty encouraged students to be open to making. The opportunities and experiences positively influenced the lives of students, exposing them to advanced class structures and techniques, increasing interest in STEM careers, and offering transferable skills for future schooling and careers. Their experiences show the possibilities for success for making in a formal learning environment, suggesting the maker movement's broader applications.

All the more, the study of student engagement in a sample making class provided many insights into how students engaged with the non-traditional learning environment of a makerspace located within a traditional school setting. Structuring the research with the components of flow opened a pathway to view engagement through a framework that divides a potential 'optimal (flow) experience' into its essential parts. Multiple flow components manifested frequently and concurrently in 11 of the 14 students throughout the period of observation. The final product quality and classroom performance reflected this outcome. Students expressed verbally and through observation, three aspects of the learning experience that were most contributed to their meaningful learning experience in this classroom: (1) the capacity to be successful on the project (a balance of the challenges to skills); (2) the capacity to have choice over aspects of what and how they learn (the paradox of control); and (3) the ability to enjoy what they are learning (autotelic experience).

CONCLUSION

The spark that ignited the TRSD's integration of making was a question to district leaders from a community partner, "Can you paint your walls?" TRSD leaders answered yes, and thus began a collaborative, district effort to reimagine and redesign key aspects of the learning environments, structure, and interactions within their schools. Over time, the district developed a broad set of classes and curricula to inspire and provide opportunities for students to design and create. The foundations of this initiative were a culture of innovation that started with constructivist, visionary leadership that allowed building principals and teachers to take the lead on implementation and explore their own areas of interest and expertise. This was embodied in the individualized and flexible PD that was offered and promoted among teachers. This constructivist approach with teachers allowed them the freedom to fail, which they then passed on to students. This culture of innovation transformed pedagogy at TRSD, relinquishing control and allowing everyone more freedom, creativity, and engagement throughout the schools.

ACKNOWLEDGMENT

This research was supported by the National Science Foundation [grant number 1623431].

The authors would like to thank Dr. Cynthia Tananis, Founder and long-time Director of the Collaborative for Evaluation and Assessment Capacity in the University of Pittsburgh, School of Education, Dr. Everett Herman, former CEAC researcher, and all the other researchers who worked on the project, and the TRSD community for their immeasurable contribution and openness to towards this work.

Parts of this work were previously presented at the American Evaluation Association (2018), American Education Research Association (2019), and the International Society for Technology in Education (2019).

REFERENCES

Barton, A. C., Tan, E., & Greenberg, D. (2017). The Makerspace Movement: Sites of Possibilities for Equitable Opportunities to Engage Underrepresented Youth in STEM. Teachers College Record, 119.

Becker, H. J., & Riel, M. M. (2000). Teacher Professional Engagement and Constructivist-Compatible Computer Use. In *Teaching, Learning, and Computing: 1998 National Survey. Report #7*. Irvine, CA: Center for Research on Information Technology and Organizations.

Becker, H. L., & Anderson, R. E. (1998). Teacher's Survey: Combined Versions 1-4. In Teacher Professional Engagement and Constructivist-Compatible Computer Use. Teaching, Learning, and Computing: 1998 National Survey. Irvine, CA: Center for Research on Information Technology and Organizations.

Bevan, B. (2017). The promise and the promises of Making in science education. *Studies in Science Education*, *53*(1), 75–103. doi:10.1080/03057267.2016.1275380

Blikstein, P. (2019). Foreword. Beyond mindsets, cultures, brands, and clichés: a possible future for equitable maker education. In P. Blikstein, S. L. Martinez, H. A. Pang, & K. Jarrett (Eds.), *Meaningful Making 2 – projects and inspirations for Fab Labs and makerspaces*. Torrance, CA: Constructing Modern Knowledge Press.

Bringuier, J.-C. (1980). *Conversations with Jean Piaget* (B. M. Gulati, Trans.). Chicago: The University of Chicago Press. doi:10.1007/BF02694634

Csikszentmihalyi, M. (1975). *Beyond boredom and anxiety*. San Francisco, CA: Jossey-Bass Publishers.

Csikszentmihalyi, M. (1990). *Flow: The Psychology of Optimal Experience*. New York: Harper and Row.

Cuban, L. (2018). *The flight of a butterfly or the path of a bullet: using technology – using technology to transform teaching and learning*. Cambridge, MA: Harvard Education Press.

Dewey, J. (1902). *The child and the curriculum*. Chicago, IL: The University of Chicago press.

Elmore, R. F. (2000). *Building a new structure for school leadership*. Washington, DC: The Albert Shanker Institute.

Fields, D., Kafai, Y., Nakajima, T., & Goode, J. (2017). Teaching practices for making e-textiles in high school computing classrooms. *ACM International Conference Proceeding Series*, 1-8. 10.1145/3141798.3141804

Freinet, C. (1993). *Education through work* (J. Sivell, Trans.). Lewiston, NY: Edwin Mellen Press. (Original work published 1943)

Halverson, E. R., & Sheridan, K. M. (2014). The maker movement in education. *Harvard Educational Review*, *84*(4), 495–504. doi:10.17763/haer.84.4.34j1g68140382063

Institute of Design at Stanford. (2010). *Introduction to design thinking - process guide*. Retrieved from https://dschool-old.stanford.edu/sandbox/groups/designresources/wiki/36873/attachments/74b3d/ModeGuideBOOTCAMP2010L.pdf

Kafai, Y., & Resnick, M. (1996). *Constructionism in practice: designing, thinking, and learning in a digital world*. Mahwah, NJ: Lawrence Erlbaum Associates.

Kier, M. W., Blanchard, M. R., Osborne, J. W., & Albert, J. L. (2014). The development of the STEM career interest survey (STEM-CIS). *Research in Science Education*, *44*(3), 461–481. doi:10.100711165-013-9389-3

Lam, T. C. M., & Bengo, P. (2003). A comparison of three retrospective self-reporting methods of measuring change in instructional practice. *The American Journal of Evaluation*, *24*(1), 65–80. doi:10.1177/109821400302400106

LearningR. (2019). Retrieved from https://remakelearning.org/

Lent, R. W., Brown, S. D., & Hackett, G. (1994). Toward a unifying social cognitive theory of career and academic interest, choice, and performance. *Journal of Vocational Behavior*, *45*(1), 79–122. doi:10.1006/jvbe.1994.1027

Martinez, S. L., & Stager, G. (2013). Invent to learn: "making", tinkering, and engineering in the classroom. Torrance, CA: Constructing Modern Knowledge Press.

Moneta, G. B. (2012). On the measurement and conceptualization of flow. In S. Engeser & A. Shipe-Tiska (Eds.), *Historical Lines and an Overview of Current Research on flow* (pp. 23–50). New York, NY: Springer. doi:10.1007/978-1-4614-2359-1_2

Montessori, M. (1912). *The Montessori Method* (A. E. George, Trans.). New York: Frederick A. Stokes Company.

Papert, S. (1980). *Mindstorms: Children, computers, and powerful ideas*. New York: BasicBooks.

Papert, S. (1986). Constructionism: a new opportunity for elementary science education. Massachusetts Institute of Technology, Media Laboratory, Epistemology and Learning Group: National Science Foundation.

Papert, S. (1991a). Situating Constructionism. In I. Harel & S. Papert (Eds.), *Constructionism* (pp. 1–11). Norwood, NJ: Ablex publishing Corporation.

Papert, S. (1991b). Perestroika and Epistemological Politics. In I. Harel & S. Papert (Eds.), *Constructionism* (pp. 13–27). Norwood, NJ: Ablex publishing Corporation.

Papert, S., & Freire, P. (1995) *Vídeo: O Futuro da escola e o impacto dos novos meios de comunicação no modelo de escola atual.* Márcia Moreno e Marco Aurélio Del Rosso. São Paulo: TV PUC de São Paulo com apoio do Jornal da Tarde e Agência Estado. 1 DVD (115 min): sonoro, colorido. DVD.

Papert, S., & Solomon, C. (1971). *Twenty Things To Do with a Computer.* Massachusetts Institute of Technology: Artificial Intelligence Laboratory.

Peppler, K., & Bender, S. (2013). Maker movement spreads innovation one project at a time. *Phi Delta Kappan, 95*(3), 22–27. doi:10.1177/003172171309500306

Piaget, J. (1964). Development and Learning. In R. Ripple & V. Rockcastle (Eds.), *Piaget Rediscovered.* Ithaca, NY: Cornel University.

Ravitz, J. L., Becker, H. J., & Wong, Y. (2000). *Constructivist-compatible beliefs and practices among U.S. teachers. Teaching, Learning, and Computing: 1998 National Survey Report #4*. Irvine, CA: Center for Research on Information Technology and Organizations.

Resnick, M. & Rosenbaum, E. (2013). *Designing for tinkerability. Design, Make, Play: growing the next generation of STEM Innovators*, 163-181. Doi:10.4324/9780203108352

Rousseau, J.-J. (1961). *Emile; Or, Education.* J.M. Dent & Sons.

Rushkoff, D. (2011). *Program or be programmed – ten commands for a digital age.* Berkeley, CA: Soft Skull Press.

Shernoff, D. J. (2013). *Optimal Learning Environments to Promote Student Engagement.* New York: Springer. doi:10.1007/978-1-4614-7089-2

Silk, Y. (2015). *STEAM Grant Program Evaluation.* Pittsburgh, PA: Allegheny Intermediate Unit.

Stevenson, M., Bower, M., Fallon, G., Forbes, A., & Hatzigianni, M. (2019). By design: Professional learning ecologies to develop primary school teachers' makerspaces pedagogical capabilities. *British Journal of Educational Technology*, *50*(3), 1260–1274. doi:10.1111/bjet.12743

The Institute for Arts Integration and STEAM. (2019, December 2). *What is STEAM Education*? Retrieved from https://educationcloset.com/what-is-steam-education-in-k-12-schools/#stemvssteam

Trahan, K., Romero, S. M., Ramos, R. A., Zollars, J., & Tananis, C. (2019). Making success: What does large-scale integration of making into a middle and high school look like? *Improving Schools*, *22*(2), 144–157. doi:10.1177/1365480219835324

Tushnet, N., Millsap, M.,Abdullah-Welsh, N., Brigham, N., Cooley, E., Elliott, J., & Rosenblum, S. (2000). *Final Report on the Evaluation of the National Science Foundation's Instructional Materials Development Program.* Academic Press.

Vossoughi, S., Hooper, P. K., & Escude, M. (2016). "Making" through the lens of culture and power: Towards transformative visions for education equity. *Harvard Educational Review*, *86*(2), 206–232. doi:10.17763/0017-8055.86.2.206

Vygotsky, L. (1978). *Mind in Society*. London: Harvard University Press.

Zollars, J. (2018) *Flow Theory and Engagement: Observing Engagement through the Lens of Flow in a Middle School Integrated Maker Space* (Doctoral Dissertation). University of Pittsburgh.

ADDITIONAL READING

Blikstein, P. (2013). Digital fabrication and making in education: The democratization of invention. In J. Walter-Herrmann & C. Buching (Eds.), *FabLabs: of machines, makers and inventors* (pp. 1–22). Bielefeld: Transcript. doi:10.14361/transcript.9783839423820.203

Kier, M. W., Blanchard, M. R., Osborne, J. W., & Albert, J. L. (2014). The development of the STEM career interest survey (STEM-CIS). *Research in Science Education*, *44*(3), 461–481. doi:10.100711165-013-9389-3

Martinez, S. L., & Stager, G. (2019). Invent to learn: “making”, tinkering, and engineering in the classroom (2nd ed.). Torrance, CA: Constructing modern knowledge press.

Papert, S. (1987). A critique of technocentrism in thinking about the school of the future. In *Children in the information age: Opportunities for creativity, innovation and new activities* (pp.1-13). Retrieved from http://www.papert.org/articles/ACritiqueofTechnocentrism.html

Peppler, K., Halverson, E. R., & Kafai, Y. B. (Eds.). (2016). *Makeology–Makerspaces as learning environments* (Vol. 1 and 2). New York: Routledge.

KEY TERMS AND DEFINITIONS

Constructionism: An epistemological theory of learning developed by Seymour Papert which, like constructivism, understands learning as “building knowledge structures” (Papert, 1991a, p. 1). Papert added to this idea the fact that building such knowledge structures is more fortuitous when a learner is constructing something to be shared. Educators who use this theory to design their environment and promote constructionist learning may be called constructionists.

Constructivism: A genetic epistemological theory of learning developed by Jean Piaget, which states that knowledge is *constructed* in continuous and incremental ways through processes of assimilation and accommodation. Knowledge, therefore, cannot be transmitted. It is constructed by learners through their personal experiences. Educators who use this theory to underline their work are called constructivists.

Flow Theory: Originated by Mihaly Csikszentmihalyi, flow theory proposes that when one is actively engaged in an activity where the skills possessed are balanced to the challenge of the activity, s/he can approach an optimal state of experience called “flow.” Several conditions contribute to this psychological state. Together, they can produce a “flow” state fostering meaningful engagement that leads to learning and increased personal satisfaction.

Maker Movement: Consist of the shared making experiences of makers on makerspaces that form a culture. Makers are the people who form the communities around the makerspaces and create a culture of creation and sharing.

Makerspace: Refers to the place where the making happens. A makerspace can be a garage, a room in a museum, library, or school.

Making: Refers to hands-on activities that pursue the solving of a personal meaningful problem that is often shared on physical and online communities.

Making Course: A course where an essential element was the creation of a new product by the students. Examples of making courses in this study are Introduction and Appreciation of Art, Graphic Technologies, Games Through the Ages, Introduction to CAD, Woodworking I, and Maker Lab.

ENDNOTES

1 This is a pseudonym.

2 "Since 2013, TRSD has used a variety of making tools and activities to promote a project-based curriculum that integrates art, technology education, and computer science at the middle and high schools. This has included: Three Rivers Academy, the Fabs Labs, Gaming, The Imagination Station, Digital Media, The Media Center, iCreate Studio, Robotics, Engineering, Entertainment Technology, etc. This section of our survey is looking at how these initiatives have impacted both schools and their teachers. How have these initiatives affected you personally or affected your teaching/classroom/curriculum?"

3 Human-centered design (HCD), defined as a "process for transforming difficult challenges and capturing unexpected insights to produce creative innovation that reflects what targets users want" (Thanapornsangsuth, 2016, p. 17), also center the students' creation of solutions. This creative process is implemented in five iterative steps (empathy, define, ideate, prototype, and test) while considering three essential elements: collaboration, optimism, and experimental iteration (fail fast, learn from it, and iterate) (Institute of Design at Stanford, 2010). HCD was the framework that inspired the leadership team to implement the maker movement at the EFSD.

Section 3
STEAM Education Beyond the Classroom

Chapter 7
A Laboratory for Creativity:
How Youth Thrive With Design Thinking and STEAM Education

Neal Barbour
Lawrence Arts Center, USA

Becky Eason
https://orcid.org/0000-0002-5167-270X
Wordcraft LLC, USA

ABSTRACT

This chapter explores how design thinking, challenge-based learning, and project-based learning have been incorporated into a STEAM program named ARTspace at the Lawrence Arts Center in Lawrence, Kansas. Using examples of curriculum and practice undertaken over the last five years, the authors share the formation of three distinct assessment tools developed to measure metacognitive awareness of the invention process, creative self-efficacy, and the acquisition and retention of standards based STEAM content. The data presented from a three-year evaluation of the ARTspace program shows statistically significant gains in awareness of process-based thinking/practice as well as confidence in practicing innovation.

INTRODUCTION

The rapid development of technology has created an uncertain future for today's students. Although it is difficult to estimate the exact percentage of future jobs that do not yet exist today, we know that today's education needs to shift in order to prepare students to be sharp, critical-thinking, adaptive, and collaborative citizens

DOI: 10.4018/978-1-7998-2517-3.ch007

(Murane & Levy, 2017). Offering students opportunities in afterschool and summer programming that focus on the practice of creation and invention helps to build skills that can translate to this uncertain future economy. Even more beneficial than this economic argument is the data indicating that arts-based, afterschool activities create more engaged citizens. According to a 2012 study by James Catterall, children who regularly study the arts are more likely to: be recognized for academic achievement, be elected to class office, show higher achievement in math and science, win academic awards, attend college, choose a professional career, regularly read the newspaper, volunteer in their community, and vote. Enhancing the present and future lives of our students by allowing them to explore their passions and apply knowledge to practice is inherent in best practices in education.

One such program, the Lawrence Arts Center's ARTspace STEAM, integrates Dance, Performing, and Visual Arts to teach Science, Technology, Engineering, and Mathematics (STEM) concepts by allowing students to create artwork that speaks to a design challenge. Combining elements of Challenge Based Learning and Project Based Learning, this program encourages risk-taking, collaboration, informed design, STEM identity-building, and awareness of the creative process. In order to measure gains students were making, three unique and distinct assessment tools were created and a three-year evaluation of the program ran from 2015 through 2017. This chapter traces the development of the ARTspace STEAM program, examines the pedagogical aspects of its curriculum using applicable, real-life examples, and shares the assessment tools and results of the three-year evaluation.

BACKGROUND

The Lawrence Arts Center (LAC) was established in 1975 as a public/private partnership between the city of Lawrence, Kansas and local residents. The center moved to expanded 40,000 square-foot facilities in 2002, continuing community arts programs and services for individuals of all ages, interests, and backgrounds through three exhibition galleries, a 300-seat proscenium theater, two dance studios, two arts-based preschool classrooms, a half day kindergarten, a black box theater, and eight visual arts studios. In addition to being a center for visual and performing arts, modern art exhibitions, film, and lectures, the LAC provides a full complement of educational programming for ballet, modern dance, theatre, and visual arts, employing over 200 teaching artists to work with over 10,000 students and over 200,000 visitors annually *(Lawrence Art Center, 2015)*. The mission of LAC is to create meaningful arts experiences with and for the community through education, exhibitions, and performances.

In 2017-2018, 8,000 students took part in visual and performing arts programming at LAC through workshops, classes, and camps, many receiving financial aid. Through LAC's support of Head Start, Boys & Girls Club, Little Nations Academic Center at Haskell Indian Nations University, and other social service organizations, over 1,000 students received financial aid to participate in arts education at LAC (*Lawrence Arts Center, 2017*). The goal of all visual arts education programs at LAC is to provide an environment where youth can develop original, methodical, and progressive thinking skills with the help of artists and art instructors, essential skills for future achievement and leadership. LAC's art education programs are founded on the understanding that art education impacts all areas of academic achievement, building the capacity of students to think boldly and creatively while motivating themselves to learn *(Kálly, 2012).* Of the LAC'S 2017-2018 $3,100,489 budget, the largest portion went to arts education (including preschool and dance) at 48%, followed by 27% for performing arts, 16% for exhibitions, 5% for scholarships and financial aid, and 4% for the visiting and resident artist program *(Lawrence Arts Center, 2017).* Each year, almost half of the budget funds sequenced education classes for children, adolescents, and adults in ceramics, creative writing, digital media, drama, drawing, film, jewelry-making, metalsmithing, painting, photography, printmaking, and textiles, plus an arts-based preschool with 500 children enrolled each year. LAC employs 28 full-time staff, 15 part-time staff, and more than 100 contract teaching artists. An additional 100 people volunteer annually.

PEDAGOGY AND PRACTICE OF ARTspace STEAM

One of the Lawrence Arts Center's many programs is a unified visual arts, movement/ performance, and STEM program for 1st through 5th grade students: ARTspace STEAM. ARTspace places itself at the vanguard of the drive to add art to initiatives promoting new and rigorous ways to teach science, technology, engineering, and math (STEM). After securing a grant from the Hearst Foundation in 2012, the LAC devoted time and resources to developing the ARTspace program. Beginning with a team of eight public school arts and sciences teachers in tandem with eight local teaching artists, the LAC crafted its first series of interdisciplinary lesson plans that focused on using the fine arts as a delivery method for STEM concepts. From the initial twelve lesson plans, the LAC has crafted over 200 additional lesson plans that use visual arts, performing arts, and dance to teach a STEM lesson. ARTspace's student-centered pedagogy uses design thinking to engage students in challenge-based learning exercises. As shown from the evaluation results, students engage at a high level as they work both collaboratively and individually toward a unified

Figure 1. Students work together in Mission to Mars to build a NASA mission control that communicates with another group of students who have built and installed a Mars habitat. Students create the installation from artwork generated from their Design Challenge lessons and then work through different problem-solving scenarios within the installation.

goal; all results are different, yet they are celebrated equally as they integrate in a final installation, performance, or sharing as seen in Figure 1.

Since 2012, ARTspace has reached over 5000 elementary students in rural and urban settings. The program was named as a Finalist for the National Arts and Humanities Youth Program Awards in 2017.

In each team-taught course, students experience learning in a holistic way, finding themselves in a place where there is time, space, support, and encouragement to try new things, expand their imagination, practice innovation, and generally be creative. ARTspace staff believe learning to be innovative and creative begins in early childhood, when kids need liberty and support to explore their distinct interests, abilities, and thoughts. Innovation and invention are both mindsets and disciplines that take practice, and, in doing so, lead to a life-long habit of creativity *(Costa.& Kallick, 2008).*

The learning goals of the ARTspace program expect that students will:

- Exercise creativity and innovation
- Expand communication skills
- Exercise critical thinking and problem solving abilities
- Improve their intellectual flexibility and adaptability
- Practice initiative and self-direction
- Improve productivity and responsibility
- Practice the design and invention process
(Lawrence Arts Center, n.d., p. 4)

Each interactive and fundamental element or principle of ARTspace's multisensory courses ties art to a STEM concept framed in a challenge-based project. For example, the *Mission to Mars* camp helps students study the physics of space travel, build and launch a rocket, design and build a habitat suitable for survival, and work through several challenging scenarios within this immersive art installation. Another example can be found in the *Prairie Power* camp, where students study the Prairie/Grassland ecosystem, then design functioning windmills to harness energy, seedpods that use their sculptural forms to survive, and habitats for the wildlife of the prairie. Guest scientists work with art educators and students to make connections between art principles and scientific concepts, often bringing in supplemental materials and experiences that enhance their STEAM projects.

In the development of the curriculum for ARTspace, teaching artists were guided through both challenge-based learning and design thinking principles from *Learning and Leading with Habits of Mind*, *Design Thinking for Educators*, and the Institute for Design at Stanford's *Design Thinking Toolkit*. Teaching artists within ARTspace crafted their lessons according to these principles with each lesson including a guiding question linked to standards, supplies needed, background information, a challenge with parameters, discipline-specific process instructions, and methods for reflection.

By integrating the A for art and converting STEM to STEAM, the ARTspace program broadcasts the importance of teaching creativity through art (a) to develop habits of mind that positively impact long term achievement *(Costa. & Kallick, 2008)*, (b) as fundamental to an innovative workforce *(Tierney & Farmer, 2002)*, and (c) as "essential to our nation's ability to develop the new products and businesses that will drive our economy in the future" *(Lawrence Arts Center, n.d.)*. It follows that arts education is required for practicing creativity, just as creativity is vital to innovation and innovation is essential for creating new industries and jobs that are the cornerstone of economic security *(Murnane & Levy, 1998)*. In short, the goal of the ARTspace program is to instill in its learners the design and invention process used by artists, designers, scientists, and inventors as seen in Figure 2.

Figure 2. The "design wheel" illustrating the design and invention process used by teaching artists and students at the Lawrence Arts Center. Specific actions have been assigned to each stage to help teachers help students engage in metacognitive practices.

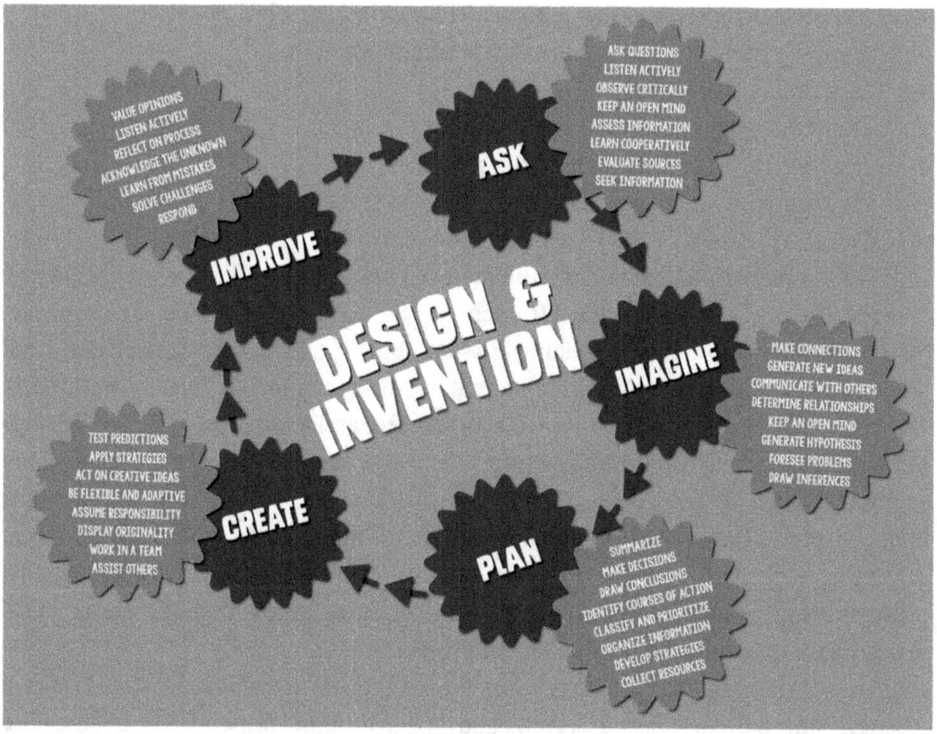

The Arts Center intentionally defined a clearly codified process that young artists can use to craft creative solutions to design challenges in order to teach more than just artistic practices and mediums. The Design Wheel (see Figure 2) is a tool to teach innovation, creativity, and metacognition in learners. This tool was developed by Neal Barbour, Director of Youth Education at the Lawrence Arts Center and graphic designer Amy Albright based off the initial discussions and development of ARTspace curriculum. The Design Wheel identifies the stages of creation and expands each step to include specific actions that can be utilized by young inventors going through the creative process. By articulating the stages of the creative process with accompanying actions, both instructors and students are able to communicate the stage they are on as they work through an ongoing process. The Design Wheel has distinct, sequential and cyclical phases, with no beginning nor end, signifying the ongoing nature of creation. Each section of any given lesson attaches to the stages of the Design Wheel and multiple lessons can be linked together to create a full ARTspace camp. For example, a teacher who creates a lesson on seedpod structures could find other lessons on plant identification, species adaptation, wind, and ecosystems in order to create an entire unit that focuses on biology.

In order to give the reader a clear picture of what a challenge-based lesson from ARTspace looks like in practice, the following explanations of the Design Wheel stages include how the *Seedpod Sculpture* lesson is presented:

- **Ask**: At the bedrock of each lesson is a STEM concept, linked to the Common Core standards for Math, Science, and Art, that is asked of the students at the beginning of each design challenge in the form of a guiding question. This question is intended to spark inquiry in the students, and also guide their creative process through the Design Wheel. Every lesson crafted within the ARTspace curriculum uses a standards-based question as an anchor point for a STEM concept. For the example lesson on seedpod structures, the lesson would have a guiding question such as: "How does surface area affect wind resistance?" or "How do plants ensure their survival?" Students interact with both the guiding question and contextual information presented by the instructor. This is an important moment where teachers can use high interest teaching aides such as short videos, physical objects, art pieces, and other materials to generate engagement with the content. For the seedpod structure lesson, teachers could present actual seedpods or footage from David Attenborough's mesmerizing documentary, *The Secret Lives of Plants*. At this point the students should be speaking with one another, asking questions, answering questions, and generally interacting with the content information provided in order to establish a base level understanding of the STEM concept.
- **Imagine**: The Imagine stage presents a challenge for the students to solve using the information they've discussed in the Ask phase. Teachers set the parameters for the challenge based on physical materials, size, time, space, and/or other controlled variables, but try to keep their parameters to three or fewer in order to allow creativity without an overwhelming number of restrictions. In the *Seedpod Sculpture* lesson, students are challenged to create a form of a seedpod that has qualities which ensure the propagation of the seeds within. The parameters given are: 1) you can use no more than one sheet of tissue and printer paper, 2) it must contain seeds, 3) it will be dropped from a high place. The solution to how a plant uses seedpods to propagate is open-ended and can take many different forms: the seedpod can transport a great number of seeds, remain aloft long enough to catch wind, have a dispersal method upon impact, or some unknown solution young learners will put forth when challenged.

- **Plan**: Once the challenge is set, students are encouraged to plan their project through short, active exercises including, but not limited to: making lists, drawing a chart, illustrating their plan, making a maquette, discussing with a neighbor, or other strategies facilitated by the instructor. In the case of our example lesson, students begin to familiarize themselves with materials and building techniques using small rectangles of pre-cut paper, essentially creating a maquette of their larger seedpod sculpture. Or, if using a pencil and paper to plan potential structures, students begin to think about how they would translate a 2D rendering into a 3D form. This planning phase can be used as a reflective tool during both the Create and Improve stages of the Design Wheel.
- **Create**: Once planned, students move to the Create stage, diving deep into a self-guided practice of experimentation, discovery, and risk taking. In the out of school time setting, teachers devote the largest amount of time to this stage. Students may move through the Design Wheel multiple times during this stage, continuously testing and improving ideas as they work toward the end of the class time. Having a testing area set up helps students to gather feedback and improve their work as they go. Initially, this is the most difficult step for teachers and students to become comfortable with because it strays from conventional teaching practices of having students work on a uniform task. If the challenge parameters are constructed and explained effectively by the teacher, this open, divergent creation phase becomes easier with practice. One of the greatest disservices we can do to Design Thinking practice is to not fully scaffold the challenges and not devote enough open time for experimentation *(Schell, 2017)*. For the Create phase of *Seedpod Sculpture,* students quickly dive into trying to create their plan in reality. This comes with lots of struggle, testing, modification, and creative risk-taking, but students throw themselves at the challenge. Some students use the entirety of their paper and others just a fraction—each sculpture comes out unique, purposeful, and ready to test.
- **Improve:** This stage is intended to celebrate both success and failure, to allow time and practice for young learners to reflect, to help students become comfortable with feedback, and to initiate jumping right back into the process of invention. When asked how many times they went through the Design Wheel in a single lesson, many students answer multiple times. They are aware of their own process, it is the challenge of the teacher to help them share with and learn from others in an efficient way. Frequently, opportunities for reflection can be self-guided, including modifying their plan to get different results the next time, creating a list of lessons learned, displaying evidence with a brief summary, sharing surprises with a neighbor, etc. In the case of

Seedpod Sculpture, students created a mini gallery of their pieces with a brief statement of how the structure failed or succeeded. They then walked around the "gallery" observing and speaking with others about their sculptures.

PROGRAM STRUCTURE

The ARTspace program takes place both during the academic year and the summer in two very different formats. During the summer, students are engaged in two to four, 90-minute design challenge sessions, daily relating to a specific theme, for five days of the week. In this programming format, students receive either fifteen or thirty hours of design thinking programming, depending on their selection of half day or full day camp. Lessons in the different disciplines of dance, performing, and visual arts inform one another thematically and typically build to several small art pieces or performances, or one large installation or performance.

During the school year, ARTspace takes place for 90 minutes daily, Monday through Friday, for three to four weeks, with students alternating between visual arts and dance/performance art. The lessons are still linked and the overall classroom time is similar to summer programming (either 15 or 20 total hours depending on the length of the camp), yet the lessons take place over the course of a longer time period with a shorter concentration of time daily afforded to design thinking lessons.

In order to engage elementary students and provide lessons that build on each other, LAC instructors across disciplines integrate their challenge-based lessons into one overall theme. For example, one ARTspace camp is called *Break the Internet* and centers around the theme of technology, computer science, and the collective sharing power of the internet. In this camp, students are challenged in all disciplines to create artwork based on computer science principles. One cross disciplinary lesson between dance and visual arts challenges students to create a short movement or series of movements that can be looped in an animated GIF. The students use background knowledge of memes, have discourse on internet trends and usage, analyze successful and non-successful examples of looped video, develop articulated movements, learn the basics of recording, discover how to use animation software to create an animated .gif file, and upload the content to their student generated website. This is just one challenge-based activity in a suite of lessons devoted to the theme. In this way, ARTspace relies on the engagement and self-chosen content creation of young learners to fill in a framework.

The ARTspace curriculum includes over 200 STEAM lessons that have been used in over fifty distinct camps. In addition to previously mentioned camps, some past ARTspace camps include:

- In *Box City*, students study the architecture, urban planning, and history of a representative city and then work together to recreate the city in cardboard. They speak with city planners, engineers, and a resident cardboard artist to ensure the buildings they create are structurally sound and can withstand the challenge of remaining in an installation that will have young children playing within it for a week.
- In *Toy Convention*, students study prototyping, industrial design, and graphic design in order to bring a toy from concept to production. They work with toy designers and graphic designers to conceive of a toy that can be displayed in a science fair type setting complete with prototype sketches, packaging, functionality, and a representation of their toy.
- In *Circus Comes to Town*, students study kinesthesiology, forces of motion, and color theory in order to design a circus performance complete with structures, games, sideshow performances, and visual accompaniments.
- In *Polar Extremes*, students study ecosystems, animal biology, and biodiversity to create an installation of a base camp that provides examples and evidence of their scientific explorations, experiments, and discoveries of the Polar Regions.
- In *Art Detectives*, students study forensics, spycraft, and engineering to work together to solve an immersive mystery. By moving through different skill building activities, students work in investigative teams to put the pieces of the mystery together and show their work through an evidence board installation.
- In *Lost in the Woods*, students study engineering basics, chemistry, and plant biology to learn how to both survive and thrive in the woods by designing and building tools and art making supplies, creating their own nature-based pigments, and forming an agreed-upon document for their new society.

Figure 3 shows students working in the design/build workshop called Box City.

INSTRUMENTS, LESSONS LEARNED, AND USE OF ASSESSMENT DATA

After three years of successful programming with over 2500 students enrolled, the Lawrence Arts Center moved to the evaluation phase for ARTspace in 2015. In collaboration with the University of Kansas, the Lawrence Arts Center originally engaged the Graduate Department of Curriculum and Instruction to develop an evaluation that would speak to four primary questions.

Figure 3. Students working in a design/build workshop called Box City. Students study the architecture, urban planning, and history of a representative city and then work together to recreate the city in cardboard

1. Does student understanding of creativity progress along Bloom's Taxonomy from the beginning of a single ARTspace camp, and from camp to camp?
2. Do students' artistic and creative abilities grow over the course of a single ARTspace camp, and from camp to camp?
3. How and to what extent do students apply the design and invention process at the beginning compared to the end of a single ARTspace camp, and from camp to camp? How does student metacognition (e.g. awareness and understanding of thinking) change throughout each camp?
4. How does student confidence in their capacity to execute creativity and innovation (e.g. creative self-efficacy) change from the beginning to end of a single ARTspace camp, and from camp to camp?

As the ARTspace curriculum was enacted, however, administrators decided to step away from Bloom's Taxonomy (Question 1) and focus more on the creative and design elements of the student experience (Questions 2 to 4). Thus, the current ARTspace student learning assessment is based on the Common Core Arts standards, the Next Generation Science Standards, and the Common Core Standards of Mathematical

Practice for K-5th graders. Because the curriculum and delivery methods are outside the norm of measured structures of pedagogy, traditional methods of measurement would not effectively capture the gains that students made through participating in ARTspace. In order to answer the three evaluation questions above, the ARTspace evaluation team created three instruments for use during the academic year, two of which are also used for summer programming.

What follows is a summary of the program evaluation conducted of ARTspace, to include the instruments and assessments used. The information has been broken down by Academic Year and Summer administration. In each, the assessment itself is briefly examined, as well as results available from 2015 – 2017.

Pre- and Post-Drawing Assessment for Metacognition and STEM Content

The first measurement tool of the ARTspace evaluation is a Pre- and Post- Drawing Assessment that was developed to evaluate students' conceptual understanding of STEAM principles linked to the Common Core Art standards, Next Generation Science Standards, and the Common Core Standards of Mathematical Practice. Because K-5th grade students are still building their literacy skills, a drawing assessment (as compared to a written assessment) of students' conceptual understanding removes any potential inconsistencies resulting from students' varying reading and writing abilities *(Nelson et al, 1998)*. Visual Arts being a form of communication dependent on fine motor control and the skill of rendering ideas into pictoral representations, each student's pre- and post-tests were solely measured against each other to account for the varying skills and developmental levels of individual students.

During the Summer ARTspace classes, the prompt for the Drawing Assessment referred specifically to the content of the particular course. For example, the prompt for a week about the environment might ask the students to draw an animal in its natural habitat, or a class on ancient Egypt might have a prompt about drawing an artifact. Thus, students were given the opportunity both to be creative and to demonstrate learning gains from participating in the class.

ARTspace seeks to not only increase students' abilities in art, but build students' understanding of their own learning processes and styles. Research has found metacognitive awareness of one's personal learning and planning approach can have a positive effect on academic success *(Kálly, 2012; Negretti, 2012)*. In the development of the ARTspace Evaluation, it was determined that the Pre- and Post- Drawing Assessment tool used during the Summer could be adapted to measure students' understanding of the process of invention, thus all of the pre- and post- drawing assessments for the Academic Year used a prompt asking specifically about the design process. Because teachers were using the Design Wheel as the primary structure for framing and executing their lessons, the Pre- and Post- Drawing Assessment was adapted to focus on the stages of design and invention as informed by the Design Wheel, and on the student's ability to retain the entire process of invention.

Administration of the Pre- and Post- Drawing Assessment was refined across the semesters it was used. Since the assessment is administered by numerous program faculty under varying circumstances, program directors determined the need for a printed set of drawing assessment administration protocols. These protocols dictated that faculty members must: use the same drawing prompt, regardless of class theme or topic; follow a printed and consistent script for delivering the pre- and post-test; administer pre- and post-tests at the same time of day in each class; and give students a consistent amount of time to respond to both the pre- and post-tests. Controlling these factors has allowed for some uniformity across the multiple iterations of the assessment.

The pre- and post- drawing assessment was administered at the beginning and end of each design camp (respectively Day 1 and Day 5 for Summer, and Day 1 and Day 15 or 20 for the Academic Year). The pre- and post-assessment measure consists of a single drawing pertaining to either a STEM concept for each weeklong summer camp, or the Design Wheel for each camp during the academic year. Each student draws to the same prompt and may use words, symbols, or any other markings to clarify their visual communication of the prompt. This data collection generated two sets of standardized information pertaining to student development: elements related to STEM subject matter during the Summer, and metacognition of the design process during the Academic Year.

Assessment of the drawings utilizes the Formal Elements Art Therapy Scale (FEATS; Gantt & Mills, 2001), a measurement method for applying numbers to comprehensive variables in two- dimensional art (see Appendix B). The elements of space, integration, logic, realism, problem-solving, details of objects and environment, and line quality will be assessed. Elements are then rated on a Likert-type scale ranging from 1 = *not apparent* to 5 = *very apparent*.

Space. Space can be used as a form of measurement in all two-dimensional art. Additionally, it can be measured on a ratio scale (that is, a scale that has the properties to permit mathematical operations). The scale utilized by Gantt & Tabone (1998) assigns points to less than 25%, between 25% and 49%, between 50% and 74%, between 75% and 99%, and 100% of the space used (see Appendix B).

Integration. This scale should have broad application to all types of arts. Most often it will be used to assess compositions consisting of two or more objects and/or people. This criterion should also be employed with non-representational compositions. Discerning the individual "parts" of such a composition should be determined prior to rating. A pilot study to assess varied nonrepresentational art may be used to create grading consensus amongst raters (Gantt & Mills, 2001).

Logic. The Logic scale addresses the inclusion of bizarre or illogical elements not part of the requested response. This scale should discriminate between intentionally humorous or satirical items, and those items that seem to have no reason for being in the picture. This scale should also be applicable to many types of drawings and paintings. Only further studies will demonstrate whether raters can use this scale in a reliable manner on nonrepresentational as well as representational art (Gantt & Mills, 2001).

Realism. This scale is used only with drawings for which there is a directive to make a specific object, or when the artist describes part of the drawing as being a specific object. Thus, if a person was asked to make a scribble and declared the image to be "a dog," then the researcher would need to relay that to the raters. This variable may also be associated with education and artistic training so it is imperative that investigators consider this in research designs (Gantt & Mills, 2001).

Problem-solving. This scale measures the degree to which the artist shows the resolution of a problem as presented in the directive. It was a serendipitous result of the single- picture assessment "Draw a person picking an apple from a tree" (PPAT; Lowenfeld, 1939) where the subject is asked to draw a person picking an apple from a tree. In a small-scale validity study the scores on this scale discriminated all five groups (four diagnostic groups and a control group) from each other (Gantt, 1990). While these results have not been replicated in other drawing measures, it still holds valuable potential application in the assessment of those pieces created as part of the ARTspace process.

Details of Objects and Environment. This scale will be used when the work requires identifiable objects, but not with abstract or non-representative art. It measures whether the person followed the directions in a concrete form, and includes the specified items in the directive, or added to the initial idea to create a comprehensive picture. For example, "abundant and inventive details" are rated 5 (Gantt & Mills, 2001).

Line Quality. This scale looks at the overall line quality and the degree of control that the student uses. Raters will be asked to "average" the lines in the drawing. For example, excessively "fluid or flowing lines" are rated 5, while lines that "appear to be drawn with a shaking hand" are rated 3. Line Quality can be used as a form of measurement in all two- dimensional art (Gantt & Mills, 2001).

Figure 4 shows an example of a post-drawing from an ARTspace student.

Figure 4. An example of a post-drawing from an ARTspace student. During the Academic Year, pre- and post- drawings are based on thinking of invention as a multi stage process that links to the Design Wheel

Pre- and Post Drawing Results for Academic Year: Metacognition Scores

The drawing prompt asked of each student (or a selection of students) at the beginning and after the completion of a project, was, "Draw or chart the steps of inventing something" *(see* Figure *4).* This prompt is geared toward determining student awareness of any process or steps he or she goes through in planning and executing a project or task. (See Appendix C for suggested questions, modified for the proper age range from the Metacognitive Awareness Inventory *(Schraw & Dennison, 1994).*) Basic metacognitive awareness is determined primarily by student answers that developed from a single drawing to a drawing comprised of stages, or a segmented drawing that moved an invention through stages, specifically those within the Design Wheel. A more specified level of cognitive awareness could be

Table 1. The scores the for the Academic Year Drawing Assessment for four sessions

Semester	Mean Increase	t (df)	p-value
Fall 2015	.49*	3.71 (41)	.001
Winter/Spring 2016	.97*	8.25 (54)	<.001
Fall 2016	.53*	3.52(27)	<.01
Winter/Spring 2017	.27	1.42(23)	.17

* Statistically Significant

determined based on additional detail provided by the student in their interpretation of what these stages could potentially encompass.

The Academic Year drawing scores demonstrated a mean positive increase in students' conceptual understanding of the invention process as shown in Table 1. The growth from pre- to post-test can show whether there are changes in how the student thinks about and processes a task over the course of the ARTspace camp. Increases in procedural awareness, such as providing more detail in description of the process or presenting it in a step- by-step format, provides support and insight into the success of teaching the Design Wheel emphasized as a goal in the ARTspace program. Results for Pre- and Post- Drawing assessments from four course sessions are summarized in Table 1. In order to better assess the different steps and where they fit on the wheel, a rubric (see Appendix D) has been created for the pre- and post-assessment measure.

Pre- and Post- Drawing Results for Summer: STEAM Content Scores

Much like the Academic Year courses, the Pre- and Post- Drawing Assessment for the Summer camps was administered to evaluate students' conceptual understanding of STEAM principles based on the Common Core Arts standards, Next Generation Science Standards, and the Common Core Standards of Mathematical Practice. Each week's pre-post student drawings were also analyzed using a modified version of the FEAT Scale (Gantt and Mills, 2001). The scale examined the students' drawings for such elements as:

- appropriate use of available space on the paper,
- capacity to draw what was asked by the simple prompt,
- the level of detail in their drawings, and
- the quality of the lines made in the drawing.

Table 2. The scores the for Summer Drawing Assessment for three sessions

Semester	Range of Mean Scores	Average
Summer 2015	.19 - .71	.43
Summer 2016	.28 – 1.03	.57
Summer 2017	.13 - .93	.69

These measures served as proxies for the students' confidence levels regarding the subject, the knowledge gained regarding the subject, and level of understanding regarding what was being asked of them. Concessions were made for the considerable age differences amongst the students (aged 5 to 10) and the concomitant differences in their fine motor skills and developmental levels.

Each student's pre- and post-drawings were paired and examined using a 5-item, 5-point scale. The difference between the score for pre- and post- was recorded for each item, and then each student received a mean pre-score, a mean post-score, and a mean difference score (see Appendix *G).* Each week of class then also received a mean pre-score, mean post-score, and a mean difference score. The range of mean differences for Summers 2015-2017 are listed in Table *2*, as well as an average of the scores for each Summer semester. It is worth noting that, in every week of all three Summers, drawing scores improved. Thus, this assessment would indicate that learning is happening consistently in the LAC's summer camps.

Over the course of assessment administration, a number of inconsistencies were addressed that may have resulted in the increase of scores over the three years. Like the Academic Year assessments, administrators needed to ensure that all students in every section have the same amount of time to complete their drawing on both the pre- and post-test.

However, while the Academic Year drawing assessments were all conducted to a prompt about the design process, the Summer assessments had specific prompts for each week's curriculum. Because of this variety of prompts, the Summer process also had to address potential inconsistencies such as:

- Ensuring that the language of the drawing prompt is written at a level that the students can understand
- Ensuring that the prompt does not unduly lead the students (one iteration of the prompt included pictures, which affected student drawings)

Creative Self-Efficacy Photograph Assessment

In the area of creative self-efficacy, the aim for ARTspace is to improve students' confidence in their own abilities to be creative and innovative. With this confidence, participants will be better prepared to formulate creative solutions to challenges in school, in the community, and later on in their careers. To accomplish this, the program presents students with design challenges throughout the weeklong camps in Summer and in the daily afterschool camps during the Academic Year. These design challenges give students a chance to struggle with a problem or creative challenge, and practice using their creativity to complete activities and solve the challenge.

The Measure of Self-Efficacy developed by Beghetto (2006) has also been adapted for use with ARTspace students. The original measure consists of three questions (see Appendix E) on which students rate how true a statement is about them on a Likert-type scale from 1 = *really not true* to 3 = *really true*. Total scores range from 3-9 with high scores indicating greater creative self-efficacy. Most research in this area requires raters to read these questions and circle answers (e.g. Beghetto, 2006; Mathisen & Bronnick, 2009; Tierney & Farmer, 2002). However, to increase the usability and feasibility of this evaluation for ARTspace, the questions were read aloud and the children posed their bodies in three areas of space to indicate their responses. Photos were taken of the children, so that their responses could be tabulated.

These self-report ratings of creative self- efficacy are measured in a pre- and post-assessment fashion, with the first rating taking place on the first day of each camp and the last rating taking place on the last day of camp. The ratings occur prior to the pre- and post- drawing assessment described previously, as it is important to capture initial attitudes not influenced by any aspect of the program.

Analyses of the data consisted of aggregate class-level comparisons between students' ratings from the beginning to the end of a camp. Growth in creative self-efficacy was also examined by looking at change in the frequencies of individual student ratings (e.g., 20 out of 28 students increased their ratings by at least one point). Evaluators also calculated the arithmetic mean of the group's creativity self-efficacy scores on the first day and compared them to the arithmetic mean of data from the last day.

One of the goals for the students who participate in ARTspace programming is to grow in their perceptions of themselves as creative individuals, whose creativity can lead to positive results. One challenge encountered was how to capture attitudinal gains with elementary students, who may be too young to take any sort of written assessment with reliability. Our solution was to administer the question verbally and capture student responses in a group photo, where "yes" responses were demonstrated through hands over the head, "maybe" through hands on the shoulders or hips, and

Table 3. Results from the Creative Self Efficacy tests for both Academic Year 2015-16 and 2016-17 that show students' responses to three questions regarding identity forming and confidence with creativity and innovation

Self-Efficacy Questions	Scores Improved	Statistically Significant	t (df)	p-value
AY 2015-16				
Fall "New Ideas"	X	X	5.22 (116)	p < .001
Fall "Good Ideas"	X		1.41 (106)	
Fall "Imagination"	X	X	3.49 (117)	p = .001
Winter/Spring "New Ideas"	X		1.38 (153)	
Winter/Spring "Good Ideas"	X		.84 (157)	
AY 2016-17				
Self-Efficacy Questions	Scores Improved	Statistically Significant	t (df)	p-value
Fall "New Ideas"	X			
Fall "Good Ideas"	X			
Fall "Imagination"	X	X	2.25 (65)	.03
Winter/Spring "New Ideas"	X			
Winter/Spring "Good Ideas"	X	X	1.94 (134)	.05
Winter/Spring "Imagination"	X	X	3.22 (132)	.02
Winter/Spring "Imagination"			.47 (160)	

"no" through hands touching knees, toes or the floor The photo assessments were administered in a pre-post fashion and cumulative scores in each category were tallied.

Academic Year Self-Efficacy Results

The Self-Efficacy Pre- and Post-tests asked students whether or not they agreed with the following statements: "I am good at coming up with new ideas"; "I have lots of good ideas"; and "I have a good imagination." Students were asked this question in camps taking place in both Winter/Spring and Fall semesters in AY 2015-16 and AY 2016-17. The daily results on the three questions were combined into Fall and Winter/Spring datasets, for a total of 6 collection points. All results are shown in Table *3*.

The results are telling about how the Academic Year ARTspace programming is consistently having a positive effect on student perception of their own efficacy. It is worth noting that only one question across four semesters did not show an improvement in responses. The scores suggest that the ARTspace program helps

Table 4. As with the Academic Year programming, the scores of the Summer Creative Self-Efficacy assessment shows students are making noticeable gains in their self-efficacy regarding the generation of new/good ideas and the use of their imagination

Self-Efficacy Questions	Scores Improved	Scores Declined	Statistically Significant	t (df)	p-value
2015 "New Ideas"	X		X	2.30(383)	.006
2015 "Good Ideas"	X			1.90(390)	.036
2015 "Imagination"	X		X	3.00(370)	<.001
2016 "New Ideas"	X		X	3.12 (454)	<.001
2016 "Good Ideas"	X		X	3.76 (454)	<.001
2016 "Imagination"		*		.22(452)	.670
2017 "New Ideas"	X		X	3.33(330)	.001
2017 "Good Ideas"	X		X	3.31(319)	.001
2017 "Imagination"	X		**	.427(290)	.670

*The Pre-test scores for 2016 "Imagination" were so high that, although they declined, that drop is insignificant.

**The Pre-test scores for 2017 "Imagination" were so high, there was not enough room to increase a statistically significant amount.

students improve their self-efficacy in relation to their ability to come up with new ideas, as well as their ability to have good ideas and a good imagination.

Summer Self-Efficacy Tests

As with the Academic Year, The Summer Self-Efficacy Pre- and Post- Tests asked students whether or not they agreed with the following statements: "I am good at coming up with new ideas"; "I have lots of good ideas"; and "I have a good imagination." Students were asked this question at the beginning of class on Monday and end of class on Friday. The weekly results on the three questions were combined across all weeks for each summer, resulting in three data points per year.

Slight improvements to the administration of the assessment were made over the course of its usage. Students were given clearer instructions on how to indicate their "yes," "maybe," and "no" responses, and photos were taken against a blank wall for easier analysis. Both of these adjustments increased the reliability of the results as assessments continued. Table 4 shows student gains.

Much like the Pre-Post Drawing Assessments, inconsistencies in the administration of the Self-Efficacy Assessments also had to be addressed. Specifically, instructions for hand placement were clarified, and photos were consistently taken in front of a plain background, so that student responses could more clearly be classified. By

Table 5. An analysis of the Summer 2017 data shows how significant the gains that students make over the course of their week-long class after inconsistencies were fully addressed, making these data the ideal set for a full analysis of Creative Self Efficacy results..

Pre-test	Post-test	Difference	T	Df	Significance
3.171	3.87	.69	12.88	200	<.001

Summer 2017, these inconsistencies were addressed and implemented, making these data the ideal set for Creative Self Efficacy results. Table 5 shows the gains in self-efficacy made by students, who were able to express their gains in this unconventional assessment process.

Guiding Questions Assessment

In Spring 2016, instructors added the administration of the Guiding Questions Measurement to the assessment of the Academic Year programming. This measurement was added to determine student learning of the specific STEAM concepts being taught each day. Students were asked a Guiding Question, tailored to the day's lesson, at the beginning and end of the class period. See Figure 5. Correct answers were tallied, and mean differences analyzed. As a refinement of

Figure 5. Examples of guiding STEAM content questions from ARTspace lessons

What is an omnivore? A. A plant eater. B. A meat eater. C. An animal that eats both plants and animals. D. An animal that gets its food from sunlight. *Tied to Next Generation Science Standard 3-LS4 Biological Evolution: Unity & Diversity.*

What is the result of a long camera exposure? A. The camera records more light. B. The camera records more sound. C. The camera records more darkness. D. The camera records more space. *Tied to National Core Arts Standards, Visual Art Anchor Standard 2: Organize and develop artistic ideas and work.*

What is the strongest shape? A. Straight line. B. Triangle. C. Whatever is longest. D. Pentagon. *Tied to Common Core Standards, Math Content 2.G.A.1: Reason with shapes and their attributes.*

What are the 3 tools of an actor? A. Voice, Body, Imagination. B. Script, Lighting, Sound C. Stage Manager, Director, Actor.

Figure 6. The results from the first and second sessions of Guiding Questions administered daily with challenge-based lessons

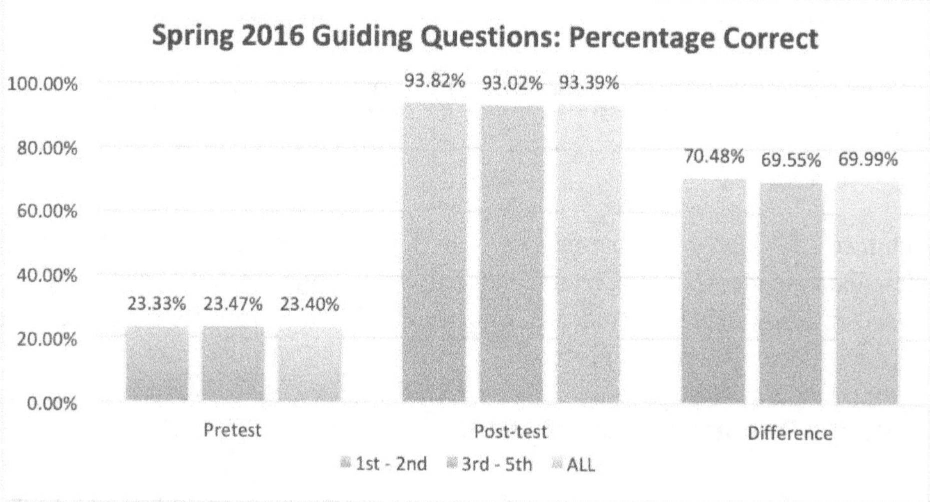

the assessment, a "Retention Question" was added to the process, where students were asked the Guiding question again after two-four weeks, to determine if their learning gains had been retained.

Because the students in every ARTspace can be at differing levels of reading and writing, the question and answer choices were written on the board and also read aloud. A few examples of guiding questions include:

Percentages of correct answers for Spring 2016 are represented in Figure 6. Percentages of correct answers for Fall 2016 are represented in Figure 7. Results were calculated for the 1st-2nd Grade students, the 3rd -5th Grade students, and all students. As the results show, students made remarkable learning gains at each grade level, all of which are statistically significant.

Because of the significant gains shown in the Spring 2016 Guiding Questions data, the evaluation team determined that an additional measure was needed to ascertain students' long term retention of STEM concepts covered in the lessons. Thus, in Fall 2016 the "retention question," was added. Students were asked to answer one of the Guiding Questions from a lesson they attended two to four weeks prior. Interestingly, retention levels were at least as high, if not higher, than the post-test results. In other words, some of the students who did not know the answer at the post-test learned from their mistake and retained that information when asked the question two to four weeks later.

Figure 7. The results from the first and second sessions of Guiding Questions administered daily with challenge-based lessons. In Fall 2016, the Guiding Question was asked two or three weeks later depending if it was a three or four week camp.

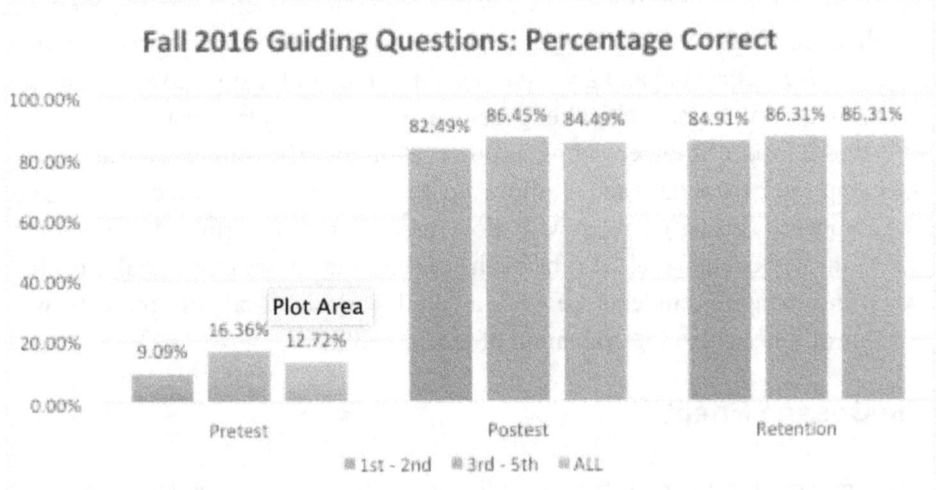

RECOMMENDATIONS FOR FUTURE STUDY

The results from the three measurements created and used to quantify the efficacy of ARTspace all integrate into a larger picture of what the program does for students. It shows that they are creating new and unique material within a design thinking structure, that they can explain both the STEM concepts and stages of invention they are using, and they are gaining confidence in identifying as innovators, and, above all, that they are engaging in learning at a deep level.

Bloom's Taxonomy

Clearly, this research only begins to examine the facets of the ARTspace program at LAC. For example, the initial plan for evaluating design challenges within ARTspace was to tie Bloom's taxonomy of instructional objectives to the effort and completion of each challenge. In Bloom's taxonomy, demonstrating mastery of higher level learning objectives requires students to first obtain knowledge and skills from lower levels. For example, as a student works through an instructional unit (cognitive domain), he or she must typically learn and comprehend basic facts before applying and analyzing, and can then engage in synthesis and evaluation.

While using Bloom's taxonomy was later abandoned for the tools used in this research study, the potential remains for its use as an assessment tool for ARTspace.

Specifically, a rubric has been created, based on a different set of objectives that better assesses the creative endeavors of ARTspace students. For this rubric (see Appendix A) several of Bloom's metacognitive knowledge objectives were distilled down to applicable sections: appropriate use, execution, construction, achievement, action, and actualization. These constructs can be rated on a scale from 0 = *no demonstration*, 1 = *some demonstration*, and 2 = *clear demonstration* of the objective, with higher scores suggesting higher student engagement at deeper levels of learning.

With ARTspace Summer camps, an observation could be administered at the start of a camp week (Monday) and a follow- up observation administered at the end of the camp week (Friday). Scores will allow researchers to determine the amount of growth throughout the week. A rubric can be used as an observational tool to capture grown and change as students are working on their design challenge projects, with portions of it used also for evaluation of the end product

The Dosage Effect

A second area for potential future studies is on the concept of a dosage effect—whether continued participation in ARTspace programming leads to additional gains for students. There is a potential dosage effect for students attending more than one ARTspace session throughout the Summer or multiple sessions during the Academic Year. Thus a future study could examine whether a dosage effect is observable in the scores of children who participate in multiple Summer course or Academic Year sessions. Products could be analyzed from a random sample of children who attend a single session, five sessions, and all 10 sessions. The rubric from the design challenges could be used, along with the modified rubrics the Formal Elements of Art Therapy Scale (FEATS), the creativity self-efficacy measure, and the addition of creativity/invention metacognition process video recordings.

Potential Research Questions that may be answered by looking at dosage effect data include (a) Is there a "tipping point" where students make the jump to next level(s) on Bloom's taxonomy? (b) Is it easier to move among the lower levels of Bloom's taxonomy? (c) How many sessions does it take to consistently demonstrate one or more of the upper levels of Bloom's taxonomy?

For those attending single sessions it will also be possible to see if a particular session had more impact on a child than another. Using the same measures mentioned previously, single session comparisons may be made. One caution to keep in mind is that single session discrepancies may have more to do with content/subject matter than with process, since the process is much the same for each of the 10 sessions.

Dosage for Creative Self Efficacy Photos

Dosage effects on Creative Self Efficacy may also be examined using the photos alone. Such a study would focus on whether dosage creates a cumulative effect on students who participate in greater numbers of ARTspace camps. Researchers could create two or more groups of students who are attending different numbers of Summer camps—one group might consist of those students who attend 1 or 2 camps, a second group could consist of those who attend 3-6 camps, and a third group of those who attend 7 or more camps. Once the groups are determined and the summer's data collection has concluded, the arithmetic mean of the ratings from the first Monday of camp for all of the children in Group 1 can be calculated and compared to the arithmetic mean of the ratings from the final Friday of camp for all children in Group 1, and so on for the rest of the groups. If a dosage effect is present, the size of the change in the average ratings from the beginning of the first session to the end of the last session should reveal that the group or groups that attended more sessions have greater change in average ratings, as compared to students who attend only one or two week-long Summer camps.

Validity and Reliability

A final area for future research would focus on the data collection instruments themselves. The three evaluation measures in the present plan (Drawing Assessment, Self-Efficacy Photos, Guiding Questions) have been designed or modified specifically for the ARTspace program. As such, they do not have established evidence for measurement validity. The creative self-efficacy measure has been adapted from a pencil-and- paper version used in research, but it has been used only a few times in published research, and no convergent or discriminant validity data could be found for the original instrument. For the original paper-and-pencil version of the creative self-efficacy measure, Beghetto (2006) reported an internal consistency alpha of .86 when used with middle school and high school students. Evaluators may wish to calculate the internal consistency reliability alpha of the adapted measure of creative self-efficacy used in this evaluation plan.

Along similar lines, future evaluators may also consider calculating interrater reliability for the evaluation measures that include observations or coding (FEATS-based rubric). Calculating interrater reliability for these components will help ensure ratings are not dependent upon which rater is coding, and also that all future coders would use the rubrics similarly. Similarly, the Self-Efficacy Photo evaluation would benefit from calculating interrater reliability, as not all of the students are consistent in their body positions. This research could result in a scoring code for one or both of these evaluation tools.

A final possibility for future research would be to examine the need to have different types of raters trained to criterion before beginning a large-scale study. For example, while a study by Gantt & Tabone (1998) found no differences in rating between art therapists, recreational therapists, and social workers, this might not be the case with tools involving abstract or nonrepresentational art. For the ARTspace program, this distinction may mean considering rating differences between teachers, volunteers, and other helpers assisting with coding or observation.

CONCLUSION

The Lawrence Arts Center developed the ARTspace STEAM program to help students exercise creativity and innovation in an environment where risk and exploration are encouraged. Additionally, teaching young learners the value of practice, commitment, and learning from mistakes is inherent to the process of art-making and STEM fields of both today and the future. Not only are STEM subjects enhanced with the inclusion of art, but with the development of the Design Wheel, ARTspace hopes to elucidate the link between the scientific process and the artistic one in a young learner's mind. This awareness of process, iteration, and improvement based off feedback is taught by practice; the more students engage in challenge based learning, the more comfortable they become with naming the stage of a process, taking design risks, and engaging in creating work not based on models. The assessment tools used to measure the outcomes of ARTspace were developed much how the program functions: in order to respond to the challenge of measuring creativity, facilitators needed to create new models of measurement. The data collected in the three year evaluation of ARTspace show students are making gains in believing in themselves as innovators and creators, in being aware of invention as a multi-stage process, and in retaining STEM concepts at a high level all by being afforded the time and structure to create freely in response to a design challenge. Learning is doing, but it is also more complex: learning is doing and making and listening and adapting and planning and communicating and developing and doing it over and over again while making connections to the world.

ACKNOWLEDGMENT

This research was supported by the Hearst Foundation and Enel Greenpower.

REFERENCES

Beghetto, R. A. (2006). *Creative self-efficacy: Correlates in middle and secondary students.* Academic Press.

Burk, C., Leahy, M., McGovern, J., Ramler, P., & Royer, D. (2015). *ARTspace summer camp program evaluation plan (Unpublished evaluation plan).* Lawrence, KS: University of Kansas.

Burnaford, G., Brown, S., Doherty, J., & McLaughlin, H. J. (2007). *Arts Integration Frameworks, Research Practice*. Washington, DC: Arts Education Partnership.

Catterall, J. S. (2012). *The Arts and Achievement in At-Risk Youth: Findings from Four Longitudinal Studies. Research Report# 55.* National Endowment for the Arts.

Cooper, J. O., Heron, T. E., & Heward, W. L. (2007). Applied behavior analysis. Creativity Research Journal, 18(4), 447-457.

Costa, A. L., & Kallick, B. (Eds.). (2008). *Learning and leading with habits of mind: 16 essential characteristics for success*. Alexandria, VA: ASCD.

Gantt, L. (1990). A validity study of the Formal Elements Art Therapy Scale (FEATS) for diagnostic information in patients' drawings (Unpublished doctoral dissertation). University of Pittsburgh, Pittsburgh, PA.

Gantt, L., & Mills, B. (2001). The Formal Elements Art Therapy Scale: A measurement system for global variables in art. *Brief Report*, *18*(1), 50–55.

Gantt, L., & Tabone, C. (1998). *The Formal Elements Art Therapy Scale: The rating manual*. Morgantown, WV: Gargoyle Press.

Jamrisko, M., Miller, L. J., & Lu, W. (2019). These Are the World's Most Innovative Countries. *Bloomberg News*. Retrieved from https://www.bloomberg.com/news/articles/2019-01-22/germany-nearly-catches-korea-as-innovation-champ-u-s-rebounds

Kálly, É. (2012). Learning strategies and metacognitive awareness as predictors of academic achievement in a sample of Romanian second-year students. *Cognition, Brain, Behavior: An Interdisciplinary Journal*, *16*(3), 369–385.

Lawrence Arts Center. (2015, May 9). *About us: History*. Retrieved from http://lawrenceartscenter.org

Lawrence Arts Center. (2018). *2017-2018 Annual Report*. Retrieved from http://lawrenceartcenter.org

Lawrence Arts Center. (n.d.). *ARTspace: Explore. Imagine. Invent.* Lawrence, KS: Lawrence Arts Center.

Lichtenberg, J., Wright, M., & Woock, C. (2008). *Ready to Innovate: Are Art Educators and Executives Aligned on the Creative Readiness of the US Workforce?* The Conference Board, Inc.

Lowenfeld, V. (1939). *The nature of creative activity*. New York: Harcourt Brace.

Luka, I. (2014). Design Thinking in Pedagogy. *Journal of Education Culture and Society.*, *2*, 63–74. doi:10.15503/jecs20142.63.74

Mathisen, G. E., & Bronnick, K. S. (2009). Creative self-efficacy: An intervention study. *International Journal of Educational Research*, *48*(1), 21–29. doi:10.1016/j.ijer.2009.02.009

Murnane, R. J., & Levy, F. (1996). *Teaching the New Basic Skills: Principles for Educating Children to Thrive in a Changing Economy*. New York: Free Press.

Negretti, R. (2012). Metacognition in student academic writing: A longitudinal study of metacognitive awareness and its relation to task perception, self-regulation, and evaluation of performance. *Written Communication*, *29*(2), 142–179. doi:10.1177/0741088312438529

Nelson, P., Martin, S. S., & Baldwin, V. G. (1998). Drawing skills and science concepts in young children: A study of relationships. *Studies in Art Education*, *39*(3), 262–269. doi:10.2307/1320368

Oregon State University. (2011). Retrieved from http://oregonstate.edu/instruct/coursedev/models/id/taxonomy/

Schell, J. (2018, March). *Design thinking has a pedagogy problem.* Future20 Session presented at the South By Southwest EDU Conference, Austin, TX.

Schraw, G., & Dennison, R. S. (1994). *Metacognitive awareness inventory.* doi:10.1037/t21885- 000

Tierney, P., & Farmer, S. M. (2002). Creative self-efficacy: Its potential antecedents and relationship to creative performance. *Academy of Management Journal*, *45*(6), 1137–1148.

KEY TERMS AND DEFINITIONS

21st Century Learning Skills: Skills and learning dispositions identified by educators, business leaders, academics, and governmental agencies as being required for success in 21st century society and workplaces; this includes a shift from traditional practices of memorization of science, math, and reading content to communication, collaboration, critical thinking, and creativity.

Bloom's Taxonomy: A hierarchal framework for educational practice along a continuum from simple to complex and concrete to abstract first developed by educational psychologist Benjamin Bloom in 1956.

Challenge-Based Learning: An educational framework that focuses learning and organizational behavior around authentic learning experiences moving through three distinct phases: engage, investigate, and act.

Creative Self-Efficacy: The belief one has the ability to produce and be successful with creative outcomes.

Design Thinking: An educational framework of learning using collaboration and problem solving to identify challenges, gather information, conceive of and test possible solutions, and gain knowledge from results.

Metacognition: The cognitive processes used to plan, monitor, and assess one's knowledge, understanding, and performance.

Pedagogy: The theory and practice of teaching regarding how knowledge and skills are shared in an educational setting.

Project-Based Learning: An educational framework in which learners acquire knowledge and skills by working for an extended period of time to investigate and respond to an authentic, engaging, and complex question or problem.

Standards-Based Learning: A framework of instruction and assessment based on students demonstrating mastery of the knowledge and skills as set by a defined set of criteria.

APPENDIX 1

Formal Elements of Art Therapy Scale, modified (Gantt & Mills, 2001)

Table 6 shows the Formal Elements of Art Therapy Scale.

Table 6. Formal Elements of Art Therapy Scale

Scale	Observation	Rating
Space	1 = less than 25% used 2 = between 25% and 49% used 3 = between 50% and 74% used 4 = between 75% and 99% used 5 = 100% used	12345
Integration*	1 = No required elements are not integrated 2 = Less than half of required elements are integrated 3 = Half of required elements are integrated 4 = More than half of required elements are integrated 5 = All required elements are integrated	12345
Logic*	1 = Product is not logical to required task 2= 3 = Product is somewhat logical 4= 5 = Product is logical to required task	12345
Realism*	1 = Product does not demonstrate realism 2= 3 = Product demonstrates some realism 4= 5 = Product demonstrates clear realism	12345
Problem-Solving*	1 = Problem-Solving is not demonstrated 2= 3 = Some Problem-Solving is demonstrated 4= 5 = Problem-Solving is clearly demonstrated	12345
Details of Objects and Environment	1 = Details do not match criteria 2 = Details somewhat match criteria 3 = Details match criteria, but are not inventive 4 = Details match criteria; some inventive qualities 5 = Details are abundant and inventive	12345
Line Quality	1 = All lines are shaky 2 = Most lines are shaky 3 = Lines vary (about 50%) between shaky and fluid 4 = Most lines are fluid 5 = All lines are fluid	12345

* Definition of constructs will vary based on task

Source: Gantt, L., & Mills B. (2001). The formal elements art therapy scale: a measurement system for global variables in art. *Brief Report 18*(1), 50-55.

APPENDIX 2

Sample Metacognitive Awareness Questions

- When we ask you to draw something, do you think about it first or do you just start and see what happens?

- When you are asked to build something, and the first thing you try does not work, do you try the same thing again or change what you are doing?
- What other ways could you have [insert daily task or project here]?
- When you are asked to do something, do you do it right away or do you do other things first?
- Do you do things a certain way because that is what you were told to do, or taught to do?

Adapted from: Schraw, G., & Dennison, R. S. (1994). *Metacognitive awareness inventory*. doi: 10.1037/t21885-000

APPENDIX 3

Rubric for Design and Invention Process Analysis

Table 7 shows a rubric for design and invention process analysis.

Table 7. Rubric for Design and Invention Process Analysis

Wheel Stage	Description	Example Keywords	**Scoring** (*based on number of keywords and details expressed*) **0 – no keywords or details** **1 – one to two keywords or details** **2 – three to four keywords or details** **3 – five or more keywords or details**
Ask	Exploring, examining & studying	- Looking up information/ reference pictures - Research - Ask others - Make sure I understand what I am supposed to do	0123
Imagine	Forming mental image, idea or concept	- Imagine possible options - Getting an idea	0123
Plan	Awareness and recognition of details and logistics	- Make a plan - Gather materials - Best option	0123
Create	Making & forming	- Start drawing - Testing a technique - Changing as I go	0123
Improve	Thinking, reflecting & evaluating	- Ask others what they think about what I made - Figure out what I could have done better - Acknowledging what did and did not work	0123

APPENDIX 4

Creative Self-Efficacy Measure

Instructions: "Campers, I'm going to tell you three different things. Your job is to think about what I say and how true it is for you. I'm going to ask you to rate how true it is about YOU on a scale from 1 to 5. [Gesture to the poster with smiley faces.] One will mean what I said is REALLY not true at all about you. Two will mean that it is just not true about you. Three will mean that what I said is not really true or false about you—you feel kind of in the middle about it. Four will mean that it *is* true about you. Five will mean that what I said is VERY true about you."

"Let's practice one: The first one is: I like pizza. Think about how true that is for you—it might be REALLY not true about you—for that hold up a 1, but it might be VERY true—for that hold up a 5. The more it is true, the more fingers you hold up. Now put your heads down to give everyone privacy, and hold up your number: 1, 2, 3, 4, or 5."

[Correct and teach as needed, answer questions]

1. "I am good at coming up with new ideas." "Think about it. Ok, heads down, hold up your number."
2. "I have a lot of good ideas." "Think about it. Ok, heads down, hold up your number"
3. "I have a good imagination." "Think about it. Ok, heads down, hold up your number."

Adapted from: Beghetto, R. A. (2006). Creative self-efficacy: Correlates in middle and secondary students. *Creativity Research Journal, 18*(4), 447-457.

APPENDIX 5

Creative Self-Efficacy Visual Aid

1/2/3/4/5: Really Not True/ Not True/ Somewhat True/ True/ Really True

Note. This is just a sample. The visual aid should be on a large scale, so that children can see it. It could be drawn on a whiteboard, printed on a poster-size piece of paper, or made into a banner. Feel free to augment as necessary to make it more visually appealing and help with student understanding.

APPENDIX 6

Scoring Rubric for Pre/Post Drawing Assessment

Table 8 shows a scoring rubric for pre/post drawing assessment.

Student #: ____________________

Course Week #: ________________

Mean Pre:_____

Mean Post:_____

Mean Difference:_____

Table 8. Scoring Rubric for Pre/Post Drawing Assessment

Scale	Observation	Rating	Diff
Space (Appropriate use of)	1= Product does not use an appropriate amount of space 2= 3= Product uses reasonable amount of space 4= 5= Product uses appropriate space	Pre: 1 2 3 4 5 Post: 1 2 3 4 5	
Logic (Did they draw what they were asked to draw?)	1= Product is not logical for the task 2= 3= Product is somewhat logical for the task 4= 5= Product is logical for the task	Pre: 1 2 3 4 5 Post: 1 2 3 4 5	
Realism (Does it look real? Is it identifiable?)	1= Product does not demonstrate realism 2= 3= Product demonstrates some realism 4= 5= Product demonstrates clear realism	Pre: 1 2 3 4 5 Post: 1 2 3 4 5	
Details (Of Objects and Environment)	1= details do not match criteria 2= details somewhat match criteria 3= details match criteria, but are not inventive 4= details match criteria; somewhat inventive 5= details are abundant and inventive	Pre: 1 2 3 4 5 Post: 1 2 3 4 5	
Line Quality (Confidence in Line Drawing)	1= all lines are shaky 2= most lines are shaky 3= lines vary between shaky and fluid 4= most lines are fluid 5= all lines are fluid	Pre: 1 2 3 4 5 Post: 1 2 3 4 5	

Chapter 8
Black Girls STEAMing Through Dance:
Inspiring STEAM Literacies, STEAM Identities, and Positive Self-Concept

Ayana Allen-Handy
https://orcid.org/0000-0002-8910-6957
Drexel University, USA

Valerie Ifill
University of Akron, USA

Raja Y. Schaar
Drexel University, USA

Michelle Rogers
https://orcid.org/0000-0001-9021-7809
Drexel University, USA

Monique Woodard
Drexel University, USA

ABSTRACT

Black Girls STEAMing through Dance (BGSD) leverages a transdisciplinary partnership among four Black women professors in urban education, dance, industrial/product design, and computing to engage Black girls in a STEAM-infused program to inspire STEAM literacies, STEAM identities, and positive self-concept. BGSD is in its third year of existence and operates across several contexts, including an

DOI: 10.4018/978-1-7998-2517-3.ch008

after-school program for 7- to 12-year-old Black girls, a co-curricular mini course program for 5th and 6th grade girls, and a professional development course for teachers. This chapter highlights how the program was developed and how the use of dance to integrate STEAM is a promising platform to encourage engaged STEAM participation amongst underrepresented Black girls.

INTRODUCTION

Black women and girls are severely underrepresented in Science, Technology, Engineering and Mathematics (STEM) fields across both educational and career continuums (Marra, Rogers, Shen, & Bogue 2009; Pinkard, Erete, Martin, & McKinney de Royston, 2017; Towns, 2010). Their access to STEM pipeline opportunities which could lead them on a path towards future STEM majors and career trajectories lags significantly behind their White female and male counterparts (Farinde & Lewis, 2012). Meanwhile, the United States has an urgent need for a highly skilled and innovative STEM workforce and STEM leaders in order to stay competitive in a global marketplace (Diekman, Steinberg, Brown, Belanger, & Clark, 2017). Nearly 25% of all STEM fields are comprised of women (Simon, Wagner, & Killion, 2017), and the numbers are even more alarming for Black women. According to the National Science Foundation's National Center for Science and Engineering Statistics (2019), Black women graduates represent 4.5% of the biological sciences, 2.2% of computer science, 2.5% of physical sciences, 2.1% of mathematics and statistics, and 1.0% of engineering conferred degrees. Although there have been gains in access to STEM education earlier in students' schooling experiences (Lindeman, Jabot, & Berkley, 2014; Murray, 2019), there remains grave inequities across socio-economic, racial/ethnic and gender lines, particularly as it relates to Black girls (Ireland, et al., 2018; King, 2017; King & Pringle, 2018; Marra, et al., 2009; Young, Tolliver, Young, & Ford, 2017). Some scholars have even deemed this lack of access to STEM opportunities at the early onset and throughout Black girls' education, a civil rights issue (Espinosa, 2011; Mensah, & Jackson, 2018; Tate, 2001), and there has been little progress toward comprehensively and systematically eradicating these gaps and challenges (Ireland, et al.,2018).

STEM education access in formal and informal learning spaces is limited for Black girls as a whole and often representative of compounding socio-political issues of racism, sexism, and generational poverty which keep Black girls pushed out of transformative learning experiences (Morris, 2015). Even if STEM education is available, these opportunities are often disconnected from the historical, cultural, gendered, and daily lives of Black girls, and the cultivation of a STEM identity frequently lacks intersectional, socio-cultural relevance for them (Charleston, Adserias, Lang, & Jackson, 2014; Ireland, et al., 2018). According to Brown, Mangram, Sun, Cross, and Raab (2017), "there are numerous ways to be African American, selecting an identity that allows you to integrate being Black and being scientific can be empowering" (p. 179). However, the neoliberal agendas of schooling in urban education (Lipman, 2013) create additional hurdles to bridging these gaps for Black girls' culturally congruent STEM identity formation.

While these lingering challenges have been well articulated and substantiated in the extant literature, there is however, a burgeoning body of scholarship that is examining ways in which to best engage Black girls in STEM education, in a format they can appreciate and in a way that leverages their funds of knowledge and cultural wealth (Amanti, González, & Moll, 2005; Khalifa, Gooden, & Davis, 2016; Yosso, 2005). Several asset-based pedagogies and programs are being employed to cultivate Black girls STEM learning throughout their PK-12 education, so that they can envision the possibilities of pursuing STEM majors and careers in the future (Morales-Chicas, Castillo, Bernal, Ramos, & Guzman, 2019; Scott, Sheridan, & Clark, 2015). Particularly, several informal learning programs such as Digital Youth Divas (Pinkard, et al., 2017), Black Girls Dive: STREAMS Program (Black Girls Dive Foundation, 2019), Black Girls CODE (Scott & White, 2013), and DanceExcel (Champion, 2018) are integrating various elements of STEM, while also utilizing concepts of project based learning and culturally relevant and sustaining pedagogies (Ladson-Billings, 1995; Paris, 2012). Such approaches have demonstrated Black girls increased engagement and motivation in STEM (Daily, et al., 2014; Parmar, et al., 2016; Pinkard, 2005), and there is a growing interest in the intersections of Art and STEM, which has resulted in the evolution to Science, Technology, Engineering, Art, and Mathematics (STEAM) education.

STEAM has been noted to support girls' ability to make connections across disciplines through creative processes, innovation, and problem solving (Lam, 2013; Liao, Motter, & Patton, 2016). Dance as a platform to integrate STEM is an early candidate to encourage STEAM engagement, particularly for Black girls, for dance has been and continues to be woven into the fabric of African American culture (Malone, 1996; Stovall, 2015). Considering that dance touches practically every aspect of Black life, unfortunately it (and the arts in general) is often absent from schools that serve majority Black students in urban environments. According to Parsad and Spiegelman (2012), there is a downward trend in access to arts education and dance instruction specifically. Particularly, they indicate that 95% of elementary school and 77% of secondary school children that attend schools where 75% or more of students receive free and reduced lunch do not have access to dance instruction. Bearing in mind, the grave disparities in STEM and Arts access and education, ***Black Girls STEAMing Through Dance*** (BGSD) was created to disrupt pervasive gaps in the underrepresentation of Black girls in STEM and STEAM, and to inspire a new generation of Black girls with strong positive self- concept and crystalizing STEAM identities.

BLACK GIRLS STEAMING THROUGH DANCE

BGSD launched in the Fall of 2017 and operates across several contexts. First, BGSD serves as an after-school program for 7-12 year old Black girls at the Dornsife Center for Neighborhood Partnerships in West Philadelphia (**BGSD Dornsife**). BGSD is also offered as a co-curricular mini course program for 5th and 6th grade girls at the Science Leadership Academy Middle School (**BGSD SLAMS**), as well as a professional development course for teachers (**BGSD Teach**). Our program is led by four African American women professors, an African American woman graduate assistant, and three African American women undergraduate assistants. BGSD represents a transdisciplinary partnership amongst our various fields in urban education, dance, product/industrial design, and computing.

BGSD Dornsife engages girls in a dance class consisting of jazz, modern, ballet, and Hip Hop taught by Val and our undergraduate assistants, while integrating various elements of STEAM taught by Raja, Michelle, and our graduate research assistant Monique. BGSD follows the Drexel University academic calendar divided into three eight-week sessions across the fall, winter, and spring quarters. Each week, girls rotate between the *The Coding Lab* and *The Design/Making Lab*. In *The Coding Lab*, they learn to code and integrate virtual reality (VR). In *The Making Lab*, they learn about design and circuits and create the wearable technologies for their dance costumes. Our weekly activities draw inspiration from and bring inspiration to their dance instruction, where girls also "make" with their bodies and design their own choreography. In **BGSD SLAMS**, the students participate in an 8-week, 80-minute weekly program that combines instruction in choreography, design, and coding through the use of the Chibitronics Make-code application and a blend of Technokio and Chibitronics hardware. The girls engage in choreographing group dance sequences and design illuminated costumes for a final showcase music video. **BGSD Teach** supports 2nd-8th grade teachers from the West Philadelphia Promise Neighborhood. During the hands-on, day-long workshop, teachers participate in standards driven STEAM activities that connect movement, coding, and design. At the conclusion of the workshop, teachers plan lessons to take back to their classrooms, with support from the authors throughout the school year.

Ultimately BGSD is purposed to create an ecology of STEAM experiences to inspire Black girls' STEAM literacies, STEAM identities (wherein the participants internalize that they are dancers, scientists, engineers, artists, designers, etc.), and the formation of positive self-concept. Another objective of BGSD is to support STEAM teachers' development of self-efficacy, cultural competence, and culturally sustaining pedagogies (McCarthy-Brown, 2017; Paris, 2012). Throughout this chapter, the authors will highlight BGSD and share how our program's use of dance to integrate STEAM is a promising platform to encourage engaged STEAM participation amongst underrepresented Black girls (Daily, et al., 2014, Pinkard, et al., 2017). Furthermore, the authors will demonstrate its promises for the field and the emerging body of literature about the transformative integration of the arts in STEM learning for historically marginalized Black girls. Figure 1 shows a chart of a BGSD unit plan on energy.

Figure 1. Chart of BGSD energy unit plan

DISCIPLINE	DEFINITIONS	CLASS THEME	CLASS EXPERIENCE	LESSONS LEARNED
DANCE	Energy is the potential for force. Energy is known as a core element of dance.	Explore active and passive energy	Everyone in the class will stand spread out across the performance space and when a solo dancer makes a sharp shape, the dancers closest to the soloist will respond by changing the direction they are facing, level (high, mid, low) and make a shape with their body. Each dancer will respond with the new shape once the dancers next to them have made a new shape.	This experience will: 1. Emphasize the difference between passive and active energy and how both feel in the body. 2. Create a ripple effect within the group, visually representing how energy passes from one point to another. 3. Allow the use of multiple bodies to emphasize how energy can be amplified.
DESIGN and MAKING	Energy is how power is transmitted through materials from a power source like a battery to an output like a light.	Demonstrate power transmission through simple circuits using batteries, conductive materials, and LED lights	Everyone in class will review basic circuit diagrams and create a sample circuit using an LED conductive tape and a battery and then they will create a light up bracelet.	Through this experience students will: 1. Understand the difference between positive and negative terminals on a power sources. 2. Understand how traces (i.e. conductive thread, ribbon, wires) can be made between a battery and an output source (i.e. LED light). 3.Students will understand how to measure materials to fit the body and how to apply flexible electronic materials to e-textiles.
CODING	Energy can be modeled and demonstrated in programming characters	Explore coding movements and concepts	Everyone in class will begin by creating their characters to represent themselves and/or an idea. They will then enable that character to move	This experience will 1. Demonstrate how characters can move and express energy 2. Use paired programming to assist girls with understanding and completion of activities

DANCE

Why We STEAM through DANCE

Dance is the foundational element of BGSD and it serves as the conduit for developing STEAM literacies, STEAM identities, and positive self-concept because we use the physical body as the vessel for learning and creative expression. Black and Brown bodies, which have been historically marginalized and classified as "minorities" in the U.S., (when in fact they have been minoritized; Harper, 2012) are significantly underrepresented in mainstream STEM and STEAM opportunities. Yet, in BGSD all bodies regardless of skin tone or shape are celebrated, honored, and are a significant part of reclaiming a positive sense of self for Black girls.

How We STEAM Through Dance

Over the last three years, the model has evolved based on the structure of our partner programs. BGSD grew out of an existing dance program at the Dornsife Center for Neighborhood Partnerships at Drexel University in Philadelphia, which was led by then Dance Program Director and co-author Val Ifill. The afterschool and Saturday dance program at the Dornsife Center was robust, and provided community dance classes for children ages 4 to 18 and adults 19 to mid-90s. After three years of running a successful *Dance at Dornsife* program, Val began to look for ways to expand programming for highly engaged participants. She joined forces with Ayana, Raja, and Michelle, and BGSD was born out of a *Learning Innovations Seed Grant* from the Drexel University ExCITe Center.

The first year of the program, BGSD partnered with a concert dance class featuring movement from modern, jazz and ballet dance. The second year we made a change by connecting to hip-hop curriculum. The students took a one-hour dance class followed by a one-hour design/making or computing lab focused on design using wearable technology or coding. Throughout each session, students participated in journaling and reflective activities about their interests and experiences in the program. We also began BGSD SLAMS in year two. Now in our third year, BGSD Dornsife has implemented a few changes based on our assessment of the first two years, and growth of the program. The first major change is that dance and lab time are integrated together once a week for an hour vs. the two hours in previous years, combining dance, design and computing in one cohesive experience. In this new model, the girls are able to make connections between all three subject areas at the same time.

Dance as an Art Form

In BGSD, dance both stands on its own as an artform in addition to being utilized to support and translate other disciplines. Performance and choreography are emphasized in our program. Students invest in improving dance technique in addition to learning the elements of dance-making. Collaborative dance-making amongst the students, supported by our team of faculty, graduate student, and undergraduate students, is a key component of our work. Allowing the dancers to have agency over their bodies and make choices based on how they want to interact with the design and code elements, deepens their kinesthetic, cognitive, and psychological-social learning.

Dance as Reflection

Dance is used as a tool to reflect and respond to learning experiences in coding and design, as well as, a way for the girls to share what they've learned about the art of Dance-making as a form of self-expression. In education, verbal and written communication often hold high value. Through our program, students have the opportunity to engage in nonverbal reflection which could be helpful for many learners including kinesthetic learners and visual-spatial learners, described by Howard Gardner's Multiple Intelligences Theory (Gardner, 1983). An example of dance as reflection is when students have experience in one of the design or code labs or when they view choreography, they are asked to create a movement gesture that summarizes how they feel in response to what they have just seen. This mode of processing, allows students who have a difficult time using words to express feelings or ideas, additional options to show an understanding of content. The teacher can gather information about the girls' responses by looking at posture, shapes, dynamics and focus.

Dance as a Form of Presentation

At BGSD Dornsife, our 9-month program focuses on technique and skill-building for the first three months, creative process for the next three months and the final three months are spent building a culminating experience. Since dance is a performing art, it is the perfect form to feature the learning that has taken place throughout the year. In BGSD Dornsife, for the first two years, the dancers, in partnership with their undergraduate student teachers, created a three-minute dance involving wearable technology that responded to their choreography. The choregraphed dance was performed on a proscenium stage in the Dance at the Dornsife Center end-of-year concert for family and friends. For the third year of the program, students will showcase the process as well as the product on their learning journeys. A gallery

Figure 2. BGSD Dornsife performance dress rehearsal

space will be created to showcase mini projects and site-specific dances will be performed inspired by what they learned using the costume pieces they design. At BGSD SLAMS, the girls create a final showcase dance on film in the dark that highlights their co-constructed choreography and the designs of their illuminated costumes, which they have created throughout the program. Figure 2 and Figure 3 show BGSD Dornsife images.

DESIGN AND MAKING

Why We STEAM Through Dance Using Design and E-textiles

Through Dance and Design, we want students to understand the relationship between our bodies and what we wear on them, and to ask the question, "*How might we celebrate, sense, or show how our bodies interact with space and over time?*" By connecting what we design to coded electronic components, we can both augment dance performances and apply computer science (CS) skills to physical objects. Design is integral to our STEAM curriculum because Design influences all aspects

Figure 3. BGSD Dornsife performance

of life—from the homes we live in, the shoes we wear, the chairs we sit in, the advertisements around us, and the cell phones we use to communicate. Designers not only dictate the way these things look, but they also have the power to inform the way these things work, how they are made, what they are made from, and how they help us with everyday tasks. Without understanding the power of design to shape our everyday lives, students are merely the users of objects and tools. Once students gain an understanding that there are people behind the decisions that shape the things we own, use, and experience, they begin to understand that design can change lives, and they can imagine themselves as future designers who can use their STEAM knowledge and skills to change the world. An emerging field in Industrial Design is wearable technology. Largely, this is seen in health/fitness monitors, light up sneakers, smart watches, hearing aids, and augmented reality headsets. Integral to successful wearable products is an understanding of the body, lifestyles, and fashion. These wearable devices also require coding and wiring in order to work. So, we are using the overlap of Industrial Design, wearable technology, coding, and Dance to introduce our girls to the design process by making wearable electronics that are paired with embodied movement.

How We STEAM through Dance Using Design and E-textiles

By leading girls in creating e-textile based wearables, we're able to address skills and knowledge related to designing and making with textiles materials and sewable electronic components, controls, and sensors. Our primary output has been working with lights in the form of small LED (light emitting diode) stickers and sewable sequins. At BGSD Dornsife, we divide the students by age, 7-9 and 10-12. The program is broken into two phases, one addresses skill-building and the other is an introduction to the process. During this skill-building portion, the students create simple accessories for the body and learn about electronic circuits. We cover concepts such as series and parallel circuits and closed and open switches. The students show evidence of understanding by prototyping the circuits using conductive tape, batteries, and LEDs. We also introduce the students to various control hardware and sensors including mechanical switches, velostat (a conductive pressure sensitive plastic sheet), light sensors, and tilt sensors. The culminating exercise for the skill-building portion is to make a light-up cuff using felt and LEDS and simple switches, and ultimately the girls design their own costumes. Figure 4, 5, and 6 show examples of BGSD costume designing.

Figure 4. BGSD Dornsife cuff design

Figure 5. BGSD Dornsife costume design

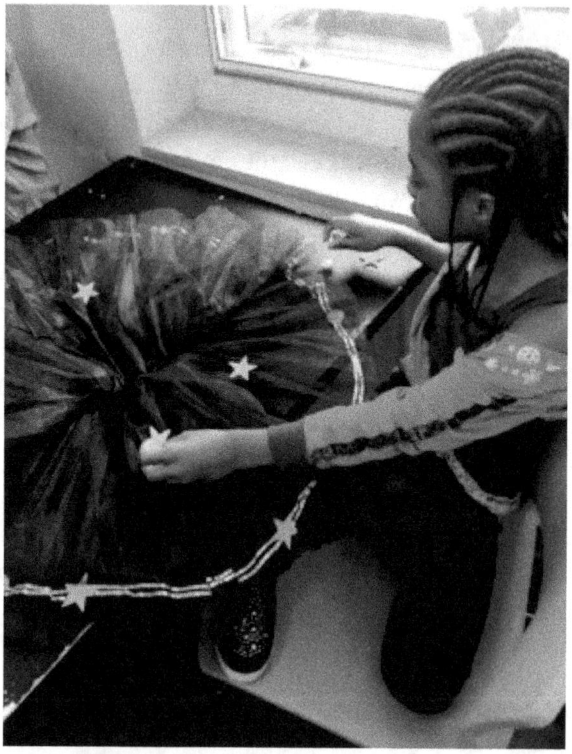

Math is used as the students use measuring tapes to measure parts of the body. We review the differences between metric and imperial measurements and the girls have to include the unit of measurement on their notes for clarity. The students also use simple addition to plan for extra material so there is an allowance for applying the battery pack and securing the cuff to the body. They use these measurements to create paper templates to sketch out design ideas and plan the circuit before transferring the design plans to felt, applying conductive tape, LED stickers, and battery packs. Through this exercise, we introduce design brainstorming through sketching, iterative development, and testing.

The two groups work with slightly different sets of electronic hardware components. This is due to age ranges of the hardware kits, the level of challenge we want the students to experience, as well as the level of support in the classroom. For the younger group of girls (ages 7-9) we largely use adhesive-based conductive components including LED stickers and conductive tape. Our older girls are introduced to hand sewing and use both the tape and conductive thread along with sew-on LEDs. In order to create stronger connections to both Dance and CS, during

Figure 6. BGSD Dornsife costume design

the second phase, the students use their basic design development skills, materials understanding, and electronic knowledge to plan light-up designs that are tied to a specific dance movement or sequence. This presents an opportunity for the dancers to think about how their embodied movements can be connected to and augmented by the lights worn on the body in strategic locations. For instance, if a dance involves a lot of stomping, a student might put a light lower on the body near the feet and make use of a tilt sensor, while a more lyrical gesture of a hand moving along with music might suggest placing lights in proximity to the hand or as an extension of the arm. Whereas turning and twisting might mean that placing lights on a skirt to accentuate the material lifting away from the body as the dancer moves. Since both choreography and coding is composed of sequences and loops, the students can use their computational understanding to control the lights as well by designing light sequences and looping patterns that align with both the music and choreography.

COMPUTING

Why We STEAM through Dance Using Computational Thinking and Coding

Coding and more specifically, computational thinking is an excellent candidate for forming a STEAM identity (Daily et al., 2015; Pinkard, 2005). While computing used to be a solo activity, it is now more focused on outcomes and co-constructing as a community of learners (Kafai, 2016). BGSD, being a transdisciplinary project between education, dance, design, and CS, allows for the girls to get direct instruction and interactive experience in coding as part of their making, and influenced by their dance.

How We STEAM through Dance Using Computational Thinking and Coding

In order to support the making and dance portion of the program, the goal of the coding content is to provide the students with the opportunity to create in technical terms. Ongoing connections are made between coding and dance and coding and design. First, between coding and dance, we use words and phrases that could demonstrate embodied movement. For design and coding, we identify coding aspects of design that allow the girls to activate their designs and make them interactive. When considering dancing, we looked at what codes would build on their knowledge. We frequently use three concepts– loops, conditional statements, and sequencing to demonstrate both dance and design. In our first year of BGSD Dornsife, the 10-12-year-old girls participated in the coding lab for three weeks, while the 7-9-year-old girls participated in the design/making lab. After that time elapsed, the groups switched. To manage the different age groups, it was decided that the younger girls would use scratch-base characters to demonstrate movements and actions. The older girls used the Tynker® platform with more mature looking characters that also used block coding to control their movements. In Figure 7 and Figure 8, the students are working in the coding lab.

The block coding platform allows for novice users to control characters without being able to write actual computer syntax (Kaplancali & Demirkol, 2017). Through the design and making components of the project, the girls participated in the application of the CS content. The CS component supported, and reinforced computational concepts observed in the dance and making. Specifically, students showed increased knowledge as they learned to correlate topics such as loops and conditional statements in coding language to their dance moves. So, if they did a dance that involved the repetition of a move, they were able to see that exemplified

Figure 7. BGSD coding lab

in the use of loops in the code they wrote. Advances to this work look to understand how the girls envision creating together as a group to produce desired outcomes. In addition, understanding is needed about how we continue to make the CS connections using actual programming syntax (e.g. Python, C++) rather than block coding (Weintrop, & Wilensky, 2015).

BGSD ON THE MOVE FORWARD

As the program enters its third year, we are constantly considering how to cater to the interests and strengths of each BGSD cohort. We do have several returning girls, but also many new ones, and most of our first weeks of lessons are spent building rapport and relationships with one another. We are also excited to launch BGSD Teach this spring. We are currently recruiting our first "class" of STEAM teachers to participate in our teacher workshop. We are also ever aware of the evolving nature of technology and the fields of industrial design and CS. As technology continuously develops, it is important for our students to be aware of current and

Figure 8. BGSD coding lab

future advancements. Therefore, virtual reality (VR) is being implemented into BGSD this year to introduce our girls to advanced technologies and to cultivate a STEAM identity in which they have agency over the technology. It must be noted that there are multiple definitions of virtual reality (VR), and for the purpose of BGSD, VR is defined as a computer-generated immersive environment that is viewed with a head-mounted display (Papanastasiou, Drigas, Skianis, Lytras, & Papanastasiou, 2018). When using VR devices in education, students are motivated and engaged in the topic they are learning (Chen, 2010). Furthermore, this technology also enables learners to transfer the information they learn from the virtual environment to reality with ease (Curcio, Dipace, & Norlund, 2016). In BGSD, VR provides an opportunity for the girls to utilize their newly gained skills in design and coding to construct a cohesive VR experience. The girls will use CoSpaces Edu, an online software, to design and code their VR environments. They will design their own dance characters in any setting they desire, such as outer space or a city. Once the characters are designed, students will then code their characters' movements and interactions. The addition of VR will also expand Math learnings. For example, in CoSpaces Edu, users only have the capability to enter the distance and time it takes for an object to move from one place to another. Therefore, as the girls begin to code their VR environments, they will learn how to calculate speed, which is distance divided by time. Using this software, the students will have the opportunity to apply their skills to create one cohesive environment that showcases their newly acquired knowledge.

In summary, ***Black Girls STEAMing through Dance*** continues to bridge the racial and gender disparities in STEM by incorporating the arts into STEM activities to develop and maintain their STEM interests. By integrating dance throughout all aspects of BGSD, dance serves as an artform and entry point to leverage the cultural wealth of Black girls. STEAM activities are scaffolded from learning to dance (Art), to making and designing costumes (Art), and lastly coding to enhance costumes and to create VR dance environments (CS, Technology, Engineering, Mathematics). BGSD is a model of transitioning from STEM to STEAM that leverages the cultural wealth and daily lives of Black girls. BGSD is an unapologetically Black girl program for and implemented by Black girls, and as such, BGSD demonstrates that African American girls can develop positive STEAM identities and positive self-concept through the transdisciplinary nature (education, dance, design/making, CS) of BGSD.

ACKNOWLEDGMENT

We would like to extend our sincere gratitude to Dr. Youngmoo Kim and the Drexel University ExCITe Center for the generous seed funding to start Black Girls STEAMing through Dance. Without your continued support for the project, we would not be able to impact all of the beautiful Black girls that are STEAMing through dance in West Philadelphia. We would also like to extend our sincere thanks to the Dornsife Center for Neighborhood Partnerships for your continued support of BGSD. It is an honor to meet with our girls every week at Dornsife. We would also like to thank The Science Leadership Academy Middle School for allowing us to share BGSD with your students, as well as Dean Penny Hammrich, Dr. Vera Lee, and the West Philly Promise Neighborhood team for your support of BGSD.

REFERENCES

Brown, B. A., Mangram, C., Sun, K., Cross, K., & Raab, E. (2017). Representing racial identity: Identity, Race, the construction of the African American STEM students. *Urban Education*, *52*(2), 170–206. doi:10.1177/0042085916661385

Champion, D. N. (2018). *The STEAM Dance Makerspace: A Context for Integration: An Investigation of Learning at the Intersections of STEM, Art, Making and Embodiment* (Doctoral dissertation). Northwestern University.

Charleston, L. J., Adserias, R. P., Lang, N. M., & Jackson, J. F. L. (2014). Intersectionality and STEM: The role of race and gender in the academic pursuits of African American women in STEM. *Journal of Progressive Policy & Practice, 2*(3), 273–293.

Chen, C. J. (2010). Theoretical bases for using virtual reality in education. *Themes in Science and Technology Education, 2*(1-2), 71–90.

Curcio, I. D. D., Dipace, A., & Norlund, A. (2016). Virtual realities and education. *REM -Research on Education and Media, 8*(2). doi:10.1515/rem-2016-0019

Daily, S. B., Leonard, A. E., Jörg, S., Babu, S., Gundersen, K., & Parmar, D. (2015). Embodying computational thinking: Initial design of an emerging technological learning tool. *Technology. Knowledge and Learning, 20*(1), 79–84. doi:10.100710758-014-9237-1

Diekman, A. B., Steinberg, M., Brown, E. R., Belanger, A. L., & Clark, E. K. (2017). A goal congruity model of role entry, engagement, and exit: Understanding communal goal processes in STEM gender gaps. *Personality and Social Psychology Review, 21*(2), 142–175. doi:10.1177/1088868316642141 PMID:27052431

Espinosa, L. L. (2011). Pipelines and pathways: Women of color in undergraduate STEM majors and the college experiences that contribute to persistence. *Harvard Educational Review, 81*(2), 209–241. doi:10.17763/haer.81.2.92315ww157656k3u

Farinde, A. A., & Lewis, C. W. (2012). *The underrepresentation of African American female students in STEM fields: Implications for classroom teachers.* Online Submission.

Gardner, H. (1983). *Frames of mind: The theory of multiple intelligences.* New York: Basic Books.

Harper, S. R. (2012). Race without racism: How higher education researchers minimize racist institutional norms. *The Review of Higher Education, 36*(1), 9–29. doi:10.1353/rhe.2012.0047

Ireland, D. T., Freeman, K. E., Winston-Proctor, C. E., DeLaine, K. D., Lowe, S. M., & Woodson, K. (2018). (Un)hidden figures: A synthesis of research examining the intersectional experiences of Black women and girls in STEM education. *Review of Research in Education, 42*(1), 226–254. doi:10.3102/0091732X18759072

Kafai, Y. (2016). From computational thinking to computational participation in K--12 education. *Communications of the ACM, 59*(8), 26–27. doi:10.1145/2955114

Kaplancali, U. T., & Demirkol, Z. (2017). Teaching coding to children: A methodology for kids 5+. *International Journal of Elementary Education, 6*(4), 32–37. doi:10.11648/j.ijeedu.20170604.11

Khalifa, M. A., Gooden, M. A., & Davis, J. E. (2016). Culturally Responsive School Leadership: A Synthesis of the Literature. *Review of Educational Research, 86*(4), 1272–1311. doi:10.3102/0034654316630383

King, N. S., & Pringle, R. M. (2019). Black girls speak STEM: Counterstories of informal and formal learning experiences. *Journal of Research in Science Teaching, 56*(5), 539–569. doi:10.1002/tea.21513

Liao, C., Motter, J. L., & Patton, R. M. (2016). Learning 21st-century skills through STEAM digital artmaking. *Art Education, 69*(4), 29–35. doi:10.1080/00043125.2016.1176492

Lindeman, K., Jabot, M., & Berkley, M. (2014). The Role of STEM (or STEAM) in the early childhood setting: Learning across the early childhood curriculum. *Advances in Early Education and Day Care, 17*, 95–114. doi:10.1108/S0270-4021(2013)0000017009

Lipman, P. (2013). *The new political economy of urban education: Neoliberalism, race, and the right to the city*. Taylor & Francis. doi:10.4324/9780203821800

Malone, J. (1996). *Steppin'on the blues: The visible rhythms of African American dance* (Vol. 21). University of Illinois Press.

Marra, R. M., Rodgers, K. A., Shen, D., & Bogue, B. (2009). Women engineering students and Self-Efficacy: A Multi-Year, Multi-Institution study of women engineering student Self-Efficacy. *Journal of Engineering Education, 98*(1), 27–38. doi:10.1002/j.2168-9830.2009.tb01003.x

McCarthy-Brown, N. (2017). *Dance pedagogy for a diverse world: Culturally relevant teaching in theory, research, and practice*. Jefferson, NC: McFarland & Company.

McCarthy-Brown, N. (2017). *Dance pedagogy for a diverse world: Culturally relevant teaching in theory, research and practice*. McFarland.

Mensah, F. M., & Jackson, I. (2018). Whiteness as property in science teacher education. *Teachers College Record, 120*(1), 1–38.

Moll, L. C., Amanti, C., & Gonzalez, N. (2005). *Funds of knowledge: Theorizing practices in households and classrooms*. Mahwah, NJ: Lawrence Erlbaum Associates.

Morales-Chicas, J., Castillo, M., Bernal, I., Ramos, P., & Guzman, B. L. (2019). Computing with Relevance and Purpose: A Review of Culturally Relevant Education in Computing. *International Journal of Multicultural Education*, *21*(1), 125–155. doi:10.18251/ijme.v21i1.1745

Morris, M. W. (2014). *Black stats: African Americans by the numbers in the twenty-first century*. New York, NY: The New Press.

Murray, J. (2019). Routes to STEM: Nurturing Science, Technology, Engineering, and Mathematics in early years education. *International Journal of Early Years Education*, *3*(1), 219–221. doi:10.1080/09669760.2019.1653508

Papanastasiou, G., Drigas, A., Skianis, C., Lytras, M., & Papanastasiou, E. (2018). Virtual and augmented reality effects on K12, higher and tertiary education students' -twentyfirst- century skills. *Virtual Reality*.

Parsad, B., Spiegelman, M., & Coopersmith, J. (2012). *Arts education in public elementary and secondary schools, 1999-2000 and 2009-10*. Washington, DC: National Center for Education Statistics, Institute of Education Sciences, U.S. Department of Education.

Pinkard, N. (2005). How the perceived masculinity and/or femininity of software applications influences students' software preferences. *Journal of Educational Computing Research*, *32*(1), 57–78. doi:10.2190/3LEE-MLCE-NK0Y-RUEP

Pinkard, N., Erete, S., Martin, C. K., & McKinney de Royston, M. (2017). Digital Youth Divas: Exploring narrative-driven curriculum to spark middle school girls' interest in computational activities. *Journal of the Learning Sciences*, *26*(3), 477–516. doi:10.1080/10508406.2017.1307199

Scott, K. A., Sheridan, K. M., & Clark, K. (2015). Culturally responsive computing: A theory revisited. *Learning, Media and Technology*, *40*(4), 412–436. doi:10.1080/17439884.2014.924966

Scott, K. A., & White, M. A. (2013). COMPUGIRLS'standpoint: Culturally responsive computing and its effect on girls of color. *Urban Education*, *48*(5), 657–681. doi:10.1177/0042085913491219

Simon, R. M., Wagner, A., & Killion, B. (2017). Gender and choosing a STEM major in college: Femininity, masculinity, chilly climate, and occupational values. *Journal of Research in Science Teaching*, *54*(3), 299–323. doi:10.1002/tea.21345

Stovall, M. (2015). African American cultural technology: The lindy hop, the king of pop, and the factory worker's experience. *Transforming Anthropology*, *23*(1), 1–13. doi:10.1111/traa.12042

Tate, W. (2001). Science education as a civil right: Urban schools and opportunity-to-learn considerations. *Journal of Research in Science Teaching*, *38*(9), 1015–1028. doi:10.1002/tea.1045

Towns, M. H. (2010). Where are the women of color? Data on African American, Hispanic, and Native American faculty in STEM. *Journal of College Science Teaching*, *39*(4), 8.

Weintrop, D., & Wilensky, U. (2015, June). To block or not to block, that is the question: students' perceptions of blocks-based programming. In *Proceedings of the 14th International Conference on Interaction Design and Children* (pp. 199-208). ACM. 10.1145/2771839.2771860

Women, Minorities, and Persons with Disabilities in Science and Engineering: 2019 | NSF - National Science Foundation. (2019). Retrieved 23 November 2019, from https://ncses.nsf.gov/pubs/nsf19304/digest/employment#employment-status

Yosso, T. J. (2005). Whose culture has capital? A critical race theory discussion of community cultural wealth. *Race, Ethnicity and Education*, *8*(1), 69–91. doi:10.1080/1361332052000341006

Young, J. L., Ero-Tolliver, I., Young, J. R., & Ford, D. Y. (2017). Maximizing opportunities to enroll in advanced high school science courses: Examining the scientific dispositions of Black girls. *Journal of Urban Learning, Teaching, and Research*, *13*, 174–183.

ADDITIONAL READING

Allen-Handy, A., & Thomas-EL, S. L. (2018). Be(com)ing critical scholars: The emergence of urban youth scholar identities through research and critical civic praxis. *Urban Education*. doi:10.1177/0042085918814589

Evans-Winters, V. E., & Esposito, J. (2010). Other people's daughters: Critical race feminism and Black girls' education. *Educational Foundations*, *24*, 11–24.

Johnston, E., Olivas, G., Steele, P., Smith, C., & Baily, L. (2018). Exploring pedagogical foundations of existing virtual reality educational applications: A content analysis study. *Journal of Educational Technology Systems*, *46*(4), 414–439. doi:10.1177/0047239517745560

Ladson-Billings, G. (2006). From the achievement gap to the education debt: Understanding achievement in US schools. *Educational Researcher*, *35*(7), 3–12. doi:10.3102/0013189X035007003

Leandro, C. R., Monteiro, E., & Melo, F. (2018). Interdisciplinary working practices: Can creative dance improve math? *Research in Dance Education*, *19*(1), 74–90. doi:10.1080/14647893.2017.1354838

Liao, C., Motter, J. L., & Patton, R. M. (2016). Tech-savvy girls: Learning 21st-century skills through STEAM digital artmaking. *Art Education*, *69*(4), 29–35. doi:10.1080/00043125.2016.1176492

Melchior, E. (2011). Culturally responsive dance pedagogy in the primary classroom. *Research in Dance Education*, *12*(2), 119–135. doi:10.1080/14647893.2011.575223

Scott, K. A., & White, M. A. (2013). COMPUGIRLS'standpoint: Culturally responsive computing and its effect on girls of color. *Urban Education*, *48*(5), 657–681. doi:10.1177/0042085913491219

Wright, C. G., Wendell, K. B., & Paugh, P. P. (2018). "Just put it together to make no commotion:" Re-imagining Urban Elementary Students' Participation in Engineering Design Practices. *International Journal of Education in Mathematics Science and Technology*, *6*(3), 285–301. doi:10.18404/ijemst.428192

KEY TERMS AND DEFINITIONS

Block Coding: A method of visual programming that allows one to program without having to write syntax in the form of a specific programming language. It is often used as an introduction to coding to beginners.

Community Cultural Wealth: The various assets or capitals that are embedded in historically marginalized communities such as social capital and linguistic capital, yet they are often overlooked by schools and institutions.

Computational Thinking: The development of thought processes that reflect the steps that one would execute to instruct computers to perform tasks.

Culturally Sustaining Pedagogy: Fostering, centering, and sustaining the ways of knowing and being of historically marginalized students throughout their educational experiences.

E-Textiles: Textiles with electronic circuits (including a power source and output like a light) embedded in or applied to the textiles.

Embodied Learning: Involving the movement and the physical body in the process of learning.

Immersive Environments: Simulation that provides the user with a sense of physical presence.

Making: Freely exploring endless possibilities through creating, designing, experimenting and producing, supporting one's innovation, creativity, and freedom of expression.

Programming: The process of writing source code that instructs computers to perform tasks.

STEAM: Science, technology, engineering, arts, and mathematics.

Virtual Reality (VR): A computer-generated immersive environment that is viewed with a head-mounted display.

Section 4
STEAM in Teacher Education

Chapter 9
The Transdisciplinary Nature of STEAM Education:
Integrating STEAM in Pre-Service Teacher Education

Douglas Huffman
University of Kansas, USA

Kelli Thomas
University of Kansas, USA

James D. Basham
University of Kansas, USA

ABSTRACT

The purpose of this chapter is to describe and explain multiple methods of integrating STEAM into the curriculum for pre-service teachers. The chapter includes both stand-alone modular methods of integration and continuous integration methods that attempt to merge STEAM concepts throughout the curriculum. The advantages and disadvantages of each method are discussed, along with the challenges teachers faces as they attempt to integrate STEAM. Recommendations are made along with suggestions for the future direction of STEAM integration as the field becomes more transdisciplinary.

DOI: 10.4018/978-1-7998-2517-3.ch009

INTRODUCTION

The movement to shift from STEM to STEAM is based on the idea that science, technology, engineering, mathematics and the arts are inter-disciplinary, and that teachers should integrate ideas from each of these fields in meaningful ways. There are many examples where one or two fields are integrated in a course. We see examples where technology or mathematics is added onto a science lesson, or technology is added onto science; however, the STEAM movement is attempting to go beyond add-on models of integration. The STEAM movement is attempting to push the field further and integrate all five areas in meaningful ways.

However, integrating science, technology, engineering, arts, and mathematics is a complex challenge for the field of education. One of the challenges is that the concepts in STEAM are constantly evolving. At times STEAM appears more multidisciplinary, meaning that various fields such as science, technology and engineering are overlapping. At other times the field is interdisciplinary in that STEAM concepts interrelate in new and ever increasing ways. Some would even take this idea further and describe STEAM as "transdisciplinary." Transdisciplinary can be defined as the merging of distinct and diverse fields into a unified whole to foster new paradigms or domains and converge on a target (NSF, 2017). The new field is larger than the sum of the parts. Thinking about STEAM as transdisciplinary is a new and potentially transformative way to use STEAM ideas. The concept of "unified whole" is important in transdisciplinary fields such as STEAM. The goal is to truly unify all areas of STEAM such that the sum of the parts creates a more meaningful experience for all students. The objective of this chapter is to describe various ways of integrating STEAM in the classroom.

BACKGROUND

According to a report by the American Academy of Arts and Science, "The lines among disciplines are blurring, leading to the emergence of new fields of study that span disciplinary boundaries and allow dramatic advances that no one field could have achieved in isolation" (AAAS, 2013). The evolving multidisciplinary, interdisciplinary and transdisciplinary nature of STEAM makes it difficult to understand how to integrate concepts into the curriculum. However, as the field evolves it is important to explore new ways of achieving the transformative potential of STEAM.

One of the challenges for the field of education is to understand how to integrate STEAM into the elementary, middle school or high school curriculum (Briener, Harkness, Johnson & Koehler, 2012; Margot & Kettler, 2019; National Research

Council, 2001). There may not actually be one best way, but rather there are likely multiple ways to effectively integrating STEAM. The interdisciplinary and multidisciplinary nature of STEAM creates a situation where at times the concepts appear to be stand-alone ideas that fit into one course; however, at other times the STEAM concepts are interdisciplinary and appear to integrate throughout a courses or across multiple courses. Stevens, Sutherland and Krajcik (2009) point out that the interdisciplinary nature of fields can impact society and provide leaps in scientific knowledge, and thus it may be desirable to emphasize the interdisciplinary nature of the field using a multiple-course, integration approach.

Previous literature on integration in the classroom indicates that the field has tried to understand how to integrate. According to Newberry (2012), the integration of pedagogical tools into the discipline provides: 1) new ways to teach traditional concepts, 2) the introduction of new and recent discoveries into classroom work, and 3) the capability to enhance student experience in critical thinking (p.4). There is also the issue of whether or not to integrate as a stand-alone unit in the curriculum. Schank and colleagues (2009) recommend concepts be integrated throughout the curriculum rather than serve as stand-alone units. They claim that this would allow teachers to help students understand the basic science concepts they need while also learning other important concepts. The various ways to integrate STEAM are not clear. There are vague views of defining STEAM integration. According to the National Research Council,

Rather than a single, well-defined experience, it involves a range of experiences with some degree of connection. The experiences may occur in one or several class periods, or throughout a curriculum; they may be reflected in the organization of a single course or an entire school, or they may be presented in an after or out-of-school activity (p. 39)

The range of different integration methods, results in a need to try and better understand the various STEAM integration methods. The purpose of this chapter is to describe and explain multiple methods of integrating STEAM into the curriculum. The chapter includes both stand-alone modular methods of integration, and continuous integration methods that attempt to merge concepts throughout the curriculum. The advantages and disadvantages of each method are discussed, along with the challenges teachers faces as they attempt to integrate STEAM. Recommendations are made along with suggestions for the future direction of STEAM integration as the field becomes more transdisciplinary.

One of STEAM's origins as an educational movement is the Rhode Island School of Design. According to Martinez (2017):

Art + Design are poised to transform our economy in the 21st century just as science and technology did in the last century. We need to add Art + Design to the equation — to transform STEM into STEAM.

Martinez does not just add the arts to STEM. Rather, he transforms the way teaching and learning are done. Martinez does not focus on the products of learning alone, but he focuses on the artistic environments people need in order to learn. The roots of Martinez practice spread wider than STEM or STEAM education and have implications even beyond schools. Martinez share these roots in his introduction to *The Search for Method in STEAM Education* (2017). Martinez writes about creating new kinds of developmental interdisciplinary learning environments that will be necessary if STEAM education reforms are going to amount to more than just another educational fad.

ISSUES AND PROBLEMS

Methods of Classroom Integration

There are multiple ways teachers can integrate STEAM in the classroom. Three of the most common methods found in the literature include multidisciplinary, interdisciplinary and transdisciplinary integration. Each of these models of integration are conceptualize by educators in different ways (Vasquez, Sneider & Comer, 2013). First, STEM used to mean science with some technology, engineering, or mathematics. Art is the least likely to be integrated. The first letter in STEAM is "S" and this implies that STEAM is often thought of primarily as science. However, as we move towards a more transdisciplinary view of STEAM, it is imperative for educators to understand ways to fully integrate all aspects of STEAM.

Multidisciplinary: STEAM is often conceptualized as a multidisciplinary practice. Multidisciplinary is viewed as instruction where students learn concepts and ideas from each discipline separately. For example, a science teacher might have students study Newton's Laws of Motion. The mathematics teacher might use laws of motion to learn about quantitative ideas, while the computer science teacher uses technology to study motion. Engineering and art are also taught in separate courses. The point is that multidisciplinary is typically separate instruction for each discipline. Graphically multidisciplinary can be viewed as in Figure 1.

Interdisciplinary is where students learn concepts and ideas from each discipline in a way that attempts to loosely link the concepts. For example, a physics teacher might have a unit on Newton's Laws of Motion. During the unit students learn about motion concepts, but they also use technology to study motion. The teacher might

Figure 1. Multidisciplinary views of STEAM

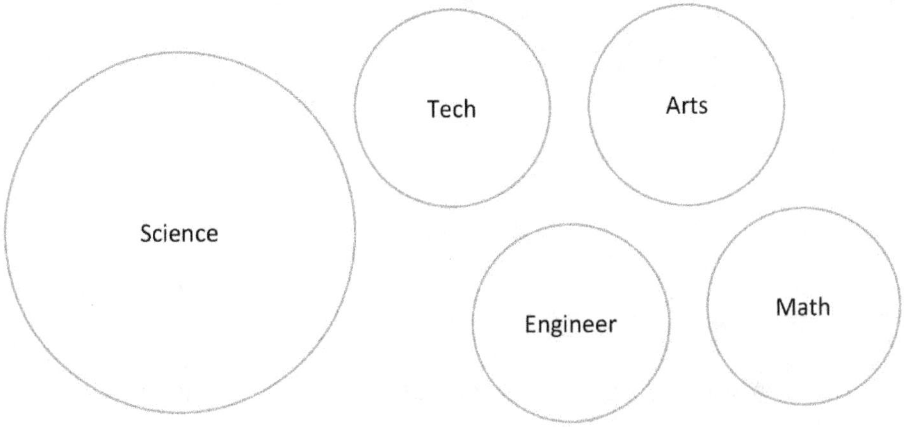

use motion meters and other technology to graphically describe laws of motion. Mathematics is also used to calculate numerical values and better understand the quantitative aspects of Newton's Laws. Engineering practices are typically less common in a science unit, but the teacher might have students design a moving device that can represent all three laws of motion. Art is even less common to find integrated into a science unit. The teacher might have the students use their creativity when designing their motion object. Students might be asked to design a creative motion object that represents all three laws of motion as in Figure 2.

Transdisciplinary is where students learn concepts and ideas in a coherent, tightly woven manner in which all the disciplines are combined to become larger than the combination of each discipline. Ideally all five areas are combined and inter-twinned in such a way that all are represented in important ways. For example, a teacher might design a unit around the concept of consumer science. The challenge is to design a piece of clothing that is both fashionable and stain resistant. In this unit students will study which brand of washer detergent cleans the best. In this unit students will use chemistry, social studies, technology, mathematics, engineering and the arts as in Figure 3.

Designing Courses to Integrate STEAM

Another important way to think about STEAM integration is in course design. There are two typical ways of integrating STEAM into courses; Stand-Alone Modules and Continuous Integration. In the continuous approach, STEAM is woven throughout the entire course, while in a stand-alone approach STEAM modules are added onto

Figure 2. Interdisciplinary views of STEAM

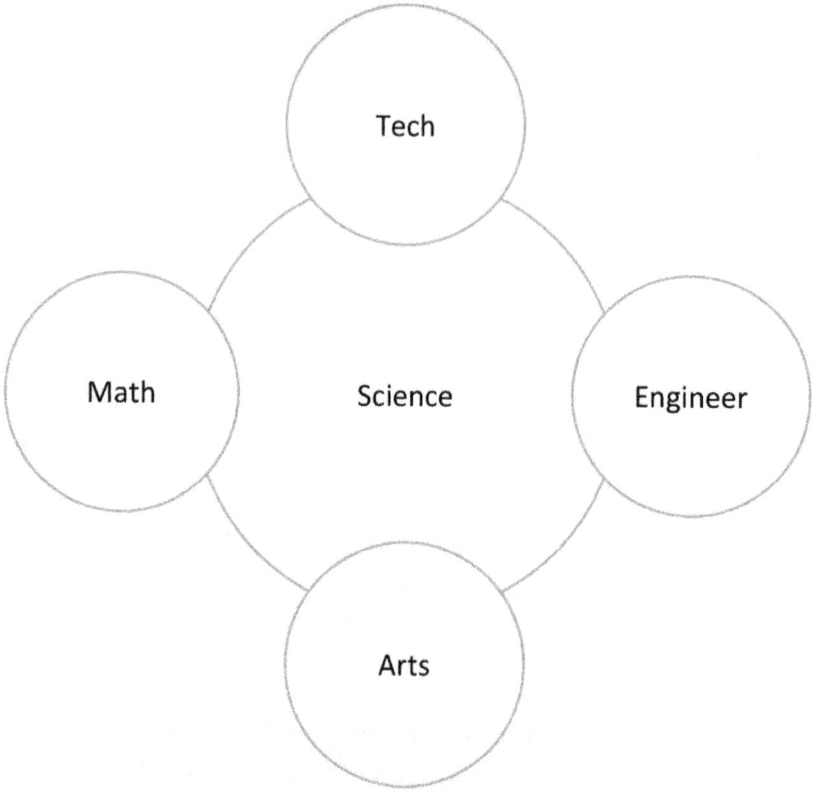

an existing course. The stand-alone modular approach is one of the more common forms of integration. The stand-alone approach is often easier to implement than trying to seamlessly weave STEAM throughout a course. With the stand-alone modular approach, teachers can add a STEAM module onto their course. In the following example, a stand-alone unit is added onto an existing science methods course for pre-service elementary teachers.

Example: Stand-Alone Modular STEAM Integration

In this university science laboratory, the room is filled with undergraduate pre-service teachers who are studying to become elementary teachers. The pre-service teachers are part of a university level methods course focused on learning how to teach science, mathematics and social studies. In this activity they are trying to figure out which brand of paper towel is "best." To begin the paper towel unit, students discuss what they think makes the "best" paper towel. Various ideas are considered

Figure 3. Transdisciplinary views of STEAM

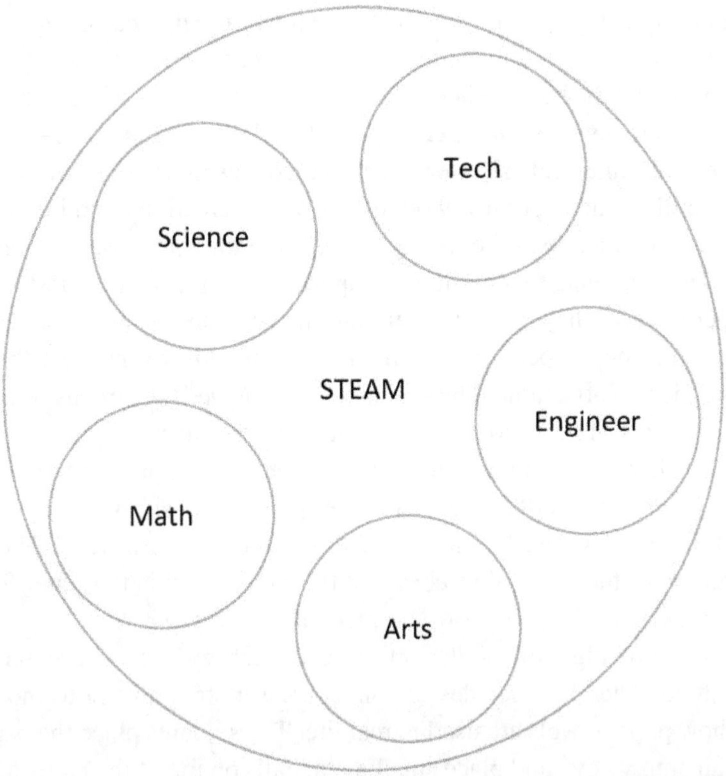

including absorbency, strength, softness, and cost. Some students simply choose the least expensive paper towel, while others choose based on the physical properties of the paper towel. The choice often depends on how one might use the paper towel. If the paper towel is used to hold food, strength might be the most important property. If the paper towel is used to clean up spills, absorbency might be most important. Through the discussion, the students realize the choice is a cost-benefit decision. How much extra strength are consumers willing to pay for? If the more expensive paper towel is twice as absorbent as the cheaper brand would you pay extra for absorbency? Some brands claim to be the "quicker picker upper." This implies they are fast at absorbing spills. Do you value how fast the paper towel can absorb water or do you value the total amount of water the paper towel can hold? These are also important variables to test in this activity.

To begin the laboratory investigation of paper towels, students were given four different brands of paper towel. The brand names of the paper towels were hidden from students to provide a blind test. First, the students designed an investigation to determine how much total amount of water different types of paper towel can hold. There are multiple ways to carry out this investigation. Students are given freedom to choose whatever method they prefer, and then discussion after the investigation will help to sort out the best methods. Some students put drops of water on the paper towel to see how many drops of water the paper towel can hold. Other students dunked their paper towels into a tub of water and measured how much water was in the paper towel after pulling the paper towel out of the water. Creativity and innovation was encouraged given there is not one way to measure how much water a paper towel can hold. Some students thought it was important to replicate the most natural use of the paper towels. They poured water onto the table and placed a paper towel on top of the spill. They wiped up the spill and measured how much water the paper towel could pick up off a table. The students have watched the famous Bounty paper towel commercials on television. In the Bounty advertisement, the paper towel claims to be "the quicker picker upper." Bounty does not claim to pick up the most water; they claim to clean up spills faster than other paper towels. This leads students to natural test how quickly each brand of paper towel can absorb water. The students place water on the table and dip a corner of the paper towel in the water. They time how long it takes for the paper towel to pick up all the water.

In the next investigation, students test how much weight each brand of paper towel can hold. The class decides to conduct a wet strength test to most closely replicate how paper towels are used in real life. The students place the paper towel on top of an empty bowl and place small metal balls on top of the paper towel until it breaks. Then they measure the weight of the metal balls.

Next, students are asked to calculate the cost of each paper towel. Different brands of paper towel have different sized towels, and different number of sheets in each roll. The students calculate the cost of one square foot of paper towel to help conduct a cost benefit analysis.

Finally, the students are asked to design a television commercial to try and sell one of the brands of paper towel. Using an iPad, the students create a television commercial based on different methods of consumerism. Some create a commercial based on celebrity endorsement. Other students create a commercial based on the scientific properties of the paper towel. Students are encouraged to use the data from their own investigations to develop their television commercial.

The goal of this paper towel unit is to help students use Science, Technology, Engineering, Art and Mathematics to fully investigate different features of paper towel. STEAM is used to provide a holistic view of the paper towel features. Science is used to study the wicking ability of paper towels. Technology is used to help create the television commercials on the iPad. Engineering process is used to evaluate the different paper towels for physical properties of strength and absorbency. Art

Figure 4. Paper towel testing

is used in the creation of the television commercial. Finally, mathematics is used in the cost-benefit analysis of the paper towels. Each aspect of STEAM is used to provide a holistic view of the paper towels. The goal is to have each aspect of STEAM equally represented in the investigation to provide an interdisciplinary perspective on what makes the best paper towel. Students leave this activity with a sense that all aspects of STEAM can be integrated into an investigation. They understand that each aspect of STEAM is actually essential to understanding different aspects of what makes the best paper towel as in Figure 4.

Continuous STEAM Integration

Another method of integrating STEAM into courses is the continuous integrated approach. In this approach, STEAM is integrated throughout the entire course. The goal is to seamlessly integrate aspects of STEAM throughout the course. This approach is often viewed as more difficult than a stand-alone module because each aspects of STEAM needs to be woven into the course in a meaningful way. The ultimate goal of this approach is to seamlessly integrate STEAM in such a way that all aspects of STEAM are experienced together.

An example of continuous integration is our university-level design based STEAM course. This course is designed for pre-service teachers who are studying to become elementary teachers. The course focuses on project-based learning, and attempts to integrate science, technology, engineering, art, and mathematics into each project. In one project, students are given the challenge to design a "Rube Goldberg" type machine to pour a bowl of cereal. Rube Goldberg machines are somewhat absurd in that they take a simple task and make it more complex and inefficient. Each year the Rube Goldberg Institute sponsors a contest where student groups across the United States design Rube Goldberg projects. One recent contest had students design a method of pouring a bowl of cereal. Students used a ten step process to pour the bowl of cereal. https://www.rubegoldberg.com/contest/pour-a-bowl-of-cereal/ The following course description provides an overview of the university STEAM course and the ways STEAM concepts are infused throughout the course as in Figure 5.

Course Description: STEAM for All Elementary School Learners

In this course science, technology, engineering, arts and mathematics (STEAM) are explored to solve complex real world problems. We will engage in "problem-based learning" (PBL) to understand disciplinary knowledge, solve problems, and

Figure 5. Rube Goldberg design challenge

develop skills in science, engineering and technology. PBL is a pedagogy that simultaneously develops problem solving strategies, disciplinary knowledge bases, and science, technology and engineering skills by placing students in the active role of problem-solvers confronted with an open-ended problem situation. Together we will collaboratively solve complex problems to learn science, technology, engineering, arts and mathematics.

STUDENT LEARNING OUTCOMES

By of end of this course, students will…

- Understand the importance and be able to demonstrate how to use a variety of assessment instruments, procedures and technologies for supporting instructional planning decisions and gathering formative assessment data in STEAM content areas.
- Use a variety of assessment instruments, procedures, and technologies for instructional planning in STEAM content areas.
- Create a STEAM learning environment that fosters an inclusive setting for all learners.
- Begin to understand and learn how to personalize and intensify instruction by applying the principles of universal design for learning and using technology to personalize and improve access to instruction and learning for all learners.
- Begin to understand and learn to account for personalized learning needs and supports through application of the principles of universal design for learning.
- Begin to understand and use scientific disciplinary core ideas, cross-cutting concepts, and science and engineering practices and create learning experiences that make these aspects of the discipline accessible and meaningful for students to assure mastery of the content.
- Begin to understand and use a variety of assessment strategies appropriate to science and engineering fields to engage learners in their own growth, monitor learning progress, communicate meaningful feedback to relevant stakeholders, evaluate instructional effectiveness, and guide instructional decisions.
- Plan and implement instruction using a variety of instructional strategies that supports all learners to engage with curiosity, creativity, and increasing skill in science and engineering practices; that supports learners in developing increasingly more sophisticated science and engineering core ideas and cross-cutting concepts; and that integrates other disciplines.

- Support personalized learning needs and supports through application of the principles of universal design for learning in STEAM areas.
- Use technology to meet the needs of individual learners.
- The STEAM course outline Table 1.

Table 1. Sample of STEAM Course Outline

Date	Concepts	Challenge & Driving Question(s)
Week #1	Course introduction and importance of STEAM in elementary learning environment for all learners. Initial introduction to authentic learning environments via Project Based Learning.	• Thinking about future: Initial Reflection. • Driving questions: Why is STEAM important to society? How do educators integrate STEAM into elementary environments? What variables are critical in planning for success in transdisciplinary or cross-cutting STEAM inclusive and equitable elementary environments?
Week #2	Concepts: Inclusion, Equity, and Personalized Learning, wrapped into applied understanding of concepts of Motion and Forces.	• Rube Goldberg Machines • Driving question(s): Can you pour a bowl of cereal using concepts in force and motion, technology, engineering, arts, and mathematics? How can RGMs support inclusion, equity, and personalized learning? What do RGMs to support elementary learners understanding of motion and forces?
Week #3	Concept: Considering and Design and Engineering in Elementary Learning Environments for all Learners.	• Learning Space and Activity Design Challenge • Driving questions: How do common learning design paradigms impact practices? What are considerations for future learning design paradigms that are inclusive and equitable in STEAM learning environments?
Week #4	Concept: Understanding Basics of Universal Design, Universal Design for Learning, & Design Thinking for STEAM Learning Environments. Integrated understanding of engineering in the inclusive elementary learning environment.	• Engineering for All Design Challenge • Driving Question: How could you build an engineering design challenge into an elementary learning environment? (e.g., CS for All design challenge)
Week #5	Concepts: Transdisciplinary understanding of STEAM in Society, Lessons from the Past and Looking Toward the Future.	• Educating Stakeholders Design Challenge. • Driving Question(s): Why is it necessary to consider ethics, creativity, and curiosity when encouraging STEAM? Can you develop a PSA that depicts why/how educators might design equitable and inclusive STEAM learning environments while considering lessons learned and future challenges of society? How can reflection enhance both your life as a professional and lives of your learners?
Week #6	Concept: Understanding the use and assessment of technology and assistive technology in supporting individual access & betterment.	• AT Data Design Challenge. • Driving Question(s): Why is technology useful in learning environment? How might it support both group and individual betterment? What would a data plan look likely for implementing an ecologically based AT SETT framework?
Week #7	Concept: Selecting and implementing individualized solutions technology and assistive technology in supporting individual access & betterment.	• Enabling and Assistive Technology Design Challenge. • Driving Question: Based on a case study (or real-life learner) can you develop an enabling or assistive technology for supporting betterment in STEM/STEAM learning environments? How can you integrate this same challenge in your learning environment?

Major Course Assignments

Throughout this course, students will integrate various concepts and standards in problem-based learning projects. The following projects are the Major Course Projects.

BIG IDEA: A primary goal of this course is for you to understand and demonstrate transdisciplinary STEAM thinking, planning, and implementation across equitable and inclusive learning environments. Aligned to this big goal are various educator preparation standards for planning and implementing STEAM content as well as effectively engaging with community stakeholders. This project has been designed for you to demonstrate your abilities in this area.

By the end of this assignment you should be able to:

- Identify and plan STEAM cross-cutting and transdisciplinary content that support scientific and engineering practices, while encouraging creativity, curiosity, and increasing STEAM subject area knowledge and skills in an equitable and inclusive learning environment.
- Begin to understand how to assess strategies appropriate to STEAM fields to engage learners in their own growth, monitor learning progress, communicate meaningful feedback to relevant stakeholders, evaluate instructional effectiveness, and guide instructional decisions.
- Design and implement activities that personalize and intensify instruction by applying the principles of Universal Design for Learning and using technology to personalize and improve access to instruction and learning for all learners including individuals with various abilities and disabilities.
- Demonstrate to stakeholders the importance of equitable and inclusive STEAM transdisciplinary thinking

Visioning Your Future: Reflections Across Class & the STEAM Field

BIG IDEA: Professional growth and understanding merges through exposure and reflection on information, activities, and overall experiences. This project has been designed to support information seeking, thoughtful experiences, and ongoing reflection.

By the end of this project you should be able to:

- Better understand your own STEAM experiences, conceptions, and misconceptions.
- Discuss/show what and how STEAM learning experiences shape both current and the emerging STEAM professionals, workforce, and citizenry.

- Discuss/show how to design equitable and inclusive STEAM learning experiences that shape the curiosity, discovery, and creativity of all elementary learners.
- Discuss how specific design solutions, strategies, practices, technologies and assessments might be integrated to encourage and support STEM/STEAM for All in an elementary learning environment.
- Discuss how the implementation of STEAM elementary learning environments potentially impact the future of society.

Overall, this university-level methods course for pre-service teachers is designed to engage teachers in STEAM through problem-based learning tasks. The goal is to seamlessly integrate all the STEAM areas into the projects in such a way that all areas are equally represented. In some ways, this best represents what is meant by the "transdisciplinary" nature of STEAM. Transdisciplinary is the merging of distinct fields in such a way that they converge on a new and evolving concept. The new transdisciplinary STEAM field is greater than the sum of the parts. Students experience something more significant than if they just studied science, technology, engineering, arts and mathematics separately. The seamless integration creates a transdisciplinary STEAM experience that is unique and far beyond what could be experienced through separate disciplines.

CONCLUSION

STEAM is still an emerging concept for the field of education. Traditionally the field has focused on the four STEM areas – Science, Technology, Engineering and Mathematics; however, today the field of education is beginning to find ways to integrate the Arts with STEM. It is still relatively rare to see all aspects of STEAM implemented in practice. It is more common to find teachers who integrate science and technology, with some mathematics, or add an engineering module onto a science course. Truly integrating engineering and the arts is less common.

Another place with recommendations for seamless continuous integration is the Next Generation Science Standards (National Research Council, 2013). The NGSS emphasize the concept of 3-Dimensional Learning. The goal of 3D learning is to integrate all three aspects of the standards, namely the Science and Engineering Practices, Disciplinary Core Ideas, and Cross-Cutting Concepts. Adding a stand-alone engineering unit to a science course does not truly meet the standards. According to the idea of 3D learning all three aspects of the standards should be woven together in a meaningful way.

One of the reasons we find less continuous integration of STEAM is that it is difficult to combine all five aspects of STEAM in meaningful ways. Each discipline is continually evolving and changing. This makes it difficult for teachers to fully understand how one might integrate all aspects of STEAM. These are transdisciplinary fields that are changing, and the challenge for teachers today is to embrace the ever changing nature of STEAM and work to find ways to integrate all aspects of STEAM in ways that can help all students learn.

REFERENCES

American Academy of Art and Sciences Transdisciplinary Opportunities. (2013). ARISE 2: Advancing Research in Science and Engineering: Unleashing America's Research & Innovation Enterprise. Cambridge: American Academy of Art and Sciences.

Breiner, J. M., Harkness, S. S., Johnson, C. C., & Koehler, C. M. (2012). What is STEM? A discussion about conceptions of STEM in education and partnerships. *School Science and Mathematics, 112*(1), 3–11. doi:10.1111/j.1949-8594.2011.00109.x

Butera, G., Horn, E. M., Palmer, S. B., Friesen, A., & Lieber, J. (2016). Understanding Science, Technology, Engineering, Arts, and Mathematics (STEAM). In B. Reichow, B. Boyd, E. Barton, & S. Odom (Eds.), *Handbook of Early Childhood Special Education*. Cham: Springer. doi:10.1007/978-3-319-28492-7_9

Bybee, R. W. (2013). *The Case for STEM Education: Challenges and Opportunities*. NSTA Press.

Estapa, A. & Tank, K. (2017). *Supporting integrated STEM in the elementary classroom: a professional development approach centered on an engineering design challenge*. Academic Press.

From STEM to STEAM. Developmental Learning. (n.d.). Retrieved fromhttps://www.psychologytoday.com/us/blog/conceptual-revolution/201708/stem-steam-developmental-learning

Margot, K. C., & Kettler, T. (2019). Teachers' perception of STEM integration and education: A systematic literature review. *International Journal of STEM Education, 6*(2). doi:10.118640594-018-0151-2

Martinez, J. (2017). *The search for method in STEAM education*. London, UK: Palgrave Macmillian. doi:10.1007/978-3-319-55822-6

National Research Council. (2011). *Successful K-12 STEM Education: Identifying Effective Approaches in Science, Technology, Engineering and Mathematics.* Washington, DC: The National Academic Press.

National Research Council. (2013). *Next Generation Science Standards: For States, By States.* Washington, DC: The National Academies Press. doi:10.17226/18290

National Science Foundation. (2017). *Convergence Research at NSF.* Convergence Press Release (17- 082). Retrieved from https://www.nsf.gov/od/oia/convergence/index.jsp

Newberry, D. M. (2012). Nanotechnology: A platform for education change. *Nanotechnology (IEEE- NANO), 12th IEEE Conference*, 20-23.

Schank, P., Wise, A., Stanford, T., & Rosenquist, A. (2009). *Can High School Students Learn Nanoscience? An Evaluation of the Viability and Impact of the NanoSense Curriculum.* SRI International.

Vasquez, J., Sneider, C., & Comer, M. (2013). *STEM Lesson Essentials: Integrating Science, Technology, Engineering and Mathematics.* Heinemann.

KEY TERMS AND DEFINITIONS

Crosscutting Concepts: These concepts in the *NGSS* and are described as "concepts that bridge disciplinary boundaries, having explanatory value throughout much of science and engineering. These crosscutting concepts were selected for their value across the sciences and in engineering. These concepts help provide students with an organizational framework for connecting knowledge from the various disciplines into a coherent and scientifically based view of the world" (NRC, 2012, p. 83).

Disciplinary Core Idea: The committee developed its small set of core ideas in science and engineering. We grouped disciplinary ideas into four major domains: the physical sciences; the life sciences; the earth and space sciences; and engineering, technology, and applications of science" (NRC, 2012, p. 31).

Interdisciplinary: Where concepts and ideas from each discipline are overlapping, interlinked, and woven together.

Multidisciplinary: Where concepts and ideas from each discipline are studied as separate distinct areas.

NGSS: The Next Generation Science Standards (NGSS) were completed in April of 2013. The NGSS represent a change in how states have traditionally approached their science standards. In embracing science education research, the NGSS represent performance expectations (PEs) that require all students have a deep understanding of a smaller number of disciplinary core ideas (DCIs), are able to show evidence of that knowledge through scientific and engineering practices, and connect crosscutting concepts across disciplines" (Pruitt, 2014, p. 145).

Transdisciplinary: Transdisciplinary is defined as the merging of distinct and diverse fields into a unified whole to foster new paradigms or domains and converge on a target (NSF, 2017). The new field is larger than the sum of the parts.

Chapter 10
Using Arts Education in STEM With the Science and Engineering Practice of Developing and Using Models

Lizette A. Burks
University of Arkansas, USA

ABSTRACT

Since 2013 more than three-quarters of the United States has adopted science education standards based on the Next Generation Science Standards (NGSS). Science education is often integrated with multiple disciplines including technology, engineering, and mathematics (STEM) and in more recent movements integrated with the arts (STEAM). This chapter examined preservice teachers' preconceptions about the practice of developing and using models in science education and practical integration of the arts through this central practice. The results of the study indicated preservice elementary preconception survey scores were higher when describing the practice as a social endeavor than any other aspect of the practice. Using social endeavors as a lever in elementary teacher education can help preservice teachers utilize this critical practice in more expansive ways (investigatory, sensemaking, critiquing). Examining the way the arts manifest in the practice of developing and using models within the NGSS serves as a first step to finding meaningful ways for integration.

DOI: 10.4018/978-1-7998-2517-3.ch010

INTRODUCTION

"The fundamental creative practices of imagination, investigation, construction, and reflection…are essential in the arts but equally important for science and mathematics learning" (National Coalition for Core Arts Standards, 2014, p. 19). The term STEAM (Science, Technology, Engineering, Arts, and Mathematics) was coined by the Rhode Island School of Design and was created to potentially enhance STEM (Science, Technology, Engineering, and Mathematics) (Allina, 2018). The acronym STEM is not a new term and has had challenges as it continues to evolve from being used primarily as an educational slogan and into more meaningful and practical opportunities for reforms (Bybee, 2013). Just as stating STEAM stands for science, technology, engineering, arts, and mathematics only clarifies the acronym, Bybee (2018) states doing the same for STEM over the last decade has not provided a definition with implications for coherence in K-12 education programs. Although slogans can be rallying symbols for potential unity in educational movements, capitalizing on the support before the term "loses the power to rally and instead becomes the subject of criticism" (Bybee, 2018, p. 6).

Since 2013 more than three-quarters of the United States has adopted science education standards based on the *Next Generation Science Standards (NGSS)* (NGSS Lead States, 2013) and/or *A Framework for K-12 Science Education: Practices, Crosscutting Concepts, and Core Ideas* (the *Framework)* (NRC, 2012) (NSTA, n.d.). This new vision for science education has changed the landscape with a central focus on student sensemaking of science phenomena and has moved away from rote memorization of scientific knowledge. Central to sensemaking of phenomena in the *NGSS* (NGSS Lead States, 2013) is the science and engineering practice of developing and using models, a new practice in science standards and is the focus of the study in this chapter. With the implementation of the *NGSS* (NGSS Lead States, 2013), specifically the science and engineering practices, shifts in science preservice professional learning will be needed. In a period of time when science education is changing, integration with other disciplines can be criticized if meaningful integration is not created from a practical view point.

One way to move towards capitalizing on the support for STEAM education is to find ways the arts inherently manifest in the *NGSS* (NGSS Lead States, 2013) as a way to move towards more meaningful and practical opportunities for reform. Elementary teacher education is changing due to changes in science education standards. Teacher preparation time is limited as elementary teachers currently "take a limited number of science courses and a single science methods course" (NRC, 2012, p. 259). This chapter focuses on the way the arts manifest in the practice of developing and using models within the *NGSS* (NGSS Lead States, 2013) and seeks to find first steps in finding meaningful ways to integrate the arts in science education.

The study focused on preservice elementary teachers' preconceptions of the *NGSS* science and engineering practice of developing and using models to guide translate of this practice into teacher education. The following questions guided the study:

1. What are elementary preservice teachers' preconceptions about developing and using models in the classroom?
2. What student-student and student-teacher interactions are identified by elementary preservice teachers as critical to developing and using models in the classroom?
3. What teaching strategies do preservice elementary teachers identify as critical to developing and using models in the classroom?

The findings from the first research question were previously explored and more than half of participants in the study were not able to identify over seventy percent of the aspects described for this practice in the *NGSS* (NGSS Lead States, 2013) (Burks & Huffman, 2019).

In science, models are used to represent a system (or parts of a system) understudy, to aid in the development of questions and explanations, to generate data that can be used to make predictions, and to communicate ideas to others. Students can be expected to evaluate and refine models through an iterative cycle of comparing their predictions with real world and then adjusting them to gain insights into the phenomenon being modeled. As such, models are based upon evidence. When new evidence is uncovered that models can't explain, models are modified. (NGSS Lead States, 2013, Appendix F p. 6)

"Considering that preservice elementary teachers have little to no experience participating in a scientific community, it is not surprising that their knowledge of scientific practices is often limited" (Ricketts, 2014, p. 2110). Although preservice elementary teachers were not able to describe the different aspects of the practice as defined by the *NGSS* (NGSS Lead States, 2013), analysis of the second and third research questions provide insight about what could be used as a lever to help support this practice through teacher education. Using social endeavors as a lever for this practice in teacher education can help elementary preservice teachers use the practice in more expansive ways described within the *NGSS* (NGSS Lead States, 2013). Examining the way the arts manifest in this practice can help support this high-leverage practice in science education.

BACKGROUND

In the U.S. the *NGSS* (NGSS Lead States, 2013) were released in 2013, and their precursor, the *Framework* (NRC, 2012), was released in 2012. The *Framework* (NRC, 2012) was the blueprint for the *NGSS* and "expresses a vision in science education that requires students to operate at the nexus of three dimensions of learning: Science and Engineering Practices, Crosscutting Concepts, and Disciplinary Core Ideas" to achieve scientific literacy (NGSS Lead States, 2013, Appendix F p.1). One dimension of the standards, the science and engineering practices, builds on prior reforms of inquiry in science classrooms and better articulate "what successful inquiry looks like when it results in building scientific knowledge" as a "kind of Inquiry 2.0-not a replacement for inquiry" (Schwarz, Passmore, & Reiser, 2017, p.5).

The *Framework* (NRC, 2012) describes the following eight science and engineering practices: asking questions and defining problems, developing and using models, planning and carrying out investigations, analyzing and interpreting data, using mathematics and computational thinking, constructing explanations and designing solutions, engaging in argument from evidence, obtaining evaluating and communicating information. Science learning "is an inherently social and cultural process that requires mastery of specialized forms of discourse and comfort with norms of participation in the scientific community of the classroom (NRC, 2007, p. 203). Engagement in practices is language intensive and requires students to participate in classroom science discourse. The practices offer rich opportunities and demands for language learning while advancing science learning for all students (Lee, Quinn, & Valdés, 2013). "Each of the eight practices, as it is introduced and elaborated and experienced in the classroom, requires that students externalize their reasoning" (Schwarz et al., 2017, p. 311). Talk and discursive practices are fundamental to all the science practices in the classroom and mimic that of experienced scientists and engineers (Schwarz et al., 2017). "Researchers in a variety of fields relating to education have begun to converge on the finding that when teachers open up the conversation and get students to participate actively in reasoning with evidence and building and critiquing academic arguments, students make dramatic learning gains" (Schwarz et al., 2017, p. 312). In *Taking Science to School: Learning and Teaching Science in Grades K-8 (NRC, 2007)*, authors describe scientific discourse as different from that of everyday life and support is needed to engage in this type of discourse. Ogborn, Kress, Martins, and McGillicuddy (1996) found that most of the talk in science classrooms comes from teachers and is mostly persuading talk about science topics. Eichinger, Anderson, Palinscar, and David (1991) found that students are more likely to engage in academically productive talk like argumentation when they are able to do this by working directly with each other instead of through teacher mediated talk. This talk needed to be supported by the teacher working to

Table 1. Grouping the Eight Science Practices into Investigating, Sensemaking, and Critiquing

	Investigating Practices	Sensemaking Practices	Critiquing Practices
Science Practices	Asking Questions	Developing and Using Models	Engaging in Argument from Evidence
	Planning and Carrying Out Investigations	Analyzing and Interpreting Data	Obtaining, Evaluating, and Communicating Information
	Using Mathematics and Computational Thinking	Constructing Explanations	

Source: (McNeill et al., 2015)

create classroom norms for this student mediated talk to occur and students were not successful without teacher intervention (Eichinger et al., 1991). In *Helping Students Make Sense of the World Using Next Generation Science and Engineering Practices*, Schwarz et al. (2017) devote a whole chapter to science discourse and base many of their recommendations on the work of Michaels and O'Connor (2012). Michaels and O'Connor (2012) describe four teacher goals for academically productive talk through their findings from research.

McNeill et al. (2015) group the practices into three categories (investigating practices, sensemaking practices, and critiquing practices) based on how they were presented in the *Framework* (McNeill et al., 2015, p. 23). "The idea is that science is fundamentally about making sense of the natural world" (McNeill et al., 2015, p. 22).

The practice of developing and using models is one of the few practices that fits into all three categories depending on how it is used in instruction (McNeill et al., 2015). "Increasingly, more science education researchers and U.S. national standards documents have noted the importance of models in science and engineering and have subsequently called for an increased role for models in K-12 science teaching and learning" (Campbell et al., 2014, p. 159-160). Developing and using models is a high-leverage practice in teacher preparation because of the use of discourse in its implementation that is also used in other practices utilized within the *NGSS* (NGSS Lead States, 2013) science classroom. Additionally, the overlap between the other seven identified *NGSS* (NGSS Lead States, 2013) practices and the development and use of models along with the use of models represented in two of the overall three dimensions of the new vision for science education (NRC, 2012) contribute to its high leverage nature.

The science and engineering practice of developing and using models has become prominent in a recent Open Education Resource (OER) initiative, entitled OpenSciEd, which seeks to improve the supply and demand for high quality K-12 science instructional materials (OpenSciEd, n.d.). Just prior to the creation of this initiative and responding to the "need for more coordination across efforts underway to support the design and implementation of instructional materials for science education in ways that build from the *Framework*" the Board on Science Education at the National Academies of Sciences, Engineering, and Medicine convened a public workshop that focused on the development of instructional materials (National Academy of Science, Engineering, and Medicine, 2018, p.4). "OpenSciEd brings together multiple partners, including ten states, a consortium of curriculum developers, and many other science education leaders and experts, to create a complete set of robust, research-based, open-source, K-12 science instructional materials" (OpenSciEd, n.d.) and is the first of its' kind aiming for coherent year-long OER resources in alignment with the *Framework* (NRC, 2012). In a recent presentation OpenSciEd revealed their instructional model for curricular units and this instructional model follows a storyline instructional framework informed by the work of Reiser (2013) that make the science and engineering practice of developing models prominent in the instructional model for each unit (Edelson et al., 2019).

Science and art have been regarded as belonging to different worlds and the education of scientists has ignored forms of cognition relating to artistry in scientific research (Eisner & Powell, 2002). "Integration of subject areas is an avenue that strengthens science learning for all students, particularly for students who have been traditionally underserved" (NGSS Lead States, 2013, Appendix D pg. 4). It does so by creating a convergence of core ideas and practices across subject areas, offering multiple entry points for students (NGSS Lead States, 2013). Integration of key pedagogical practices across subject areas can help support programs preparing teachers for this new vision in science education. The primary goals of the arts in education are the creation of experience, meaning, and mind (Eisner, 2002b). In a study of artistry in the practice of scientific research, conducted at Stanford University, Eisner and Powell (2002) found five major themes of artistic modes of thought in interviews conducted with scientists. "Art in science inspires, motivates, and enriches the pursuit of inquiry; indeed, for good work to be done, artistry appears inevitable" (Eisner & Powell, 2002, p.157).

Meaning is not limited to what words can express. Among the most important ideas is the idea that humans are meaning-making creatures. All of us wish to create meaningful experience. Some meanings are readable and expressible through literal language; other meanings require literary forms of language; still others demand other forms through which meanings can be represented and shared. The arts provide a spectrum of such forms we call them visual arts, music, dance, theater through which meanings are made, revised, shared, and discovered. (Eisner, 2002a, p.230)

Table 2. Number of Instances of Alignment Recorded for Each Framework Component (Possible 48 Connections)

	Philosophical Foundations and Lifelong Goals					Creative Practices			
	Arts as Communication	Arts as Creative Personal Realization	Arts as Culture, History, and Connectors	Art as a Means to Well-Being	Arts as Community Engagement	Imagine	Investigate	Construct	Reflect
# of Connections to Common Core	39	2	6	1	0	27	36	28	43

Source: (The College Board, 2012, p.15)

Integration can also serve as a means to increase instructional time for science in elementary education. Instructional time in elementary school settings is often lacking science education and is dominated by English Language Arts and Mathematics (NGSS Lead States, 2013; Banilower et al., 2018). The development of national common core education standards for English language arts and math occurred in 2009 and yielded a similar focus for science in 2013 as well as the arts following in 2014. In preparation for this chapter a formal review of connections between the *NGSS* (NGSS Lead States, 2013) and the 2014 National Core Arts Standards (National Coalition for Core Arts Standards, 2012) could not be found, but a formal review of connections between the Common Core State Standards (Math and English Language Arts) and the National Core Standards Conceptual Framework took place in 2012 prior to the creation of The 2014 National Core Arts Standards by the National Coalition for Core Arts Standards (National Coalition for Core Arts Standards, 2012). This review revealed instances of alignment in the Common Core State Standards to two core areas in the National Core Standards Conceptual Framework: Philosophical Foundations and Lifelong Goals, and the four Creative Practices (The College Board, 2012). Although there were instances of alignment to philosophical foundations and lifelong goals, there were very high number of instances of alignment (rate of more than 50 percent) for the four Creative Practices.

The review found that every Standard for Mathematical Practice referenced all four of the Creative practice numerous times though "popular belief holds that artists and mathematicians operate in fundamentally different ways" (The College Board, 2012, p. 15). This review commissioned by the National Coalition for Core Arts Standards stated since the release of the Common Core Standards in 2010 "arts educators and advocates have publicly discussed the relationship between this set of expectations and teaching and learning in the arts" (The College Board, 2012, p. 4). This review can serve as first steps to find connections between the science and engineering practices within the NGSS (NGSS Lead States, 2013) and the 2014 National Core Arts Standards (National Coalition for Core Arts Standards, 2012).

ASSUMPTIONS, LIMITATIONS, AND DELIMITATIONS

It was assumed that the participants in this study had little to no knowledge of the *NGSS* (NGSS Lead States, 2013) and specifically the practice of developing and using models. Students in this study were only in the first weeks of their first elementary science methods course at the university. It was assumed that the *NGSS* (NGSS Lead States, 2013) and the practice of developing and using models was not used for instruction during the first two weeks of class.

The science and engineering practice of developing and using models described in the *Framework* (NRC, 2012) and the *NGSS* (NGSS Lead States, 2013) was the focus of this study. This practice is one of the three main dimensions (disciplinary core ideas, science and engineering practices, and crosscutting concepts) students use to make sense of natural phenomena described in the new vision for science education (NGSS Lead States, 2013). This study revealed the preconceptions of the science and engineering practice of developing and using models for a sample of elementary preservice teachers.

There were several limitations to this study. First, teachers were asked about their preconceptions of this practice and a video was analyzed during one phase of the study. This video was not meant to be an intervention, rather it was only shown to teachers to solicit responses to the survey questions. Second, analysis of written responses and interviews was conducted by one researcher. A more robust study would include the use of multiple evaluators to establish inter-rater reliability. An effort was made to use tools developed in previous research studies and a pilot test was conducted to improve reliability and validity. Lastly, the results are from two small groups of elementary preservice science teachers at a Midwestern U.S. university. Due to this relatively small sample size implications of this study cannot be generalized to a larger context, however, they can provide further insight about the practice of developing and using models and the support needed to achieve critical thinking by students described in the *Framework* (NRC, 2012).

METHOD

This was a descriptive non-experimental study and used a mixed methods design. Thirty-six preservice elementary teachers at a large mid-western public university in the U.S. were surveyed to discover the state of their knowledge about the science practice of developing and using models in the *NGSS (NGSS Lead States, 2013)*. The preservice teachers were shown a video of an experienced teacher using modeling in the science classroom. They were then asked to describe each of the following: student discourse, student-teacher discourse, and teaching strategies around the use of models. The preservice teachers' responses to this survey were coded and themes on the use of discourse and teaching strategies during modeling instruction began to emerge. As a follow-up, a sample of six preservice teachers were interviewed. The interview data were qualitatively coded using the themes that emerged from the survey responses. The conglomeration of both types of data provided an in-depth examination of preservice teachers conceptions of student discourse, student-teacher discourse, and teaching strategies around the use of models.

PRESERVICE TEACHERS' DISCOURSE PRECONCEPTIONS

A portion of this study analyzed elementary preservice teachers' preconceptions about social discourse in the science and engineering practice of developing and using models as described in The *Framework* (NRC, 2012) and *NGSS* (NGSS Lead States, 2013). A pilot study was conducted and preservice teachers had limited ideas about this practice as a social endeavor and the viewing of the video was not used as an intervention, but instead to provide minimal context in order to respond to survey questions. Teachers responded to the survey questions below while watching a video of modeling being used in the classroom: What communication patterns between students do you think led to groups successfully developing and using their model?; What communication patterns between the students and teacher do you think led to the groups successfully developing and using their model?

In a pilot study two discrete scales were adapted from the web-based tools created for professional learning to improve academic productive talk in the science classroom (TERC, 2012a; TERC, 2012b) to code responses to the survey questions. The responses to the survey received a score of zero (0) beginner, one (1) novice, two (2) transitional, or three (3) skilled by the author to interpret their preconceptions for student-student and student-teacher discourse. Tables 3 and 4 include example responses (one category from each scale) that relate to the scores for each of the scales to further clarify how these instruments were used.

Table 3. Discrete scale student-student communication: example survey responses

Example Survey Responses Discrete Scale: Student-Student Communication		
Category 2: Critiquing Their Own and Others' Answers/Ideas (Represents a Sample of One Category Only)		
Score	**Example Response**	**Coding Rationale**
A Score of 0 (Novice)	"They were talking between each other, throwing out ideas and information and would write down the one they felt was most suitable" (Participant #10)	Describes students generally having discussions in a group, but no distinctions made to show there were differences in answers/ideas of the students
A Score of 1 (Emerging)	"Some students in the group asked other group members questions. The students shared their own ideas and collaborated to come up with a master idea. " (Participants #8)	Describes students generally discussing answers/ideas of their own and others, but does not include specific descriptions of students critiquing those answers/ideas (agree/disagree, build on each other's answers, distinguish evidence from opinion, identify questions, ask if we have enough evidence)
A Score of 2 (Transitional)	"Asking one another questions and making short statements to see if beliefs were supported. The students would restate the student's ideas/belief to make sure they understood. They would confirm support for stated belief or they would build onto it. This became a pattern of communication as the student built their model. " (Participant #16)	Describes students generally discussing answers/ideas of their own and others, but includes little description of students critiquing those answers/ideas (agree/disagree, build on each other's answers, distinguish evidence from opinion, identify questions, ask if we have enough evidence)
A Score of 3 (Skilled)	No responses at this level	Describes students critiquing their own and others' answers/ideas (students agree/disagree, build on each other's answers, distinguish evidence from opinion, identify questions, ask if we have enough evidence)

The frequency and percentage statistics when coding preservice teacher survey response about critical student-student communication patterns, regarding the four categories of the discrete scale, are in Table 5.

Criteria from categories one (use of evidence to support answers), two (critiquing their own or others' answers or ideas), and three (merging ideas to develop and explanation) were used by more than 90 percent of participants when describing what student-student communication patterns led to the students successfully developing and using their models. Category two (critiquing their own or others' answers or ideas), and three (merging ideas to develop and explanation) had more participants score at an emerging (score of 1) level than at a novice (score of 0) level. All three categories where more than 90 percent of participants identified criteria from were still at the novice (score of 0) or the emerging (score of 1) level. There were

Table 4. Discrete scale student-teacher communication: example survey responses

Example Survey Responses Discrete Scale: Student-Teacher Communication (See Appendix J)		
Category 1: Helping Individual Students Share, Expand, and Clarify Their Own Thinking (represents a sample of one category only)		
Score	**Example Response**	**Coding Rationale**
A Score of 0 (Novice)	"Walk around and ask guiding questions that got students thinking about whether their model would function correctly. Give feedback to each individual group so they can develop their models even more. " (Participant #23)	Describes the teacher asking students general questions or providing support with no specific descriptions of the types of questions or support
A Score of 1 (Emerging)	"The teacher used open ended questions to make the students think about their ideas. The teacher avoided telling them the facts and instead led them to the answers. He connected it to other subjects. He encouraged students when if they didn't come to the right conclusion. " (Participants #5)	Describes the teacher asking students questions that fall under one of the following categories: share, expand, or clarify using general descriptions
A Score of 2 (Transitional)	"Teacher came around and individually clarified questions and probed for questions. Challenged students to come up with alternate ways. "If you think that's what happens, I want you to write it down". " (Participant #30)	Describes the teacher helping students to share, expand, and clarify their own thinking using some descriptions of strategies like the following: -Giving time to think (partner talk, writing as think time, wait time) -Asking students to say more (elaborate, clarify, and ask for examples) -Asking students to validate teacher summary of their thinking (giving space for the original student to agree/disagree or say more)
A Score of 3 (Skilled)	No responses at this level	Describes the teacher helping students to share, expand, and clarify their own thinking using descriptions of strategies like the following: -Giving time to think (partner talk, writing as think time, wait time) -Asking students to say more (elaborate, clarify, and ask for examples) -Asking students to validate teacher summary of their thinking (giving space for the original student to agree/disagree or say more)

few that could describe the criteria in this category at higher levels. Criteria from category four (apply learning to a new or different context) was not used by more than 90 percent of participants. The frequency and percentage statistics when coding

Table 5. Discrete scale student-student communication: frequency and percentage statistics

Frequency and Percentage Statistics Discrete Scale: Student-Student Communication						
Category	Statistics	Score				
		0-Novice	1-Emerging	2-Transitional	3-Skilled	No Use
1	Frequency	16	16	1	0	3
	Percentage (n=36)	44.4%	44.4%	2.8%	0%	8.3%
2	Frequency	10	23	3	0	0
	Percentage (n=36)	27.8%	63.9%	8.3%	0%	0%
3	Frequency	14	21	1	0	0
	Percentage (n=36)	38.9%	58.3%	2.8%	0%	0%
4	Frequency	0	2	0	0	34
	Percentage (n=36)	0%	5.6%	0%	0%	94.4%

preservice teacher survey response about critical student-teacher communication

Table 6. Discrete scale student-teacher communication: frequency and percentage statistics

Frequency and Percentage Statistics Discrete Scale: Student-Teacher Communication						
Category	Statistics	Score				
		0-Novice	1-Emerging	2-Transitional	3-Skilled	No Use
1	Frequency	11	20	4	0	1
	Percentage (n=36)	30.6%	55.6%	11.1%	0%	2.8%
2	Frequency	7	28	1	0	0
	Percentage (n=36)	19.4%	77.8%	2.8%	0%	0%
3	Frequency	8	19	7	0	2
	Percentage (n=36)	22.2%	52.8%	19.4%	0%	5.6%
4	Frequency	25	10	0	0	1
	Percentage (n=36)	69.4%	27.8%	0%	0%	2.8%

Table 7. Discrete scale student-student communication: statistical mean of scores

Statistical Mean of Scores Discrete Scale: Student-Student Communication		
Category	**Mean Min.=0, Max.=3**	**Percentage of Participants Who Identified Criteria in this Category**
Category 1	0.5	33 of 36 = 91.7%
Category 2	0.8	36 of 36 = 100%
Category 3	0.6	36 of 36 = 100%
Category 4	1	2 of 36 = 5.6%

patterns, regarding the four categories of the discrete scale, are in Table 6.

The student-teacher discrete scale was used by the researcher to see what preconceptions elementary preservice teachers have about important communication patterns and all four categories were used by almost all participants. Category one (helping individual students share, expand, and clarify their own thinking), two (helping students listen carefully to one another), and three (helping students deepen their reasoning) found more participant descriptions scoring at an emerging (score of 1) than at a novice (score of 0) level. Category four (helping student think with others) had more participant responses score at a novice (score of 0) level than at an emerging (score of 1) level. All four categories were used in preservice teacher descriptions for this response in the survey, but they majority were still at the novice (score of 0) or emerging (score of 1) level. The mean of the scores for each of the four categories in the two discrete scales used to elicit the preservice teachers' preconceptions about communication patterns in this practice are listed in Table 7 and 8 below. The mean only represents those answers that could be coded for each category, because the responses to each survey questions for each preservice teacher did not identify criteria for every category.

Table 8. Discrete scale student-teacher communication: statistical mean of scores

Statistical Mean of Scores Discrete Scale: Student-Teacher Communication		
Category	**Mean Min.=0, Max.=3**	**Percentage of Participants Who Identified Criteria in this Category**
Category 1	0.8	35 of 36 = 97.2%
Category 2	0.8	36 of 36 = 100%
Category 3	1.0	34 of 36 = 94.4%
Category 4	0.3	35 of 36 = 97.2%

Although more than 90 percent of participants identified criteria for three first three categories (one, two, and three) of the student-student communication pattern discrete scale, their mean scores are still below a score of one (emerging). The mean score for category four (apply learning to a new or different context) is high, but only two participants identified criteria in this category skewing the mean.

More than 90 percent of participants identified criteria for all four categories in the student-teacher communication discrete scale. The mean scores are the highest of all mean scores for this study. The mean scores for responses to survey questions about the practice of developing and using models as a social endeavor (second research question) were higher than the scores that displayed elementary preservice teacher's preconceptions about how the practice of developing and using models is used in the classroom (first research question) explored by Burks and Huffman (2019).

Six elementary participants were interviewed using a protocol that asked preservice teachers elaborate on their survey responses about this practice as a social endeavor (student-student and student-teacher communication) to probe for possible influences for higher mean scores than the first research question. Three emerging themes developed when analyzing the data that explain the higher scores; accumulated classroom management knowledge, previous practicum experience, and previous course work attending to student thinking. All participants described classroom management experiences in their descriptions. These descriptions included participants describing previous knowledge about how to facilitate group work and the intervention as the teacher during this group work as one of the primary emerging themes. One participant used the term "classroom management" to mean "good interactions and bad interactions" students were having (Participant #5, October 9, 2016). Another participant described needing "training on classroom management and also intervention" (Participant #6, October 16, 2016) when describing what would help them strengthen their ability to facilitate student-student and student-teacher communication while using the science and engineering practice of developing and using models in the classroom. Another emerging theme was previous experience observing students in the classroom (as a teacher or student) and most participants referred to this experience as practicum experience. All participants described experiences under this emerging theme. Experiences ranged from those as a student in the classroom and noticing how students interact or how the teacher helped facilitate conversations. Other experiences included practicum college experience; "I think our practicum classes in math and social studies have really helped with providing a lesson or providing an overall unit talking about how you might interject with questions for students" (Participant #4, October 9, 2016). Another participant described their previous practicum experience as "the opportunity we have had the most to observe and understand student interactions" (Participant #9, October 18, 2016).

The last emerging theme is previous course work attending to student thinking. Participants described a wide range of teacher strategies attending to student thinking and described these strategies as those learned in previous course work. One participant described student-teacher communication patterns as those you would see in an informal assessment check with students during class; "I think of it in terms of assessment that you as the teacher kind of know for sure that the students are learning the material and if they are struggling they have to recognize that they are struggling" (Participant #6, October 16, 2016). Another participant described student-student interaction as that where students can express communication using their preferred learning style. The participant described learning about learning styles in a previous teacher pedagogy course and said, "it is okay for them to choose the learning style that would appeal to them when communicating, but you just have to make sure that everyone had their own individual ideas before coming up with the same group idea in creating their model" (Participant #4, October 9, 2016). The elementary preservice teachers were probably better able to score higher mean scores when noticing the social aspects (student-student and student-teacher communication) of this practice because of previous practicum experience observing students in the classroom or observing this behavior as a student in the classroom, previous accumulated classroom management knowledge, and the attention to student thinking in other college courses. When participants described these emerging themes, almost all events were not linked to any science related experiences. Only one participant described a previous science related experience that led him to respond the way he did in the survey; "I was in anatomy lab and we were working with cadavers and there were a lot of times where all three of us just had a different idea of what was happening and so that was a time she would come and mediate to help us get to the final answer" (Participant #6, October 16, 2016). The descriptions of these emerging themes were also not described as being isolated from one experience, but instead were accumulated experiences throughout many of mostly non-science related events they learned from. The varied and numerous experiences about student-student and student-teacher communication outside of their science pedagogy work could have also led to higher mean scored on this topic. One participant described how they came to the conclusions they did on the survey about this practice as a social endeavor over many accumulated experiences and said, "I mean not one specific course I think it is kind of a variety" (Participant #4, October 9, 2016).

Table 9. Nominal scale teacher strategy responses: example survey responses

Example Survey Responses Nominal Scale: Teacher Strategy Responses		
Coding Category	**Example Response**	**Coding Rationale**
Investigating	"I would inquire as to what my students' perceptions were about a science topic. I would have them write up a hypothesis. I would then pair my students up for a lab. I would model the use of materials first. Then I would have student perform the lab with a partner. Lastly, the partnered pair would write down their findings together" (Participant #16)	Described the developing and using models practice by describing students asking questions or implementing methods of data collection to investigate the natural world
Sensemaking	"Students would be building something to visually represent a concept. Students will be using the models to understand how things work. Building models is a hands-on learning approach that involved inquiry-based learning by students. Models should be used frequently in the classroom for elementary science" (Participants #32)	Described the developing and using models practice by describing students analyzing data, looking for patterns or relationships to develop explanations and design representations based (constructing models) on data to explain how and why phenomena occur
Critiquing	"I would have the students write about what they liked about the other classmates models and what they would change if they could do it again. As a later project I would have the students go back to their original models and use that information to help them construct a 3 dimensional model of the eye. I would have the students group up with someone different from their group and compare/contrast each other's models, discuss what you liked and what you were confused on" (Participant #33)	Described the developing and using models practice by describing students evaluating, comparing, and contrasting different claims, explanations, or models as they make sense of the world around them

PRESERVICE TEACHERS' TEACHING STRATEGIES PRECONCEPTIONS

Another portion of the study analyzed elementary preservice teachers' preconceptions about the critical teacher strategies needed for successful use of the science and engineering practice of developing and using models in the classroom. Teachers responded to the survey question below while watching a video of modeling being used in a classroom to elicit teachers' preconceptions about critical teacher strategies needed for the practice. The key questions were: What three things would you do if you were the teacher at the end of the video to continue to help students develop and use their model?; What are your perceptions of how this practice of developing and using model should be used in the classroom?

Table 10. Nominal scale survey question one teacher strategy responses: frequency and percentage statistics

Survey Question One: Frequency and Percentage Statistics Nominal Scale: Teacher Strategy Responses		
Category	**Frequency Statistic**	**Percentage Statistic (n=46)**
Investigating Practice	32	69.7%
Sensemaking Practice	12	26.1%
Critiquing Practice	2	4.3%

Table 11. Nominal scale survey question two teacher strategy responses: frequency and percentage statistics

Survey Question Two: Frequency and Percentage Statistics Nominal Scale: Teacher Strategy Responses		
Category	**Frequency Statistic**	**Percentage Statistic (n=53)**
Investigating Practice	11	20.8%
Sensemaking Practice	18	33.4%
Critiquing Practice	24	45.3%

In the same pilot study a nominal scale adapted from a study conducted by McNeill et al. (2015) identifying the practice of developing and using models in three distinct ways (investigating, sensemaking, and critiquing tools). The responses to the survey questions for this portion of the study received a code of investigating, sensemaking, or critiquing to interpret the teacher preconceptions for each of the categories in the

Table 12. Nominal scale combined teacher strategy responses: frequency and percentage statistics

Total Frequency and Percentage Statistics Nominal Scale: Teacher Strategy Responses		
Category	**Frequency Statistic (used in both questions)**	**Percentage Statistic (n=99)**
Investigating Practice	43	43.4%
Sensemaking Practice	30	30.3%
Critiquing Practice	26	26.3%

nominal scale. Tables 9 includes example responses (one category from the scale) that relate to the scores for the scale to further clarify how this instruments was used.

The frequency and percentage statistics when coding preservice teacher survey responses (in both survey questions) about their preconceptions of how the practice of developing and using models is used as a tool (a critical teacher strategy), regarding the three categories of the nominal scale, are in Tables 10, 11, and 12. The frequency represents how many times in each question the practice of developing and using models was referred to as an investigating, sensemaking, or critiquing practice. The percentage statistic represents the percentage of use for that category in responses out of the total combined use of all three categories. Frequency and percentage statistics are reported for each question individually and then combined for an overall analysis of elementary preservice teachers' use of the three categories during the survey (both questions combined).

The findings from the survey responses indicated the preservice teachers have a limited understanding of the varying ways the practice of developing and using models is used in the *Framework* and *NGSS* described by McNeill et al. (2015). The practice of developing and using models was referred to more as an investigating practice than any other type of practice overall. Developing and using models was described as a critiquing practice the least when coding for all three categories (investigating, sensemaking, and critiquing) overall. Participants described the practice of developing and using models primarily as an investigating practice and a quarter of the time as a sensemaking practice. Although sensemaking was used to describe the practice, this was done to make sense of learning after the learning had already taken place. This is not in keeping with the new vision for this practice in the classroom. Describing the practice as a critiquing practice was used less than 5 percent by participants.

The same six participants were interviewed using a protocol that asked preservice teachers to elaborate on their survey responses. When coding the interviews using the three ways (investigating, sensemaking, critiquing) to think about this practice, all participants described this practice primarily as an investigating practice when elaborating about their responses and searching for further knowledge to display about this practice. One participant described the practice of developing and using models as an investigating practice by saying, "we were supposed to draw out what we think the model of the foil boats would be and then predict the number of pennies it could hold" (Participant #9, October 18, 2016). When seldom describing the practice as a sensemaking practice, participants used descriptions that implied the model was made by the teacher or student by making sense of content already previously learned. This is not the sensemaking described in the *Framework* (NRC, 2012) and *NGSS* (NGSS Lead States, 2013). Sensemaking refers to students working to develop or use a model to better make sense of a phenomena during their learning and not as a representation after learning the content has occurred. One participant described this type of sensemaking by saying, "I keep thinking of the ocean and kind of like how different parts of the ocean work together and so maybe if you're studying one part at one time you would kind of design your own model of the ocean revolving around the sea floor or whatever that could be" (Participant #6, October 16, 2016). The only time participants referred to the practice of developing and using models as a critiquing practice was in response to questions using the video as context. This mimicked the evidence in the frequency and percentage statistics tables for these findings. Any other time participants reached into their funds of knowledge to display their preconceptions about this practice, they reverted to thinking about this practice as primarily an investigatory practice.

RECOMMENDATIONS

"Because science and engineering practices are basic to science education and the change from inquiry to practices is central, this innovation for the new standards will likely be one of the most significant challenges for the successful implementation of science education standards" (Bybee, 2011, p. 34). These results indicate that preservice teachers had a better beginning knowledge base (preconceptions) about this practice as a social endeavor than they did about the reformed vision for this practice identified within the *NGSS* (NGSS Lead States, 2013) from the results of the first research question (Burks & Huffman, 2019). The elementary preservice teachers were able to score higher mean scores when noticing the social aspects (student-student and student-teacher communication) of this practice because of three emerging themes that were found in the interview data; previous practicum experience observing students in the classroom or observing this behavior as a student in the classroom, previous accumulated classroom management knowledge, and the attention to student thinking in other college courses. Emerging themes in the interview data revealed almost all events were not linked to any science related experiences. The descriptions of these emerging themes were also not described as being isolated from one experience, but instead were accumulated experiences throughout mostly non-science related events they learned from. The varied and numerous experiences about student-student and student-teacher communication outside of their limited science pedagogy work led to higher mean scores on this topic. Central use of the practice of developing and using models in recent science standards-aligned Open Education Resource curricular efforts led by a national consortium of curriculum developers, OpenSciEd (OpenSciEd, n.d.), uses this practice in the expansive ways (investigating, sensemaking, critiquing) described by McNeill et al. (2015). It is important for elementary preservice teachers to move

Table 13. Creative Practices

Imagine	To form a mental image or concept
Investigate	To observe or study through exploration or examination
Construct	To make or form by combining parts or elements
Reflect	To think deeply or carefully about

Source: (National Coalition for Core Arts Standards, 2012)

away from a focus on the investigating practice focus, results found in this study, if the practice of developing and using models is going to have the impact on the reformed science classroom as described in the *NGSS* (NGSS Lead States, 2013) to include all three categories that McNeill et al. (2015) describe. Preservice teachers had relatively better beginning understandings (preconceptions) about the science and engineering practice of developing and using models as a social endeavor and this could be used as a lever in preservice teacher education to help support expansive use of this central practice where teachers fell short per the study.

FUTURE RESEARCH DIRECTIONS

Although a formal review of the NGSS (NGSS Lead States, 2013) and the 2014 National Core Arts Standards (National Coalition for Core Arts Standards, 2012) could not be found, the review of the arts standards and national math standards yielded a high number of connections within the arts creative practices. Likewise, upon initial comparison there appear to be many correlations between the science and engineering practice of developing and using models in the *NGSS* (NGSS Lead States, 2013) and the National Core Arts Standards (National Coalition for Core Arts Standards, 2012). Appendix F of the *NGSS* further develops the progression of this practice across grades K-12 (NGSS Lead States, 2013, Appendix F p.6). Both of these defining documents use similar language.

Students can be expected to evaluate and refine models through an iterative cycle of comparing their predictions with the real world and then adjusting them to gain insights into the phenomenon being modeled. As such, models are based upon evidence. When new evidence is uncovered that the models can't explain, models are modified (NGSS Lead States, 2013, Appendix F p.6).

One effective classroom approach to elicit creative process (which is common in the arts and supported by Understanding by Design) is to encourage open-ended responses by asking essential questions and providing lessons that allow for more than one solution (National Coalition for Core Arts Standards, 2012).

These connections can also be seen in the following table from the 2014 National Core Arts Standards Framework.

A more comprehensive study of how the recent art and science standards overlap and support one another for more practical ways of integration is needed. As mentioned above, developing and using models helps to foster equitable science education by offering multiple entry points for culturally and linguistically diverse students. Integrating the arts into science education can only strengthen and expand the ability of science education to be more equitable. The creative practices listed in the table above which are fundamental in the arts, science, and mathematics, "are cognitive processes by which students not only learn within an individual discipline but also transfer their knowledge, skill, and habits to other contexts and settings" (National Coalition for Core Arts Standards, 2012, p.19).

CONCLUSION

This study examined elementary preservice teachers' preconceptions around the components of the science and engineering practice of developing and using models as a starting point in the improvement process, to think about creating preservice teacher education for this new vision. Previous analysis of the data from a portion of this study (Burks & Huffman, 2019) showed the lack of understanding about the practice of developing and using models as described in the *Framework* (NRC, 2012) and *NGSS* (NGSS Lead States, 2013). The analysis for discourse showed more promising results when preconceptions about this practice as a social endeavor were elicited from the elementary preservice teachers. When thinking about this practice in a variety of ways (investigating, sensemaking, and critiquing) as a teacher strategy, elementary preservice teachers identified this practice as an investigating practice more than any other category.

Combining preservice teacher's previous knowledge of the practices with awareness of their collective renewed meaning will help them reach the new vision for science education. This categorization proposed by McNeill et al. (2015) can help preservice teacher improve their understanding about how to implement this practice. A strong connection between these categories, as well as the definition of modeling in the *Framework* and *NGSS*, that show the range of developing and using models and the four Creative Practices from the arts can be seen. The four Creative Practices are "meta-cognitive activities" that "nurture the effective work habits of curiosity, creativity and innovation, critical thinking and problem solving, communication, and collaboration, each of which transfer to all aspects of learning and life in the 21st Century" (National Coalition for Core Arts Standards, 2012). The science and engineering practice of developing and using models is one way to move towards capitalizing on the support for STEAM education due to the way the arts manifest in the *NGSS* in this central practice. This connection may serve as a way to move towards more meaningful and practical opportunities for reform.

REFERENCES

Allina, B. (2018). The development of STEAM educational policy to promote student creativity and social empowerment. *Arts Education Policy Review, 119*(2), 77–87. doi:10.1080/10632913.2017.1296392

Banilower, E. R., Smith, P. S., Malzahn, K. A., Plumley, C. L., Gordon, E. M., & Hayes, M. L. (2018). *Report of the 2018 NSSME.* Chapel Hill, NC: Horizon Research, Inc.

Burks, L. A., & Huffman, D. (2019). Pre-Service Teachers' Critical Thinking and Developing and Using Models in Science. In S. Robinson & V. Knight (Eds.), *Handbook of Research on Critical Thinking and Teacher Education Pedagogy* (pp. 194–212). Hershey, PA: IGI Global; doi:10.4018/978-1-5225-7829-1.ch011.

Bybee, R. (2013). *The case for STEM education: Challenges and opportunities.* Arlington, VA: NSTA Press.

Bybee, R. (2018). *STEM education now more than ever*. Arlington, VA: NSTA Press.

Bybee, R. W. (2011). Scientific and engineering practices in K–12 classrooms. *Science Teacher (Normal, Ill.), 78*(9), 34–40.

Campbell, T., Oh, P. S., Maughn, M., Kiriazis, N., & Zuwallack, R. (2014). A review of modeling pedagogies: Pedagogical functions, discursive acts, and technology in modeling instruction. *Eurasia Journal of Mathematics, Science, and Technology Education, 11*, 159–176.

Edelson, D., Ryan, J., & Cowart, J. (2019, April). OpenSciEd: An Open Educational Middle School Science Curriculum Designed for NGSS. Presentation at the *annual meeting of the National Science Teachers Association*, St. Louis, MO.

Eichinger, D., Anderson, C. W., Palinscar, A. S., & David, Y. M. (1991). An illustration of the roles of content knowledge, scientific argument, and social norms in collaborative problem solving. Paper presented at the annual meeting of the American Educational Research Association, Chicago, IL.

Eisner, E. (2002a). *The arts and the creation of the mind*. New Haven, CT: Yale University Press.

Eisner, E. (2002b). What the arts do for the young. *School Arts, 102*, 16–17.

Eisner, E., & Powell, K. (2002). Art in Science? *Curriculum Inquiry, 32*(2), 131–159. doi:10.1111/1467-873X.00219

Lee, O., Quinn, H., & Valdés, G. (2013). Science and language for English language learners in relation to Next Generation Science Standards and with implications for Common Core State Standards for English language arts and mathematics. *Educational Researcher, 42*(4), 223–233. doi:10.3102/0013189X13480524

McNeill, K. L., Katsh-Singer, R., & Pelletier, P. (2015). Assessing science practices – Moving your class along a continuum. *Science Scope, 39*(4), 21–28. doi:10.2505/4/ss15_039_04_21

Michaels, S., & O'Connor, C. (2012). *Talk science primer*. Cambridge, MA: Technical Education Research Centers; Retrieved from http://inquiryproject.terc.edu/shared/pd/TalkScience_Primer.PDF

National Academies of Sciences, Engineering, and Medicine. (2018). Design, Selection, and Implementation of Instructional Materials for the Next Generation Science Standards: Proceedings of a Workshop. Washington, DC: The National Academies Press. doi:10.17226/25001

National Coalition for Core Arts Standards. (2012). The 2014 National Core Arts Standards. New York, NY: The National Coalition for Core Arts Standards. Retrieved from https://www.nationalartsstandards.org/content/national-core-arts-standards

National Research Council (NRC). (2007). *Taking science to school: Learning and teaching science in grades K–8*. Washington, DC: National Academies Press.

National Research Council (NRC). (2012). *A Framework for K-12 science education: Practices, crosscutting concepts, and core ideas*. Washington, DC: The National Academies Press.

National Science Teachers Association (NSTA). (n.d.). About the Next Generation Science Standards. Retrieved May 21, 2019, from https://ngss.nsta.org/About.aspx

NGSS Lead States. (2013). *Next Generation Science Standards: For states, by states*. Washington, DC: The National Academies Press.

Ogborn, J., Kress, G., Martins, I., & McGillicuddy, K. (1996). *Explaining science in the classroom*. Buckingham, UK: Open University Press.

OpenSciEd. (n.d.). OpenSciEd. Retrieved from https://www.openscied.org/openscied/?

Reiser, B. J. (2013, September). What professional development strategies are needed for successful implementation of the next generation science standards. Paper presented at the International Research Symposium on Science Assessment, Washington, DC.

Ricketts, A. (2014). Preservice elementary teachers' ideas about scientific practices. *Science and Education*, *23*(10), 2119–2135. doi:10.1007/s11191-014-9709-7

Schwarz, C., Passmore, C., & Reiser, B. J. (2017). *Helping students make sense of the world using next generation science and engineering practices*. Arlington, VA: NSTA Press, National Science Teachers Association.

Technical Education Research Centers. (2012a). Goals for productive discussions and nine talk moves. The Inquiry Project: Bridging research & practice supported by the National Science Foundation.

Technical Education Research Centers. (2012b). Reflection tool: Are students progressing toward scientific understanding? The Inquiry Project: Bridging research & practice supported by the National Science Foundation.

The College Board. (2012). *The Arts and the Common Core: A Review of Connections Between the Common Core State Standards and the National Core Arts Standards Conceptual Framework*. New York, NY: The College Board.

ADDITIONAL READING

Bell, P., Shouse, A., & Peterman, T. (2014). Next generation science standards: What's different, and do they matter?: Stem Teaching Tools. Retrieved February 27, 2017, from http://stemteachingtools.org/brief/14

Campbell, T., Oh, P. S., & Neilson, D. (2013). Reification of five types of modeling pedagogies with model-based inquiry (MBI) modules for high school science classrooms. In M. S. Khine & I. M. Saleh (Eds.), *Approaches and Strategies in Next Generation Science Learning* (pp. 106–126). Hershey, Pennsylvania: IGI Global; doi:10.4018/978-1-4666-2809-0.ch006.

Lee, O., Goggins, M., Haas, A., Januszyk, R., Llosa, L., & Grapin, S. E. (forthcoming). Making everyday phenomena phenomenal: NGSS-aligned instructional materials using local phenomena with student diversity. In P. Spycher & E. Haynes (Eds.), *Culturally and linguistically diverse learners and STEAM: Teachers and researchers working in partnership to build a better path forward*. Charlotte, NC: Information Age Publishing.

McNeill, K. L., & Reiser, B. J. (2018). Open source for opening minds: New OpenSciEd materials support science standards. *The Learning Professional*, *39*(6), 44–48.

Michaels, S., & O'Connor, C. (2017). From recitation to reasoning: Supporting scientific and engineering practices through talk. In C. V. Schwarz, C. M. Passmore, & B. J. Reiser (Eds.), *Helping students make sense of the world through next generation science and engineering practices* (pp. 311–336). Arlington, VA: NSTA Press.

National Academies of Sciences, Engineering, and Medicine. (2015). Science teachers' learning: Enhancing opportunities, creating supportive contexts. Washington, DC: The National Academies Press.

National Academies of Sciences, Engineering, and Medicine. (2018). How People Learn II: Learners, Contexts, and Cultures. Washington, DC: The National Academies Press. . doi:10.17226/24783

Oh, P. S., & Oh, S. J. (2011). What teachers of science need to know about models: An overview. *International Journal of Science Education*, *33*(8), 1109–1130. doi:10.1080/09500693.2010.502191

Reiser, B. J., Novak, M., & McGill, T. A. W. (2017). *Coherence from the students' perspective: Why the vision of the framework for K-12 science requires more than simply "combining" three dimensions of science learning*. Washington, DC: The National Academies of Sciences, Engineering, and Medicine, Board on Science Education.

Schwarz, C. (2009). Developing preservice elementary teachers' knowledge and practices through modeling-centered scientific inquiry. *Science Education*, *93*(4), 720–744. doi:10.1002/sce.20324

KEY TERMS AND DEFINITIONS

Crosscutting Concepts: Described as "concepts that bridge disciplinary boundaries, having explanatory value throughout much of science and engineering. These crosscutting concepts were selected for their value across the sciences and in engineering. These concepts help provide students with an organizational framework for connecting knowledge from the various disciplines into a coherent and scientifically based view of the world" (NRC, 2012, p. 83).

Developing and Using Models: "In science, models are used to represent a system (or parts of a system) under study, to aid in the development of questions and explanations, to generate data that can be used to make predictions, and to communicate ideas to others. In engineering, models may be used to analyze a system to see where or under what conditions flaws might develop, or to test possible solutions to a problem. Models can also be used to visualize and refine a design, to communicate a design's features to others, and as prototypes for testing design performance" (NGSS Lead States, 2013, p. 386).

Disciplinary Core Idea: "The committee developed its small set of core ideas in science and engineering. We grouped disciplinary ideas into four major domains: the physical sciences; the life sciences; the earth and space sciences; and engineering, technology, and applications of science" (NRC, 2012, p. 31).

Framework: "Beginning in January of 2010, the Carnegie Corporation of New York funded a two-step process to develop a new set of state developed science standards intended to prepare students for college and career readiness in science" (Pruitt, 2014, p. 145). The first phase ended with Achieve creating *A Framework for K-12 Science Education* (the *Framework*). "The goal of the Framework was to articulate the vision for science education in the twenty first century and to articulate what students need to know in their K-12 experience to be considered scientifically literate" (Pruitt, 2014, p. 146).

Models: "Models include diagrams, physical replicas, mathematical representations, analogies, and computer simulations. Although models do not correspond exactly to the real world, they bring certain features into focus while obscuring others. All models contain approximations and assumptions that limit the range of validity and predictive power, so it is important for students to recognize their limitations" (NGSS Lead States, 2013, p. 386).

NGSS: The Next Generation Science Standards (*NGSS*) were completed in April of 2013. The *NGSS* represent a change in how states have traditionally approached their science standards. In embracing science education research, the *NGSS* represent performance expectations (PEs) that require all students have a deep understanding of a smaller number of disciplinary core ideas (DCIs), are able to show evidence of that knowledge through scientific and engineering practices, and connect crosscutting concepts across disciplines" (Pruitt, 2014, p. 145).

Phenomena: "Natural phenomena are observable events that occur in the universe and that we can use our science knowledge to explain or predict. The goal of building knowledge in science is to develop general ideas, based on evidence, that can explain and predict phenomena" (Achieve, 2016, p. 1).

Chapter 11
Learning From Place:
Developing a Relationship With the Land and Our Partners

Kevin O'Connor
Mount Royal University, Canada

Gladys Sterenberg
Mount Royal University, Canada

Norman Vaughan
Mount Royal University, Canada

ABSTRACT

This chapter investigates how teacher candidates' experiences in STEAM field studies with community partners can inform work in teacher education within an integrated practicum based on curriculum of place. The overall goal of the inquiry is to better understand and articulate the particular ways in which people value place-based knowledge. Through relationships with Indigenous communities, the team of educators has a deeply held conviction that sustained deliberations on the connections between Indigenous knowledge systems and place-based thinking can provide significant opportunities for reframing education. Learning from place emphasizes a relationship with the land, something deeply respected in Indigenous communities and something absent from much of place-based education. The research explores this tension as we come to a deeper and shared understanding of co-responsibility within Treaty 7 relationships. The project seeks to close this gap by considering varying perspectives of place as it informs STEAM teacher education pedagogy.

DOI: 10.4018/978-1-7998-2517-3.ch011

INTRODUCTION

We acknowledge the Blackfoot, Tsuu T'ina, and Stoney-Nakoda Nations on whose traditional territories we live and work. This chapter investigates how teacher candidates' experiences in STEAM (Science, Technologies/Education, Arts and Mathematics) field studies with community partners can inform our work in teacher education. As educators who teach within an integrated STEAM practicum semester, we are interested in linking STEAM courses, Indigenous knowledge systems, and place-based education. This project seeks to foster new understandings around a curriculum of place (Chambers, 2008) among Indigenous communities, not-for-profit organizations, public schools, and our Canadian post-secondary institution.

The research described in this chapter is part of a longitudinal programmatic qualitative study that investigates the impact of transformative pedagogies (described later in the chapter) on teacher educators, mentor teachers, and teacher candidates. Central to the longitudinal research is the study of how theory and practice can be integrated by attending to various perspectives of place within institutions and schools. Here, the goal of our inquiry is to better understand and articulate the particular ways in which we value place-based knowledge within a STEAM integrated practicum semester.

Many contributions to bridging theory and practice in STEM (Science, Technology, Engineering, and Mathematics) education have been made through non-Indigenous perspectives of place. The field of place-based education has strong roots in environmental education (Knapp, 1996; Orr, 1994; Raffin, 1995; van Eijck, & Roth, 2009). During the past twenty years, place-based education (Emekauwa, 2004; Greunewald, 2003; Penetito, 2009; Relph, 1992; Sobel, 2004; Wattchow & Brown, 2011) has become wide-spread as an approach to teaching that is grounded in the context of community and environment (Kincheloe, Mckinley, Lim & Barton, 2006; Raffan, 1993; Theobald & Curtiss, 2000).

However, Indigenous knowledge systems embedded in relationships with nature have informed peoples' understanding of place over many centuries (Cajete, 1999; 2000). Emerging research suggests that place-based education is limited because it does not critique colonial legacies in theoretical frameworks of place (Calderon, 2014). Indeed, many Indigenous scholars are replacing the term *place* with *land* and argue that land-based pedagogies promote the decolonization of education (Ballantyne, 2014; Wildcat, McDonald, Irlbacher-Fox, & Coulthard, 2014) by recognizing the intimate relationship that Indigenous peoples have with the land. Learning from place emphasizes a relationship with the land (Blood & Chambers, 2006), something deeply respected in Indigenous communities and something absent from much of place-based STEM education. Our project seeks to close this gap by considering varying perspectives of place as it informs STEAM teacher education pedagogy.

Our research questions are: 1) What do we learn as STEAM educators through the design and implementation of field studies that link Indigenous knowledge systems and place-based education? and 2) What are the teacher candidates' experiences of STEAM field studies framed by a curriculum of place?

BACKGROUND

A Curriculum of Place

In a Canadian context where calls to action are made in response to the *Truth and Reconciliation Commission* (Truth and Reconciliation Commission of Canada, 2012) and where the *United Nations Declaration on the Rights of Indigenous Peoples* (United Nations, 2008) has been recently adopted, it becomes important for educators to engage in the decolonization of education. Both documents acknowledge the need to rebuild relationships with First Nations, Métis, and Inuit peoples and these have important implications for educators. Calls to action made in response to the *Truth and Reconciliation Commission* (Truth and Reconciliation Commission of Canada, 2012) include co-developing strategies "to eliminate the educational and employment gaps between Aboriginal and Non-Aboriginal Canadians" that incorporated the principles of "developing culturally appropriate curricula", and "respecting and honouring treaty relationships" (p. 2). A curriculum of place that integrates the teaching and learning of children, undergraduate students, teachers and educators can support these calls to action.

Based on the current momentum with place-based education, our project to reframe an understanding of place is timely. Many contributions to education have been made through non-Indigenous perspectives of place. Place-based education emerges from the particular attributes of a place. The content is specific to the geography, ecology, sociology, politics, and other dynamics of that place (Woodhouse & Knapp, 2000). It provides a purpose to the knowledge and reasoning taught in schools, it offers a contextual framework for much of the curriculum by giving meaning to the studies and it engages learners in the conditions of their own realities (Smith, 2007). Place-based education often seeks to help not only students but also communities through employing students, school staff, educators, scientists and other interested community members in solving community problems (Denise & Harris, 1989). Place-based education differs from conventional text and classroom-based education in that it understands students' local community as one of the primary resources for learning (Sobel, 2004). Thus, place-based education promotes learning that is rooted in what is local—the unique history, environment, culture, economy, literature, and art of a particular place. Addressing a variety of community problems brings into play many different aspects of learning subjects/areas of studies (Kincheloe, 2001). The "problems" call for solutions requiring multidisciplinary approaches. When students deal with "real" issues and work with community members to find solutions, they become engaged (Dillon et al., 2013).

Drawing on a curriculum of place as a theoretical framework contributes to the objectives of this project as we study the conceptualization of place in STEAM teacher education by integrating Indigenous knowledge systems and place-based education. Integral in our work is a sense of dwelling-in-place, a stance dependent on listening as a "highly reflective and revelatory mode of communication that can open one to the mysteries of unity between the physical and spiritual, to the relationships between natural and human forms, and to the intimate links between places and persons" (Carbaugh, 1999, p. 250). It is these intimate links between places and persons that many researchers acknowledge (Blood, Chambers, Donald, Hasebe-Ludt & Big Head, 2012; Cajete, 2000; Chambers, 2006, 2008; Kissling, 2012; Kulnieks, Longboat, & Young, 2013; Little Bear, 2000). In the Blackfoot context, balance and harmony with the environment are recognized as part of the knowledge system. Bastien (2004) writes, "Ontological responsibilities of *Siksikaitsitapi* are the beginning of affirming and reconstructing ways of knowing. These fundamental responsibilities must be renewed by coming to know the natural alliances" (p. 4). She suggests that Indigenous knowledge is linked to intricate interrelationships within nature. The environment is understood as "the source from which all life originates and from which all knowledge is born" (p. 39). Battiste (2002) links Indigenous knowledge systems to particular "landscapes, landforms, and biomes where ceremonies are properly held, stories properly recited, medicines properly gathered, and transfers of knowledge properly authenticated" (p. 13). Little Bear (2000) describes the land as integral to the Indigenous mind. He writes:

Events, patterns, cycles, and happenings occur at certain places. From a human point of view, patterns, cycles, and happenings are readily observed on and from the land. Animal migrations, cycles of plant life, seasons, and cosmic movements are detected from particular spatial locations; hence, medicine wheels and other sacred observatory sites. Each tribal territory has its sacred sites, and its particular environmental and ecological combinations resulting in particular relational networks. All of this happens on the Earth; hence, the sacredness of the Earth in the Native American mind. The Earth is so sacred that it is referred to as "Mother," the source of life. (p. xi)

A curriculum of place recognizes the intimate relationship that Indigenous peoples have with the land and emphasizes relational ways of knowing. Chambers (2008) presents four dimensions of a curriculum of place as part of a conversation about how this theoretical stance might impact education: a different sense of time, enskillment, an education of attention, and wayfinding. She claims:

In a curriculum of place the activities in which we engage children are the very activities they need to dwell in this place, to be nourished by the place and to nourish it. In a curriculum of place, young people or novices grow into knowledge through engagement in hands-on activities learning side-by-side with masters of the crafts. This knowledge enables people to find their way in that place where they dwell and this knowledge and these skills endow them with identity. (p. 120)

She suggests that "a curriculum of place is no longer optional" (p. 125). Our inquiry is to develop a deeper understanding of STEAM education through Indigenous, disciplinary, interdisciplinary, and community perspectives of a curriculum of place.

Perspectives of Place in Post-Secondary Institutions And Schools

In this section, various perspectives of place within institutions and schools that have shaped teacher education are considered. A STEAM integrated practicum semester can assist teacher candidates and their faculty in supporting a more unified approach to theory and practice. This integration (Theory-Practice) will support increased learning opportunities based on better Theory informed by better Practice. We make the case that having the teacher candidates systematically integrate their STEAM pedagogies through place-based and experiential studies and an integrated practicum can lead to a continuous learning loop that then informs innovative and emerging practice (Raelin, 2007).

Within the field of teacher education, it is widely accepted that the development of teacher candidates' professional practice is of critical importance yet, as teacher educators, we struggle in our understanding of practice. We are limited in how to support its development and specifically its relationship to teacher candidates' learning through teacher education programs. It is through these gaps of understanding how teacher candidates' develop during practicum experiences that suggest we have more work to do so that education coursework (Theory) has more influence on their actual execution (Practice) as classroom teachers. We begin with addressing the traditional problematic of teacher candidates tending to be socialized into the status quo of school practice or to simply reproduce their own school experiences (Munby & Russell, 1994).

The separation of post-secondary and school places has contributed to ineffective teacher education programs where teacher candidates tend not to use the research-based guidelines offered to them in their courses when they subsequently engage in their practicum placements (Clift & Brady, 2005; Segall, 2002). Instead, they tend either to be socialized into typical practices of school or to teach as they were taught themselves, due to the powerful influence of their "apprenticeship of observation" (Lortie, 1975) as students in school. The evidence is longstanding (Zeichner & Tabatchnik, 1981) and widespread (Clift & Brady, 2005; Wideen et al., 1998). The only exceptions to this general trend appear to be programs that can provide a high degree of congruence between the content of course work and the models provided by mentor teachers in their practice (Beck & Kosnik, 2001; Darling-Hammond, 2006). We believe one approach that can have a significant impact merging the theoretical perspectives of the university with the practical experiences in the schools is through the longitudinal development of cross-curricular curricular opportunities based in a curriculum of place.

Reforms to teacher education have called for extensive practical experience for teacher candidates and many of these calls have advocated for strong school-university partnerships as a means for providing these. Clinical schools (The Carnegie Forum on Education and the Economy, 1986), professional development schools (Holmes Group, 1986) and partner schools (Goodlad, 1988) are proposed ways of increasing reciprocal collaboration in order to connect theory and practice (Teitel, 1998, p. 85). Yet despite these long-standing calls for more extensive and enhanced practical experiences for teacher candidates, recent proposals reiterate the same need. Korthagen, Loughran, and Russell's (2006) development of fundamental principles for reforming teacher education programs begins by highlighting the significance of experience in a program. Darling-Hammond's (2006) conclusions about the nature of effective teacher education programs includes as a major component "extensive and intensely supervised clinical work integrated with course work using pedagogies that link theory and practice" (p. 300).

As three teacher educators, we feel very fortunate to be active practitioners in a field that has shown a recent interest and special focus on the examination of effective teacher education programs. Research conducted by both Darling-Hammond (2006) and Beck and Kosnik (2006) have yielded considerable insight into the 'how' of better preparing teacher candidates for the changing field of education. These two studies identify some strikingly common themes: an essential need for partnerships between participating communities of practice that support social constructivism; a learning environment that is supportive and representative of all the complex areas of education and is achieved through experiential, inquiry and place-based initiatives that support teacher candidates to take risks and develop a critical and responsive pedagogy; and the need for a deliberate and extensive integration of both curriculum and pedagogy as it relates to the areas of theory and practice in teacher education programs. We are developing our program intentionally around this research. Specifically, we are examining how to enact transformative pedagogies, such as place-based pedagogies, focused on theory-and-practice integration. In our investigation we are attending to relationships as we study the impact of such pedagogies with teacher educators, teacher candidates, and beginning teachers.

Perspectives of Place Through Integrated Holistic Experiences

The primary foci of this chapter is on what our teacher candidates experience and what we learn as STEAM educators through the design and implementation of field studies that link Indigenous knowledge systems and place-based education. The common use of the terms "theory" and "practice" in teacher education contributes to the notion that they are mutually exclusive. However, we believe that they are inseparable. All teacher candidates have developed some kind of theory about all aspects of their experience and all practice is driven by those theoretical guidelines. The distinction between theory and practice in teacher education seems to highlight a disconnection between the abstract guidelines offered for teaching in teacher education courses and teacher candidates' experience of the insufficiency of these guidelines when encountering the complexity of the classroom. What we seek to explicate are the relationships of theory-to-practice, practice-to-theory, and theory-and-practice.

Schön's (1987) reflective practice theoretical stance of knowing-in-action informs his proposal that the practicum should be at the heart of teacher education programs (Dillon et al., 2013). Knowing-in-action refers to teacher candidates' intuitive wisdom that allows them to adjust their practice to the unfolding complexities of the classroom environment. Reflection becomes an integral part of a place-based experience of teacher education as teacher candidates begin to articulate their practical knowledge. Schön suggests that a reflective practicum depends on the quality of interaction between mentor teachers, teacher educators, and teacher candidates and among teacher candidates. This is enhanced by extensive opportunities for dialogue and reflection while teacher candidates are engaged in extensive practicum experiences. When addressing the need for theory-and-practice integration, experiential place-based opportunities become significant.

The design of our program reflects the importance of place. It has been co-designed within existing partnerships to honour and reflect Indigenous knowledge systems and developed so we can "grow into knowledge through engagement in hands-on activities learning side-by-side with masters of the crafts" (Chambers, 2008, p. 120).

OUR PLACE: MOUNT ROYAL UNIVERSITY BACHELOR OF EDUCATION PROGRAM

As authors, we are three non-Indigenous teacher educators with each specific expertise in the disciplines of Science, Mathematics and Educational Technologies.

We share experience, interest and passion for learning and supporting Indigenous ways of knowing through our pedagogies. Our students are predominantly female from the Calgary area, most are non-Indigenous yet we have a student population of 8-10% who self-identify as Indigenous descent. The context of our study is Mount Royal University, a four-year undergraduate institution located in Calgary, Alberta, Canada (http://www.mtroyal.ca/). In the fall of 2011, the University launched a new Bachelor of Education (B.Ed.) program, a four-year direct entry B.Ed. degree, with an emphasis on connecting theory with practice through the programmatic incorporation of the following transformative pedagogies:

- *Professional Learning Plan* (e-portfolios): Teacher candidates begin an electronic professional learning plan in the first year of the program and add artifacts and reflections in each year of the program. This program level portfolio provides an integrated platform for student learning throughout the degree and invites metacognition regarding professional growth. Teacher candidates reflect upon their progress towards the five major program competencies:
    - Professional Responsibilities
    - Planning for Learning
    - Facilitating Learning
    - Classroom Environment
    - Assessment
- *Cohort placements and in-school seminars:* Teacher candidates in practicum (and where possible in field experience) are placed in cohorts of 4-6 teacher candidates at each school. The cohort placements support the programmatic feature of in-school seminars. This structure has several benefits:
    - teacher candidates build a learning community and support network in their school
    - teacher candidates capture immediate correlations between the theory and its application to their own practice and experience
    - teachers better understand the unique features of the university program and the evaluation process
    - teachers become engaged partners in the post-secondary education of the teacher candidates
    - faculty are better able to visit every week to engage in weekly seminar discussions and build relationships with school personnel
- *Cross-curricular course integration:* teacher candidates are required to demonstrate the integration of outcomes from multiple subjects in their planning. Specific semesters are designed to focus on such integration (e.g.,

Language Arts/Music/Assessment in year 2, STEAM semester in year 3, Social Studies/Indigenous Education in year 4)

- *Community partnerships*: teacher candidates plan and support learning at various sites including Ann and Sandy Cross Conservation Area, Tim Horton Children's Ranch, Telus Spark Science Centre and Calgary Reads.
- *Extended, integrated practicum and school partnerships:* Courses are taught in schools for teacher candidates in the final practicum semester. This follows a Professional Development School model. Cohorts from 4-6 schools gather together at one school each week.
- *Faculty practicum supervision in schools*: All full-time faculty members supervise practicum in schools. This practice has been forefront from the inception of the degree. Full-time faculty are ideally positioned to build partnerships with schools, to intervene in challenging situations, and to bring awareness of the program vision and structure. Most contract faculty who are practicum supervisors are also instructors in courses so they have a strong understanding of the program and teacher candidates' course experiences. Schools repeatedly praise the regular presence and responsiveness of faculty members.
- *Peer mentorship:* teacher candidates are encouraged to provide peer mentorship to each other during practicum and field experiences. teacher candidates in year 2 are peer mentors for teacher candidates in year 1.
- *Inquiry:* Teacher candidates in their final practicum engage in a capstone inquiry in their school (EDUC 4201). This is either a curriculum inquiry with their practicum classes or an inquiry into some aspect of professional practice.
- *Reflective journals:* Teacher candidates write journal reflections in courses throughout each year of the program and include these in their professional learning plans.
- *Indigenous ways of knowing:* Place-based experiential learning opportunities take place with our Indigenous partners (e.g., Stoney Nakoda Reserve).

These transformative pedagogies seek to avoid the traditional separation and gap between theory and practice by not disintegrating them in the first place. They create a hybrid structure and organization that integrates "course work" and "student teaching," by blending and overlapping traditionally separate roles of "instructor," "supervisor," and "mentor teacher." Our work is to support teacher candidates' development of professional teaching competencies through both immersion in experience and reflection on that experience, particularly through the development of the STEAM semester as an integrating targeted experience.

Since our four-year teacher education program is in the midst of implementation, we are uniquely positioned to do self-study research. Our methodology is designed to address our research questions: 1) What do we learn as STEAM educators through the design and implementation of field studies that link Indigenous knowledge systems and place-based education? and 2) What are the teacher candidates' experiences of STEAM field studies framed by a curriculum of place?

METHODS

Drawing on the research involving self-study as a methodology for studying professional practice settings (Pinnegar, 1998), program improvement (Kosnick et al., 2006), and teacher education (Kitchen & Russell, 2012) and based on principles of self-study design (Dinkelman, 2003; LaBoskey, 2004), this research was self-initiated, focused on inquiry into our practice, collaborative, aimed at improvement of our practice, and using multiple and primarily qualitative means of inquiry.

Researchers have identified the need to decolonize research methodologies (Tuhiwai Smith, 2012) and offer insight into appropriate methods, protocols, and ethical responsibilities for Indigenous research (Wilson, 2007). Kovach (2009) presents a methodological approach that centers on a tribal epistemology. Her non-linear description of its characteristics include researcher preparation, decolonizing and ethics, gathering knowledge, making meaning, and giving back (p. 45). We drew on this research framework and culturally relational research methods (Donald, Glanfield, & Sterenberg, 2012), and Blackfoot concepts of *aoksisawaato'p* (visiting/renewal of relations), *aokakiosiit* (be wisely aware; pay attention), and *aatsimaak'ssin* (responsibility to balance giving/taking reciprocity) to inform our methods of community dialogues and sharing circles.

We have designed a STEAM semester for our teacher education program that utilizes Shulman's (2006) concept of signature pedagogies and Kuh's (2008) high impact practices (HIPs). Our STEAM semester focuses specifically on the following transformative pedagogical approaches:

1. *Cross-curricular course integration:* Teacher candidates learn through 'hands-on" experience and assessment feedback how to integrate the outcomes from Science, Educational Technology, Visual Arts, and Mathematics in their lesson and unit planning.
2. *Cohort placements and in-school seminars:* There is a 5 week practicum during this semester and our teacher candidates are placed in cohorts of 4-6 at each school. The cohort placements support weekly in-school seminars, which are co-facilitated by mentor teachers and school administrators.

3. *Community partnerships:* Teacher candidates plan and support learning at various community sites including the Ann and Sandy Cross Conservation Area, Calgary Open Minds, Tim Horton Children's Ranch and Telus Spark Science Centre.
4. *Faculty practicum supervision in schools*: All full-time faculty members involved in the STEAM semester supervise the practicum in schools. Full-time faculty are ideally positioned to build partnerships with schools, to intervene in challenging situations, and to bring awareness of the program vision and structure.
5. *Peer mentorship:* Teacher candidates are encouraged to provide peer mentorship to each other during the STEAM courses and practicum experience.
6. *Inquiry:* Teacher candidates engage in a semester long inquiry project in partnership with their mentor teacher.
7. *Reflective journals:* Teacher candidates write journal reflections throughout the semester in order to connect the theory of their courses with the practice of their practicum experience.
8. *Professional learning plan* (e-portfolio): Teacher candidates use a digital learning plan to document their growth and development throughout the semester with respect to the B.Ed. programs core competencies; planning, facilitation, assessment, environment, and professional responsibilities.
9. *Indigenous ways of knowing:* Place-based experiential learning opportunities with our First Nations partners such as the Tsuu T'ina, Stoney Nakoda, Kainai and the Siksika First Nations.

In this self- study, we investigated the possibilities for deliberate place-based pedagogical interventions.

Through a social-constructivist lens, in-school seminars that are integrated weekly within a 5-week practicum, each involving cohorts of 8-12 teacher candidates from partner schools, are facilitated by teacher educators. This is intended to develop a sense of community (Lave & Wenger, 1991; Wenger, 1998) and support teacher candidates to reflect upon their teaching and develop responsive educational practices and adaptive expertise (Beck & Kosnik, 2006; Loughran, 2002). Prior to the 5-week practicum, teacher candidates participate in STEAM courses that include daylong field studies and a 3-day intensive experiential STEAM program at a remote mountain ranch focused on environmental science land-based pedagogies. While at the mountain ranch, they also lead a daylong place-based STEAM inquiry project with cohorts of Grade 4 students from a partner school. In this self- study, we investigated the possibilities for deliberate place-based pedagogical interventions.

Data Sources

The participants in this study included the three teacher educators, a Blackfoot Elder (who acted as our critical friend), three undergraduate student research assistants and sixty-three teacher candidates. Together, we piloted integrated 7-week intensive STEAM courses in coordination with teacher candidates' practicum experiences and field studies. As researchers, we engaged in bi-monthly collaborative research conversations, exchanged numerous emails, and kept research notes about our experiences. Mid-way through the course we interviewed 9 teacher candidates and at the conclusion of the year, we interviewed 11 teacher candidates.

Multiple data sources provided trustworthiness as experiences were explicitly documented and analyzed by the researchers in various forms and sites. Data was first coded individually across these sites according to emerging themes that related to our research focus on the process of designing and implementing a curriculum of place (Strauss & Corbin, 1998). We then reviewed our analysis, collaboratively adjusted the codes, and wrote findings together.

RESULTS

Our study involved cross-sector co-creation of knowledge and understanding through engagement with Indigenous, not-for-profit, professional and academic communities. As researchers, our disciplinary and interdisciplinary relationships emphasized the integration of Indigenous and non-Indigenous knowledge systems in a STEAM integrated practicum semester. The results of this study represent a preliminary project framed around a curriculum of place. Here, we present five themes emerging from our 2015-2016 data that continue to inform our understanding of learning from place: place as pedagogy, integration through place, the importance of reflection, resistance, and reconsidering place.

Place as Pedagogy

During the planning stages of the STEAM field studies, Sterenberg reflected on her own experiences of place and how her view of mathematics has shifted because of it:

I have grappled with integrating my notions of place and mathematics education for many years...In this study, I wanted to investigate how 'place' impacted how I taught a STEAM-focused, community and field-based course. I wonder how my experiences of the land and home can be brought forth in teaching mathematics. (Sterenberg, Reflection)

As educators, we were struck by the increase in engagement level demonstrated by our teacher candidates and the K-9 students. Not only were they excited and passionate to discuss and engage in the studies, they brought a heightened level of professional conduct and engagement and expectations that required us, as facilitators to "raise our game".

There also was enhanced ability of the teacher candidates to draw on their field studies during their practicum teaching as a result of the STEAM programming.

It was going out into the field that helped me connect theory and practice because I began to realize that it doesn't have to be direct teaching. (Teacher Candidate, Interview).

I loved the integration of the [field study] experience. I thought that was very valuable. It gave us a first-hand view of inquiry learning. It gave us a first-hand experience with kids and that it is okay that it is a little bit messy...I hadn't experienced that before. (Teacher Candidate, Interview).

Integration Through Place

The interconnectedness and the understanding of the relation of things, which is a key component to both Indigenous ways of knowing and place-based education, became a fundamental component of the STEAM program design.

We developed this STEM model based on the multidisciplinary element often espoused through Indigenous ways of knowing; the notion of interconnectedness is essential. It promotes a relational aspect to knowledge... My biggest struggle, when trying to assist my students in seeing these connections is how to be authentic and respectful, especially when I am unable to facilitate an Indigenous voice in the field (O'Connor, Reflection)

Despite the dissonance that we were experiencing, some of our students seemed to understand the importance of integration, as shown by this representative quote:

It's easier to do cross-curricular work. Kids are so engaged and it's real learning and it can even inspire them for their future careers or even get them involved in other areas. So it's not just that you sit and do your math, it's you do math as you're doing water quality testing...I feel like everything just goes together well. (Teacher Candidate, Interview)

The cross-curricular focus of the STEAM program seemed to create a deeper understanding of science, mathematics, educational technologies and visual arts through the integrated experiences and assignments.

It forced you to do that reflective piece…and I have boxes in my brain. I have a Language Arts box and I am starting to realize that I can't just have boxes because things don't fit into one or the other, you can have bits of everything. (Teacher Candidate, Interview)

Other teacher candidates spoke about the challenge to integrate disciplines becoming easier because of attention to student learning:

At first, it seems a little silly because we are, 'Oh we have to integrate all this stuff and this is really difficult,' but if you look at a classroom you are integrating everything all day long (Teacher Candidate, Interview).

Importance of Reflection

The teacher candidates acknowledge the impact of reflection facilitated through the in-school seminars. Faculty supported teacher candidates to reflect upon, or learn from, their immersion in experience.

I feel that reflection time was huge, and that is something that has really helped making those ties between theory and practice…I found we would naturally start talking about it, but I feel if it were prompted a bit, too, it also really helped, like bringing up some of the topics that we worked through, reflected on, and gave examples. I think that is one of the biggest things, having that time to make those connections when they don't occur to me or to one of my peers. Sometimes somebody else brought it up and it made a lot more sense because I had a very similar example that hadn't occurred to me that also applied to that scenario, or hadn't occurred to me that I could use, that actually occurred to a kid, or they came up with something I hadn't thought of. (Teacher Candidate, Interview)

Resistance

We often ran into resistance by teacher candidates, colleagues and other participants, as this is often not a conventional or comfortable approach to teaching and learning. We even questioned our own commitment to this process based on the academic, professional and personal demands of our time, as shown by Sterenberg's writing:

I've just returned from teaching a 3-day intensive experiential STEM program at a remote ranch in the mountains. I've taken time away from my family, haven't slept well, and am now doing laundry because of a possible bed bug infestation. Is this worth it? What have my students experienced in this place-based curriculum of mathematics? (Sterenberg, Reflection)

Many of the teacher candidates felt the experience in the field was too demanding and long (i.e. 2 nights/3 days). As educators, passionate about place-based pedagogies and an integrated STEAM model, we discussed the value of experience and resiliency.

I still remember reading Ted Aoki's writings of a Japanese Canadian, who stressed the need for those key times to learn as those "moments of tension", when we are no longer experts of those microworlds and we become beginners who search for ways to understand what is foreign (through deliberation and analysis) so that we may feel comfortable and at ease with the task at hand. This speaks to me through experiential and place-based learning, by the inclusion of adversity and resilience into curricular delivery and content. (O'Connor, Reflection)

Our school and community partners are helping us respond to resistance and think more deeply about embedding and sustaining a curriculum of place.

Reconsidering Place

We all felt that our attempts to enact a curriculum of place that recognizes the intimate relationship that Indigenous people have with the land and emphasizes relational ways of knowing were impactful. However, we were disappointed that our students seemed to experience a place-based curriculum that was not linked explicitly to Indigenous ways of knowing.

A gap that is emerging is that most students are not making the link between Indigenous ways of knowing place and the importance of land-based pedagogies as something they, as a non-Indigenous educator, can incorporate into their teaching and learning. It is a perception of the "other" that seems to create dissonance. How can I help promote a culture of knowledge creation for these students? One that allows them to incorporate varying perspectives of place, including an Indigenous curriculum of place. (O'Connor, Reflection)

DISCUSSION

As educators, we were struck by the increase in engagement level demonstrated by our teacher candidates. The teacher educators spoke often of how we felt like "true" facilitators, as we would consistently be mediating the educational needs of our students (as defined by our course requirements) with the opportunities arising through community engagement and place-based field studies and data collection. The interconnectedness and the understanding of the relation of things became a fundamental component of STEAM design. The teacher candidates spoke of the difficulties of enacting their theories in practice (e.g., inquiry and field studies) and acknowledged the complexities associated with such alternative pedagogies.

In the next year of teaching, we intend to begin the STEAM semester in ceremony, learning closely with one of our partners teaching in a reserve school located in close proximity to the remote ranch. We will engage more closely within the First Nation school communities and incorporate more clinical placements and place-based field studies. We also intend to better scaffold the cross-curricular integration by modeling, reflecting and embedding in clinical settings. Lastly, we are designing a revised curriculum of readings and prompts for critical discussion for in-school seminars that better reflect the context and experience of the semester.

CONCLUSION

Throughout this study, we have wondered how we can be nourished by the place and how we can nourish it. We anticipate that we will engage in stories of the land and that our design of field studies will invite students to dwell in the place and to experience all dimensions of the place, as we look to shifts in identity needed to authentically experience a curriculum of place.

We suggest that this research has the potential to inform future STEAM teacher education courses and projects including the creation of relevant Indigenous curricula for post-secondary institutions, K-12 schools and not-for-profit organizations. This research distinguishes itself in originality from other research because it investigates new ways of co-creating knowledge about integrating Indigenous knowledge systems and place-based curriculum in Indigenous nations, schools, universities, and not-for-profit organizations and it fosters robust relationships within our community. Further investigations into children's learning in a context that integrates Indigenous knowledges and place-based curriculum shapes our longitudinal research.

REFERENCES

Ballantyne, E. F. (2014). Dechinta Bush University: Mobilizing a knowledge economy of reciprocity, resurgence and decolonization. *Decolonization, 3*(3), 67–85.

Bastien, B. (2004). *Blackfoot ways of knowing*. Calgary, Canada: University of Calgary Press.

Battiste, M. (2002). *Indigenous knowledge and pedagogy in First Nations education: A literature review with recommendations*. Ottawa, Canada: Indian and Northern Affairs.

Beck, C., & Kosnik, C. (2001). Reflection-in-action: In defence of thoughtful thinking. *Curriculum Inquiry, 31*(2), 217–227. doi:10.1111/0362-6784.00193

Beck, C., & Kosnik, C. (2006). *Innovations in preservice teacher education: A social constructivist approach*. Albany, NY: State University of New York Press.

Blood, N. & Chambers, C. (Producers/Writers/Directors). (2006). "*Kaaáhsinnooniksi: If the land could speak...*" [Digital-video documentary]. (Produced in cooperation with Alberta Community Development Branch and the University of Lethbridge.)

Blood, N., Chambers, C., Donald, D., Hasebe-Ludt, E., & Big Head, R. (2012). *Aoksisowaato'op:* Place and story as organic curriculum. In N. Ng-A-Fook & J. Rottman (Eds.), *Reconsidering Canadian curriculum studies* (pp. 47–82). New York, NY: Palgrave Macmillan. doi:10.1057/9781137008978_4

Cajete, G. (1999). *Igniting the sparkle: An Indigenous science education model*. Skyland, NC: Kivaki Press.

Cajete, G. (2000). *Native science: Natural laws of interdependence*. Santa Fe, NM: Clearlight Publishers.

Calderon, D. (2014). Speaking back to manifest destinies: A land education-based approach to critical curriculum inquiry. *Environmental Education Research, 20*(1), 24–36. doi:10.1080/13504622.2013.865114

Carbaugh, D. (1999). "Just listen": "Listening" and landscape among the blackfeet. *Western Journal of Communication, 63*(3), 250–270. doi:10.1080/10570319909374641

Carnegie forum on education and the economy, task force on teaching as a profession. (1986). *A nation prepared: Teachers for the 21st century*. New York: Carnegie Corporation of New York.

Chambers, C. (2006). "The land is the best teacher I have ever had": Places as pedagogy for precarious times. *Journal of Curriculum Theorizing, 22*(3), 27–37.

Chambers, C. (2008). Where are we? Finding Common Ground in a Curriculum of Place. *Journal of the Canadian Association for Curriculum Studies*, *6*(2), 113–128.

Clift, R., & Brady, P. (2005). Research on methods courses and field experiences. In M. Cochran-Smith & K. Zeichner (Eds.), *Studying teacher education: The report of the AERA panel on research and teacher education* (pp. 309–424). Washington, DC: American Education Research Association / Lawrence Earlbaum.

Darling-Hammond, D. (2006). *Powerful teacher education*. San Francisco, CA: Jossey-Bass.

Denise, P. S., & Harris, I. M. (Eds.). (1989). *Experiential education for community development*. Westport, CT: Greenwood Press.

Dillon, D., Bullock, S., O'Connor, K., Russell, T., Martin, A., & Thomas, L. (2013). Place-based Teacher Development: Placing Practicum Learning at the Heart of Pre-service Teacher Education. In L. Thomas (Ed.), *Becoming teacher: Sites for teacher development* (pp. 94-120). Canadian Association for Teacher Education/Association canadienne pour la formation à'enseignement. Retrieved from https://docs.google.com/file/d/0BwVGDOGBDzJdR1R2VGZvakZjUDQ/edit

Dinkelman, T. (2003). Self-study in teacher education: A means and ends tool for promoting reflective teaching. *Journal of Teacher Education*, *54*(1), 6–18. doi:10.1177/0022487102238654

Donald, D., Glanfield, F., & Sterenberg, G. (2012). Living ethically within conflicts of colonial authority and relationality. *Journal of the Canadian Association for Curriculum Studies*, *10*(1), 53–77.

Emekauwa, E. (2004). *The star with my name: The Alaska rural systemic initiative and the impact of place-based education on Native student achievement. The case for place-based. Rural trust white paper on place-based education.* Washington, DC: The Rural School and Community Trust.

Freire, P. (1970). *Pedagogy of the oppressed*. New York: Continuum.

Goodlad, J. (1988). School-university partnerships: Rationale and concepts. In K. Sirotnik & J. Goodlad (Eds.), *School-university partnerships in action: Concepts, cases, and concerns* (pp. 3–31). New York: Teachers College Press.

Grande, S. (2004). *Red pedagogy: Native American social and political thought*. Lanham, MD: Rowman & Littlefield Publishers Inc.

Gruenewald, D. (2003). The best of both worlds: A critical pedagogy of place. *Educational Researcher*, *32*(4), 3–12. doi:10.3102/0013189X032004003

Holmes Group. (1986). *Tomorrow's teachers: A report of the Holmes Group*. East Lansing, MI: Author.

Kincheloe, J. (2001). *Getting beyond the facts: Teaching social studies/social sciences in the twenty-first century*. New York: Peter Lang.

Kincheloe, J. (2005). *Critical pedagogy*. New York: Peter Lang.

Kincheloe, J. L., Mckinley, E., Lim, M., & Calabrese Barton, A. (2006). Forum: A conversation on 'sense of place' in science learning. *Cultural Studies of Science Education, 1*(1), 143–160. doi:10.100711422-005-9003-8

Kissling, M. T. (2012). A living curriculum of place(s). *Journal of Curriculum Theorizing, 28*(3), 109–127.

Kitchen, J., & Russell, T. (Eds.). (2012). *Canadian perspectives on the self-study of teacher education practices*. Ottawa, Canada: Canadian Association of Teacher Education.

Knapp, C. E. (1996). Just beyond the classroom: Community adventures for interdisciplinary learning. Charleston, WV: ERIC Clearinghouse on Rural Education and Small Schools. (ERIC Document Reproduction Service NO. ED 388485)

Korthagen, F., Loughran, J., & Russell, T. (2006). Developing fundamental principles for teacher education programs and practices. *Teaching and Teacher Education, 22*(8), 1020–1041. doi:10.1016/j.tate.2006.04.022

Kosnik, C., Freese, A., Samaras, A., & Beck, C. (Eds.). (2006). Making a difference in teacher education through self-study: Studies of personal, professional, and program renewal. Dordrecht: Springer Academic Publishers.

Kovach, M. (2009). *Indigenous methodologies: Characteristics, conversations, and contexts*. Toronto: University of Toronto Press.

Kulnieks, A., Longboat, D., & Young, K. (Eds.). (2013). *Contemporary studies in environmental and indigenous pedagogies: A curricula of stories and place*. Rotterdam, The Netherlands: Sense Publishers. doi:10.1007/978-94-6209-293-8

LaBoskey, V. K. (2004). The methodology of self-study and its theoretical underpinnings. In J. J. Loughran, M. L. Hamilton, V. K. LaBoskey, & T. Russell (Eds.), *International handbook of self-study of teaching and teacher education practices* (pp. 817–869). Dordrecht: Kluwer. doi:10.1007/978-1-4020-6545-3_21

Lave, J., & Wenger, E. (1991). *Situated learning: Legitimate peripheral participation*. Cambridge, UK: Cambridge University Press. doi:10.1017/CBO9780511815355

Little Bear, L. (2000). Jagged worldviews colliding. In M. Battiste (Ed.), *Reclaiming Indigenous voice and vision* (pp. 78–85). Vancouver, Canada: University of British Columbia Press.

Lortie, D. (1975). *Schoolteacher: A sociological study*. Chicago: University of Chicago Press.

Loughran, J. J. (2002). Effective reflective practice: In search of meaning in learning about teaching. *Journal of Teacher Education, 53*(1), 33 43.

Munby, H., & Russell, T. (1994). The authority of experience in learning to teach: Messages from a physics methods class. *Journal of Teacher Education, 45*(2), 86–95. doi:10.1177/0022487194045002002

Orr, D. W. (1994). *Earth in mind: On education, environment, and the human prospect*. Washington, DC: Island Press.

Penetito, W. (2009). Place-based education: Catering for curriculum, culture and community. *New Zealand Annual Review of Education, 18*, 5–29.

Pinnegar, S. (1998). Introduction to Part II: Methodological perspectives. In M. L. Hamilton (Ed.), *Reconceptualizing teaching practice: Self-study in teacher education* (pp. 31–33). London: Falmer Press.

Raelin, J. A. (2007). Toward an epistemology of practice. *Academy of Management Learning & Education, 6*(4), 495–519. doi:10.5465/amle.2007.27694950

Raffan, J. (1993). The experience of place: Exploring land as teacher. *Journal of Experiential Education, 16*(1), 39–45. doi:10.1177/105382599301600109

Raffan, J. (1995). Experiential education and teacher education. *Journal of Experiential Education, 18*(3), 117–119. doi:10.1177/105382599501800301

Relph, E. (1992). Modernity and the reclamation of place. In D. Seamon (Ed.), *Dwelling, Seeing, and Designing: Toward a Phenomenological Ecology* (pp. 25–40). New York: State University of New York Press.

Schön, D. A. (1987). *Educating the reflective practitioner: Toward a new design for teaching and learning in the professions*. San Francisco, CA: Jossey-Bass.

Segall, A. (2002). *Disturbing practice: Reading teacher education as text*. New York: Peter Lang.

Smith, G. A. (2007). Place-based education: Breaking through the constraining regularities of public school. *Environmental Education Research, 13*(2), 189–207. doi:10.1080/13504620701285180

Sobel, D. (2004). *Place-Based Education: Connecting Classrooms and Communities*. Great Barrington, MA: The Orion Society.

Strauss, A., & Corbin, J. (1998). *Basics of qualitative research: Grounded theory procedures and techniques* (2nd ed.). Thousand Oaks, CA: Sage.

Teitel, L. (1998). Separations, divorces, and open marriages in professional development school partnerships. *Journal of Teacher Education, 49*(2), 85–96. doi:10.1177/0022487198049002002

Theobald, P., & Curtiss, J. (2000). Communities as curricula. *Forum for Applied Research and Public Policy, 15*(1), 106–111.

Truth and Reconciliation Commission of Canada. (2012). *Truth and reconciliation commission: Calls to action*. Retrieved May 12, 2016 from http://www.trc.ca/websites/trcinstitution/File/2015/Findings/Calls_to_Action_English2.pdf

Tuhiwai Smith, L. (2012). *Decolonizing methodologies: Research and Indigenous peoples* (2nd ed.). London, UK: Zed Books.

United Nations. (2008). *United Nations declaration on the rights of Indigenous peoples*. Retrieved May 12, 2016 from http://www.un.org/esa/socdev/unpfii/documents/DRIPS_en.pdf

van Eijck, M. W., & Roth, W.-M. (2009). Towards a chronotopic theory of "place" in place-based education. *Cultural Studies of Science Education, 5*(4), 869–898. doi:10.100711422-010-9278-2

Wattchow, B., & Brown, M. (2011). *A pedagogy of place: Outdoor education for a changing world*. Clayton, Australia: Monash University Publishing.

Wenger, E. (1998). *Communities of practice*. Cambridge, UK: Cambridge University Press. doi:10.1017/CBO9780511803932

Wideen, M., Mayer-Smith, J., & Moon, B. (1998). A critical analysis of the research on learning to teach: Making the case for an ecological perspective on inquiry. *Review of Educational Research, 68*(2), 130–178. doi:10.3102/00346543068002130

Wildcat, M., McDonald, M., Irlbacher-Fox, S., & Coulthard, G. (2014). Learning from the land: Indigenous land-based pedagogy and decolonization. *Decolonization, 3*(3), i–xv.

Wilson, S. (2007). Guest editorial: What is an Indigenist research paradigm? *Canadian Journal of Native Education, 30*(2), 193–195.

Woodhouse, J. L., & Knapp, C. E. (2000). *Place-based curriculum and instruction: Outdoor and environmental education approaches*. Charleston, WV: ERIC Clearinghouse on Rural Education and Small Schools.

Zeichner, K., & Tabatchnik, B. (1981). Are the effects of university teacher education "washed out" by school experiences? *Journal of Teacher Education*, *32*(3), 7–11. doi:10.1177/002248718103200302

KEY TERMS AND DEFINITIONS

Decolonization: Involves valuing and revitalizing Indigenous knowledge and approaches and weeding out settler biases or assumptions that have impacted Indigenous ways of being. For non-Indigenous people, decolonization is the process of examining your beliefs about Indigenous Peoples and culture by learning about yourself in relationship to the communities where you live and the people with whom you interact.

Experiential: Process of learning through experience and is more specifically defined as "learning through reflection on doing."

Indigenization: A collaborative process of naturalizing Indigenous intent, interactions, and processes and making them evident to transform spaces, places, and hearts. In the context of post-secondary education, this involves including Indigenous perspectives and approaches.

Inquiry: Involves developing questions, making observations, doing research to find out what information is already recorded, developing methods for experiments, developing instruments for data collection, collecting, analyzing, and interpreting data, outlining possible explanations and creating predictions for future study.

Place-Based Education: Place-based education is interdisciplinary, it promotes learning that is rooted in what is local—the unique history, environment, culture, economy, literature, and art of a particular place.

Reconciliation: Is about addressing past wrongs done to Indigenous Peoples, making amends, and improving relationships between Indigenous and non-Indigenous people to create a better future for all.

STEAM: Is an approach to learning that uses science, educational technology, engineering, the arts, and mathematics as access points for guiding student inquiry, dialogue, and critical thinking.

Theory-Practice: The integration between the individual and society, art and science, field and classroom; connecting different aspects of the same phenomenon and not viewing these aspects in opposition to each other.

Transformative Pedagogies: An activist pedagogy combining the elements of constructivist and critical pedagogy that empowers students to examine critically their beliefs, values, and knowledge with the goal of developing a reflective knowledge base, an appreciation for multiple perspectives, and a sense of critical consciousness and agency.

Compilation of References

Adetunji, O. O. (2019). *SciToons Behind the Scenes*. Presentation, Brown University, Providence, RI. Retrieved July 19, 2019, from https://repository.library.brown.edu/studio/item/bdr:934244/

Adetunji, O. O. (2015). UMRI Alumni: From Characterization of High Temperature Solution Growth Cr 2+:CdSe to Developing Videos to Enhance Metacognition and Diversity in Materials Science and STEM (Science, Technology, Engineering, and Mathematics). *MRS Online Proceedings Library Archive*, *1762*. doi:10.1557/opl.2015.188

Adetunji, O. O., & Levine, R. (2015). Developing Effective STEM Animations: Application of a Multimedia Learning Theoretical Framework. *Journal of Research in STEM Education*, *1*(20), 106–124.

Adetunji, O. O., & Levine, R. (2016). Toward a New Model of Science Learning, Teaching, and Communication. *MRS Advances*, *1*(56), 3709–3714. doi:10.1557/adv.2016.105

Ahn, C. (2015). Ecoscience + art initiative: Designing a new paradigm for college education, scholarship, and service. *The STEAM Journal*, *2*(1), 1–10. doi:10.5642team.20150201.11

Akkerman, S. F., & Bakker, A. (2011). Learning at the boundary: An introduction. *International Journal of Educational Research*, *50*(1), 1–5. doi:10.1016/j.ijer.2011.04.002

Al-Balushi, S. M., & Al-Hajri, S. H. (2014). Associating animations with concrete models to enhance students' comprehension of different visual representations in organic chemistry. *Chemistry Education Research and Practice*, *15*(1), 47–58. doi:10.1039/c3rp00074e

Alhadad, S. S. J., & Thompson, K. (2017). Understanding the mediating role of teacher inquiry when connecting learning analytics with design for learning. *Interaction, Design, & Architecture (Washington, D.C.)*, *33*, 54–74. Retrieved from http://www.mifav.uniroma2.it/inevent/events/idea2010/doc/33_3.pdf

Allina, B. (2018). The development of STEAM educational policy to promote student creativity and social empowerment. *Arts Education Policy Review*, *119*(2), 77–87. doi:10.1080/10632913.2017.1296392

Alterator, S., & Deed, C. (2013). Teacher adaptation to open learning spaces. *Issues in Educational Research*, *23*(3), 315.

Alterator, S., & Deed, C. (2016). Reacting to 'irregular' learning environments in a senior secondary school. *Curriculum and Teaching, 31*(2), 47–69. doi:10.7459/ct/31.2.04

American Academy of Art and Sciences Transdisciplinary Opportunities. (2013). ARISE 2: Advancing Research in Science and Engineering: Unleashing America's Research & Innovation Enterprise. Cambridge: American Academy of Art and Sciences.

American Institute of Architects. (2018, July 26). *Career and Technical Education bill to recognize Architecture as STEM*. Retrieved from https://www.aia.org/articles/206266-career-and-technical-education-bill-to-reco

Americans With Disabilities Act of 1990, Pub. L. No. 101-336, 104 Stat. 328 (1990).

Andrade, H., Du, Y., & Mycek, K. (2010). Rubric-referenced self-assessment and middle school students' writing. *Assessment in Education: Principles, Policy & Practice, 17*(2), 199–214. doi:10.1080/09695941003696172

Ballantyne, E. F. (2014). Dechinta Bush University: Mobilizing a knowledge economy of reciprocity, resurgence and decolonization. *Decolonization, 3*(3), 67–85.

Ball, K. M. (2013). Meet Your meat: The argument for increasing education and public outreach regarding the regulation and safety of animal biotechnology. *Food and Drug Law Journal, 68*(3), 281–307. PMID:24640611

Bancino, R., & Zevalkink, C. (2007). Soft Skills: The New Curriculum for Hard-Core Technical Professionals. *Techniques: Connecting Education and Careers (J1), 82*(5), 20-22.

Banilower, E. R., Smith, P. S., Malzahn, K. A., Plumley, C. L., Gordon, E. M., & Hayes, M. L. (2018). *Report of the 2018 NSSME*. Chapel Hill, NC: Horizon Research, Inc.

Barak, M., Ashkar, T., & Dori, Y. J. (2011). Learning science via animated movies: Its effect on students' thinking and motivation. *Computers & Education, 56*(3), 839–846. doi:10.1016/j.compedu.2010.10.02

Barton, A. C., Tan, E., & Greenberg, D. (2017). The Makerspace Movement: Sites of Possibilities for Equitable Opportunities to Engage Underrepresented Youth in STEM. Teachers College Record, 119.

Basham, J. D., & Blackorby, J. (in press). UDL next: The future of the framework. In K. Lowrey (Ed.), *Critical Issues in Universal Design for Learning*. Oviedo, FL: Knowledge by Design.

Basham, J. D., Hall, T. E., Carter, R. A. Jr, & Stahl, W. M. (2016). An operationalized understanding of personalized learning. *Journal of Special Education Technology, 31*(3), 126–136. doi:10.1177/0162643416660835

Basham, J. D., Israel, M., & Maynard, K. (2010). An ecological model of STEM education: Operationalizing STEM for all. *Journal of Special Education Technology, 25*(3), 9–19. doi:10.1177/016264341002500303

Basham, J. D., & Marino, M. T. (2013). Understanding STEM education and supporting students through universal design for learning. *Teaching Exceptional Children, 45*(4), 8–15. doi:10.1177/004005991304500401

Basham, J. D., Meyer, H., & Perry, E. (2010). The design and application of the digital backpack. *Journal of Research on Technology in Education, 42*(4), 339–359. doi:10.1080/15391523.2010.10782555

Bastien, B. (2004). *Blackfoot ways of knowing*. Calgary, Canada: University of Calgary Press.

Battiste, M. (2002). *Indigenous knowledge and pedagogy in First Nations education: A literature review with recommendations*. Ottawa, Canada: Indian and Northern Affairs.

Beck, C., & Kosnik, C. (2001). Reflection-in-action: In defence of thoughtful thinking. *Curriculum Inquiry, 31*(2), 217–227. doi:10.1111/0362-6784.00193

Beck, C., & Kosnik, C. (2006). *Innovations in preservice teacher education: A social constructivist approach*. Albany, NY: State University of New York Press.

Becker, H. L., & Anderson, R. E. (1998). Teacher's Survey: Combined Versions 1-4. In Teacher Professional Engagement and Constructivist-Compatible Computer Use. Teaching, Learning, and Computing: 1998 National Survey. Irvine, CA: Center for Research on Information Technology and Organizations.

Becker, H. J., & Riel, M. M. (2000). Teacher Professional Engagement and Constructivist-Compatible Computer Use. In *Teaching, Learning, and Computing: 1998 National Survey. Report #7*. Irvine, CA: Center for Research on Information Technology and Organizations.

Begg, M. D., & Vaughan, R. D. (2011). Are Biostatistics Students Prepared to Succeed in the Era of Interdisciplinary Science? (And How Will We Know?). *The American Statistician, 65*(2), 71–79. doi:10.1198/tast.2011.10222

Beghetto, R. A. (2006). *Creative self-efficacy: Correlates in middle and secondary students.* Academic Press.

Benade, L. (2017). Is the classroom obsolete in the twenty-first century? *Educational Philosophy and Theory, 49*(8), 796–807. doi:10.1080/00131857.2016.1269631

Bevan, B. (2017). The promise and the promises of Making in science education. *Studies in Science Education, 53*(1), 75–103. doi:10.1080/03057267.2016.1275380

Black, P., & Wiliam, D. (2009). Developing the theory of formative assessment. *Educational Assessment, Evaluation and Accountability, 21*(1), 5–31. doi:10.100711092-008-9068-5

Blikstein, P. (2019). Foreword. Beyond mindsets, cultures, brands, and clichés: a possible future for equitable maker education. In P. Blikstein, S. L. Martinez, H. A. Pang, & K. Jarrett (Eds.), *Meaningful Making 2 – projects and inspirations for Fab Labs and makerspaces*. Torrance, CA: Constructing Modern Knowledge Press.

Blood, N. & Chambers, C. (Producers/Writers/Directors). (2006). "*Kaaáhsinnooniksi: If the land could speak...*" [Digital-video documentary]. (Produced in cooperation with Alberta Community Development Branch and the University of Lethbridge.)

Blood, N., Chambers, C., Donald, D., Hasebe-Ludt, E., & Big Head, R. (2012). *Aoksisowaato'op:* Place and story as organic curriculum. In N. Ng-A-Fook & J. Rottman (Eds.), *Reconsidering Canadian curriculum studies* (pp. 47–82). New York, NY: Palgrave Macmillan. doi:10.1057/9781137008978_4

Blumenfeld, P., Soloway, E., Marx, R., Krajcik, J., Guzdial, M., & Palincsar, A. (1991). Motivating project-based learning: Sustaining the doing, supporting the learning. *Educational Psychologist, 26*(3-4), 369–398. doi:10.1080/00461520.1991.9653139

Borrego, M., & Newswander, L. K. (2010). Definitions of Interdisciplinary Research: Toward Graduate-Level Interdisciplinary Learning Outcomes. *Review of Higher Education, 34*(1), 61–84. doi:10.1353/rhe.2010.0006

Boud, D. (1989). The role of self-assessment in student grading. *Assessment in Higher Education, 14*(1), 20–30. doi:10.1080/0260293890140103

Boud, D. (1995). *Enhancing learning through self-assessment.* London: Kogan Page.

Boud, D., & Falchikov, N. (1989). Quantitative studies of self-assessment in higher education: A critical analysis of findings. *Higher Education, 18*(5), 529–549. doi:10.1007/BF00138746

Breiner, J. M., Harkness, S. S., Johnson, C. C., & Koehler, C. M. (2012). What is STEM? A discussion about conceptions of STEM in education and partnerships. *School Science and Mathematics, 112*(1), 3–11. doi:10.1111/j.1949-8594.2011.00109.x

Bringuier, J.-C. (1980). *Conversations with Jean Piaget* (B. M. Gulati, Trans.). Chicago: The University of Chicago Press. doi:10.1007/BF02694634

Brown, B. A., Mangram, C., Sun, K., Cross, K., & Raab, E. (2017). Representing racial identity: Identity, Race, the construction of the African American STEM students. *Urban Education, 52*(2), 170–206. doi:10.1177/0042085916661385

Brown, D. C., Kemp, M. C., & Hall, J. (1998). On teaching biotechnology in Kentucky. *Journal of Industrial Teacher Education, 36*(4), 44–60.

Buchanan, R. (1992). Wicked problems in design thinking. *Design Issues, 8*(2), 5–21. doi:10.2307/1511637

Burk, C., Leahy, M., McGovern, J., Ramler, P., & Royer, D. (2015). *ARTspace summer camp program evaluation plan (Unpublished evaluation plan).* Lawrence, KS: University of Kansas.

Burks, L. A., & Huffman, D. (2019). Pre-Service Teachers' Critical Thinking and Developing and Using Models in Science. In S. Robinson & V. Knight (Eds.), *Handbook of Research on Critical Thinking and Teacher Education Pedagogy* (pp. 194–212). Hershey, PA: IGI Global; doi:10.4018/978-1-5225-7829-1.ch011.

Burnaford, G., Brown, S., Doherty, J., & McLaughlin, H. J. (2007). *Arts Integration Frameworks, Research Practice*. Washington, DC: Arts Education Partnership.

Burton, R. R., Brown, J. S., & Fischer, G. (1984). Skiing as a model of instruction. In B. Rogoff & J. Lave (Eds.), *Everyday cognition: Its development in social context*. Cambridge, MA: Harvard University Press.

Butera, G., Horn, E. M., Palmer, S. B., Friesen, A., & Lieber, J. (2016). Understanding Science, Technology, Engineering, Arts, and Mathematics (STEAM). In B. Reichow, B. Boyd, E. Barton, & S. Odom (Eds.), *Handbook of Early Childhood Special Education*. Cham: Springer. doi:10.1007/978-3-319-28492-7_9

Bybee, R. (2013). *The case for STEM education: Challenges and opportunities*. Arlington, VA: NSTA Press.

Bybee, R. (2018). *STEM education now more than ever*. Arlington, VA: NSTA Press.

Bybee, R. W. (2011). Scientific and engineering practices in K–12 classrooms. *Science Teacher (Normal, Ill.)*, *78*(9), 34–40.

Bybee, R. W. (2013). *The Case for STEM Education: Challenges and Opportunities*. NSTA Press.

Byers, F. (2017). A Critical Evaluation of the Effectiveness of Animation within Education. *Selected Computing Research Papers, 6,* 15-19.

Byers, T. (2016). *Development of an observation metric for linking pedagogy, technology and space*. Academic Press.

Byers, T., & Imms, W. (2014). *Making the space for space: The effect of the classroom layout on teacher and student usage and perception of one-to-one technology*. Paper presented at the 26th Australian Computers in Education Conference, Adelaide, Australia.

Byers, T., Hartnell-Young, E., & Imms, W. (2018). Empirical evaluation of different classroom spaces on students' perceptions of the use and effectiveness of 1-to-1 technology. *British Journal of Educational Technology*, *49*(1), 153–164. doi:10.1111/bjet.12518

Byers, T., & Imms, W. (2016). Evaluating the change in space in a technology-enabled primary years setting. In K. Fisher (Ed.), *The translation design of schools: An evidence based approach to aligning pedagogy and learning environment design* (pp. 215–236). Sense. doi:10.1007/978-94-6300-364-3_10

Byers, T., Imms, W., & Hartnell-Young, E. (2014). Making the case for space: The effect of learning spaces on teaching and learning. *Curriculum and Teaching*, *29*(1), 5–19. doi:10.7459/ct/29.1.02

Byers, T., Imms, W., & Hartnell-Young, E. (2018). Evaluating teacher and student spatial transition from a traditional classroom to an innovative learning environment. *Studies in Educational Evaluation*, *58*, 156–166. doi:10.1016/j.stueduc.2018.07.004

Cajete, G. (1999). *Igniting the sparkle: An Indigenous science education model*. Skyland, NC: Kivaki Press.

Cajete, G. (2000). *Native science: Natural laws of interdependence*. Santa Fe, NM: Clearlight Publishers.

Calderon, D. (2014). Speaking back to manifest destinies: A land education-based approach to critical curriculum inquiry. *Environmental Education Research*, *20*(1), 24–36. doi:10.1080/13504622.2013.865114

Campbell, T., Oh, P. S., Maughn, M., Kiriazis, N., & Zuwallack, R. (2014). A review of modeling pedagogies: Pedagogical functions, discursive acts, and technology in modeling instruction. *Eurasia Journal of Mathematics, Science, and Technology Education*, *11*, 159–176.

Caper, R. (1996). Play, experimentation and creativity. *The International Journal of Psycho-Analysis*, *77*, 859–869. PMID:8933214

Carbaugh, D. (1999). "Just listen": "Listening" and landscape among the blackfeet. *Western Journal of Communication*, *63*(3), 250–270. doi:10.1080/10570319909374641

Carnegie forum on education and the economy, task force on teaching as a profession. (1986). *A nation prepared: Teachers for the 21st century*. New York: Carnegie Corporation of New York.

Carvalho, L., & Goodyear, P. (2014). *The architecture of productive learning networks*. New York, NY: Routledge.

CAST. (2018). *Universal Design for Learning Guidelines version 2.2*. Retrieved from http://udlguidelines.cast.org

Catterall, J. S. (2002). The arts and the transfer of learning. In R. J. Deasy (Ed.), *Critical Links: Learning in the arts and student academic and social development*. Washington, DC: Arts Education Partnership.

Catterall, J. S. (2012). *The Arts and Achievement in At-Risk Youth: Findings from Four Longitudinal Studies. Research Report# 55*. National Endowment for the Arts.

Chambers, C. (2006). "The land is the best teacher I have ever had": Places as pedagogy for precarious times. *Journal of Curriculum Theorizing*, *22*(3), 27–37.

Chambers, C. (2008). Where are we? Finding Common Ground in a Curriculum of Place. *Journal of the Canadian Association for Curriculum Studies*, *6*(2), 113–128.

Champion, D. N. (2018). *The STEAM Dance Makerspace: A Context for Integration: An Investigation of Learning at the Intersections of STEM, Art, Making and Embodiment* (Doctoral dissertation). Northwestern University.

Charleston, L. J., Adserias, R. P., Lang, N. M., & Jackson, J. F. L. (2014). Intersectionality and STEM: The role of race and gender in the academic pursuits of African American women in STEM. *Journal of Progressive Policy & Practice*, *2*(3), 273–293.

Chen, C. J. (2010). Theoretical bases for using virtual reality in education. *Themes in Science and Technology Education*, *2*(1-2), 71–90.

Christensen, C. M., Horn, M. B., & Johnson, C. W. (2008). How' disruptive innovation' will change the way we learn. *Education Week*, *27*(39), 25–36.

Clift, R., & Brady, P. (2005). Research on methods courses and field experiences. In M. Cochran-Smith & K. Zeichner (Eds.), *Studying teacher education: The report of the AERA panel on research and teacher education* (pp. 309–424). Washington, DC: American Education Research Association / Lawrence Earlbaum.

Colucci-Gray, L., Burnard, P., Cooke, C., Davies, R., Gray, D., & Trowsdale, J. (2017). *Reviewing the potential and challenges of developing STEAM education through creative pedagogies for 21st learning: How can school curricula be broadened towards a more responsive, dynamic, and inclusive form of education?* London: British Educational Research Association.

Connor, A., Karmokar, S., & Whittington, C. (2015). From STEM to STEAM: Strategies for enhancing engineering & technology education. *International Journal of Engineering Pedagogy*, *5*(2), 37–47. doi:10.3991/ijep.v5i2.4458

Cook, K. L., & Bush, S. B. (2017). Design thinking in integrated STEAM learning: Surveying the landscape and exploring exemplars in elementary grades. *School Science and Mathematics*, *118*(3-4), 93–103. doi:10.1111sm.12268

Cook, S. C., & Rao, K. (2018). Systematically applying UDL to effective practices for students with learning disabilities. *Learning Disability Quarterly*, *41*(3), 179–191. doi:10.1177/0731948717749936

Cooper, J. O., Heron, T. E., & Heward, W. L. (2007). Applied behavior analysis. Creativity Research Journal, 18(4), 447-457.

Costa, A. L., & Kallick, B. (Eds.). (2008). *Learning and leading with habits of mind: 16 essential characteristics for success*. Alexandria, VA: ASCD.

Costin, D., Thompson, K., & Chapman, S. (2018). *Productive collaborative design of a STEAM unit of work in a multi-age class. Integrated education for the real world.* Presented at the 5th International STEM in Education Conference, Brisbane, Australia.

Csikszentmihalyi, M. (1975). *Beyond boredom and anxiety*. San Francisco, CA: Jossey-Bass Publishers.

Csikszentmihalyi, M. (1990). *Flow: The Psychology of Optimal Experience*. New York: Harper and Row.

Cuban, L. (2018). *The flight of a butterfly or the path of a bullet: using technology – using technology to transform teaching and learning*. Cambridge, MA: Harvard Education Press.

Curcio, I. D. D., Dipace, A., & Norlund, A. (2016). Virtual realities and education. *REM -Research on Education and Media*, *8*(2). doi:10.1515/rem-2016-0019

Curry, C., Cohen, L., & Lightbody, N. (2006). Universal design in science learning. *Science Teacher (Normal, Ill.)*, *73*(3), 32.

Dail, W. (2013). On cultural polymathy: How visual thinking, culture, and community create a platform for progress. *The STEAM Journal, 1*(1). Available at: http://scholarship.claremont.edu/steam/vol1/iss1/7

Daily, S. B., Leonard, A. E., Jörg, S., Babu, S., Gundersen, K., & Parmar, D. (2015). Embodying computational thinking: Initial design of an emerging technological learning tool. *Technology. Knowledge and Learning*, *20*(1), 79–84. doi:10.100710758-014-9237-1

Darling-Hammond, D. (2006). *Powerful teacher education*. San Francisco, CA: Jossey-Bass.

Deed, C., & Lesko, T. (2015). 'Unwalling' the classroom: Teacher reaction and adaptation. *Learning Environments Research*, *18*(2), 217–231. doi:10.100710984-015-9181-6

Denise, P. S., & Harris, I. M. (Eds.). (1989). *Experiential education for community development*. Westport, CT: Greenwood Press.

Dewey, J. (1902). *The child and the curriculum*. Chicago, IL: The University of Chicago press.

Diekman, A. B., Steinberg, M., Brown, E. R., Belanger, A. L., & Clark, E. K. (2017). A goal congruity model of role entry, engagement, and exit: Understanding communal goal processes in STEM gender gaps. *Personality and Social Psychology Review*, *21*(2), 142–175. doi:10.1177/1088868316642141 PMID:27052431

Dillenbourg, P. (1999). *What do you mean by collaborative learning?* Oxford, UK: Elsevier.

Dillon, D., Bullock, S., O'Connor, K., Russell, T., Martin, A., & Thomas, L. (2013). Place-based Teacher Development: Placing Practicum Learning at the Heart of Pre-service Teacher Education. In L. Thomas (Ed.), *Becoming teacher: Sites for teacher development* (pp. 94-120). Canadian Association for Teacher Education/Association canadienne pour la formation à'enseignement. Retrieved from https://docs.google.com/file/d/0BwVGDOGBDzJdR1R2VGZvakZjUDQ/edit

Dinkelman, T. (2003). Self-study in teacher education: A means and ends tool for promoting reflective teaching. *Journal of Teacher Education*, *54*(1), 6–18. doi:10.1177/0022487102238654

Dixon, R. A., & Brown, R. A. (2012). Transfer of Learning: Connecting Concepts during Problem Solving. *Journal of Technology Education*, *24*(1), 2–17. doi:10.21061/jte.v24i1.a.1

Donald, D., Glanfield, F., & Sterenberg, G. (2012). Living ethically within conflicts of colonial authority and relationality. *Journal of the Canadian Association for Curriculum Studies*, *10*(1), 53–77.

Dovey, K., & Fisher, K. (2014). Designing for adaptation: The school as socio-spatial assemblage. *The Journal of Architecture*, *19*(1), 43–63. doi:10.1080/13602365.2014.882376

Drake, S. M., & Burns, R. C. (2004). *Meeting standards through integrated curriculum*. Alexandria, VA: Association for Supervision and Curriculum Development.

Dumont, H., & Istance, D. (2010). Analyzing and designing learning environments for the 21st century. *The nature of learning: Using research to inspire practice*, 19-34.

Dumont, H., Instance, D., & Benavides, F. (Eds.). (2010). The nature of learning: Using research to inspire practice. Organization for Economic Cooperation and Development.

Dym, C. L., Agogino, A. M., Eris, O., Frey, D. D., & Leifer, L. J. (2005). Engineering design thinking, teaching, and learning. *Journal of Engineering Education, 94*(1), 104–120. doi:10.1002/j.2168-9830.2005.tb00832.x

Eagleman, D. M. (2013). Why public dissemination of science matters: A manifesto. *The Journal of Neuroscience, 33*(30), 12147–12149. doi:10.1523/JNEUROSCI.2556-13.2013

Edelson, D., Ryan, J., & Cowart, J. (2019, April). OpenSciEd: An Open Educational Middle School Science Curriculum Designed for NGSS. Presentation at the *annual meeting of the National Science Teachers Association*, St. Louis, MO.

Eger, J. (2013). STEAM...Now! *The STEAM Journal, 1*(1). Available at: http://scholarship.claremont.edu/steam/vol1/iss1/8

Eichinger, D., Anderson, C. W., Palinscar, A. S., & David, Y. M. (1991). An illustration of the roles of content knowledge, scientific argument, and social norms in collaborative problem solving. Paper presented at the annual meeting of the American Educational Research Association, Chicago, IL.

Eisner, E. (2002a). *The arts and the creation of the mind*. New Haven, CT: Yale University Press.

Eisner, E. (2002b). What the arts do for the young. *School Arts, 102*, 16–17.

Eisner, E., & Powell, K. (2002). Art in Science? *Curriculum Inquiry, 32*(2), 131–159. doi:10.1111/1467-873X.00219

Elmore, R. F. (2000). *Building a new structure for school leadership*. Washington, DC: The Albert Shanker Institute.

Elmore, R. F. (2004). *School reform from the inside out: Policy, practice, and performance*. Cambridge, MA: Harvard Education Press.

Emekauwa, E. (2004). *The star with my name: The Alaska rural systemic initiative and the impact of place-based education on Native student achievement. The case for place-based. Rural trust white paper on place-based education*. Washington, DC: The Rural School and Community Trust.

Erakhrumen, A. A. (2007). Phytoremediation: An environmentally sound technology for pollution prevention, control and remediation in developing countries. *Educational Research Review, 2*(7), 151–156.

Espinosa, L. L. (2011). Pipelines and pathways: Women of color in undergraduate STEM majors and the college experiences that contribute to persistence. *Harvard Educational Review, 81*(2), 209–241. doi:10.17763/haer.81.2.92315ww157656k3u

Estapa, A. & Tank, K. (2017). *Supporting integrated STEM in the elementary classroom: a professional development approach centered on an engineering design challenge*. Academic Press.

Farinde, A. A., & Lewis, C. W. (2012). *The underrepresentation of African American female students in STEM fields: Implications for classroom teachers*. Online Submission.

FCCSET Committee on Life Sciences and Health. (1993). *Biotechnology for the 21st century: Realizing the promise*. Washington, DC: Federal Coordinating Council for Science, Engineering, and Technology.

Fields, D., Kafai, Y., Nakajima, T., & Goode, J. (2017). Teaching practices for making e-textiles in high school computing classrooms. *ACM International Conference Proceeding Series*, 1-8. 10.1145/3141798.3141804

Fisher, K. (2006). The new learning environment: Hybrid designs for hybrid learning. *Education Futures Public, 2*.

Flowers, N., Mertens, S. B., & Mulhall, P. F. (2000). What makes interdisciplinary teams effective? *Middle School Journal, 31*(4), 53–56. doi:10.1080/00940771.2000.11494640

Freinet, C. (1993). *Education through work* (J. Sivell, Trans.). Lewiston, NY: Edwin Mellen Press. (Original work published 1943)

Freire, P. (1970). *Pedagogy of the oppressed.* New York: Continuum.

From STEM to STEAM. Developmental Learning. (n.d.). Retrieved fromhttps://www.psychologytoday.com/us/blog/conceptual-revolution/201708/stem-steam-developmental-learning

Gantt, L. (1990). A validity study of the Formal Elements Art Therapy Scale (FEATS) for diagnostic information in patients' drawings (Unpublished doctoral dissertation). University of Pittsburgh, Pittsburgh, PA.

Gantt, L., & Mills, B. (2001). The Formal Elements Art Therapy Scale: A measurement system for global variables in art. *Brief Report, 18*(1), 50–55.

Gantt, L., & Tabone, C. (1998). *The Formal Elements Art Therapy Scale: The rating manual.* Morgantown, WV: Gargoyle Press.

Gardner, H. (1983). *Frames of mind: The theory of multiple intelligences*. New York: Basic Books.

Gelamdin, R. B., Alias, N., & Attaran, M. (2013). Students' and teachers' perspectives on biotechnology education: A review on publications in selected journals. *Life Science Journal, 10*(1), 1210–1221.

Goldin, C., & Katz, L. F. (2008). *The Race between Education and Technology*. Cambridge, MA: Harvard University Press.

Goodlad, J. (1988). School-university partnerships: Rationale and concepts. In K. Sirotnik & J. Goodlad (Eds.), *School-university partnerships in action: Concepts, cases, and concerns* (pp. 3–31). New York: Teachers College Press.

Goodyear, P., & Carvalho, L. (2013). The analysis of complex learning environments. *Re-thinking pedagogy for a digital age: Designing for 21st century learning*, 49-63.

Goodyear, P., Jones, C., & Thompson, K. (2014). Computer-supported collaborative learning: Instructional approaches, group processes and educational designs. In J. M. Spector, M. D. Merrill, J. Elen, & M. J. Bishop (Eds.), *Handbook of Research on Educational Communications and Technology* (pp. 439–452). New York, NY: Springer. doi:10.1007/978-1-4614-3185-5_35

Grande, S. (2004). *Red pedagogy: Native American social and political thought*. Lanham, MD: Rowman & Littlefield Publishers Inc.

Gruenewald, D. (2003). The best of both worlds: A critical pedagogy of place. *Educational Researcher, 32*(4), 3–12. doi:10.3102/0013189X032004003

Guo, P. J., Kim, J., & Rubin, R. (2014). How Video Production Affects Student Engagement: An Empirical Study of MOOC Videos. *Proceedings of the First ACM Conference on Learning @ Scale Conference*, 41–50. 10.1145/2556325.2566239

Halverson, E. R., & Sheridan, K. M. (2014). The maker movement in education. *Harvard Educational Review, 84*(4), 495–504. doi:10.17763/haer.84.4.34j1g68140382063

Harper, S. R. (2012). Race without racism: How higher education researchers minimize racist institutional norms. *The Review of Higher Education, 36*(1), 9–29. doi:10.1353/rhe.2012.0047

Harris, K. S., & Rogers, G. E. (2008). Soft Skills in the Technology Education Classroom: What Do Students Need? *Technology Teacher, 68*(3), 19–24.

Hattie, J. A., & Donoghue, G. M. (2016). Learning strategies: A synthesis and conceptual model. *NPJ Science of Learning, 1*, 16013.

Henriksen, D. (2014). Full STEAM ahead: Creativity in excellent STEM teaching practices. *The STEAM Journal, 1*(2), 15.

Henriksen, D. (2017). Creating STEAM with design thinking: Beyond STEM and arts integration. *The STEAM Journal, 3*(1), 1-11. Retrieved from https://doin.org/10.5642/steam.20170301.11

Herro, D., & Quigley, C. (2017). Exploring teachers' perceptions of STEAM teaching through professional development: Implications for teacher educators. *Professional Development in Education, 43*(3), 416–438. doi:10.1080/19415257.2016.1205507

Hoban, T. J. (1998). Trends in consumer attitudes about agricultural biotechnology. *AgBioForum, 1*(1), 3–7. Retrieved from http://www.agbioforum.org/v1n1/v1n1a02-hoban.htm

Holmes Group. (1986). *Tomorrow's teachers: A report of the Holmes Group*. East Lansing, MI: Author.

Householder, D. L., & Hailey, C. E. (2012). *Incorporating Engineering Design Challenges into STEM Courses*. National Center for Engineering and Technology Education. Retrieved from http://eric.ed.gov/?id=ED537386

Howland, J. L., Jonassen, D. H., & Marra, R. M. (2012). *Meaningful learning with technology* (4th ed.). Boston, MA: Pearson.

Huffman, D., Thomas, K., & Lawrenz, F. (2003). Relationship between professional development, teachers' instructional practices, and the achievement of students in science and mathematics. *School Science and Mathematics, 103*(8), 378–387. doi:10.1111/j.1949-8594.2003.tb18123.x

Imms, W., & Byers, T. (2016). Impact of classroom design on teacher pedagogy and student engagement and performance in mathematics. *Learning Environments Research, 19*(2), 1–14.

Institute of Design at Stanford. (2010). *Introduction to design thinking - process guide.* Retrieved from https://dschool-old.stanford.edu/sandbox/groups/designresources/wiki/36873/attachments/74b3d/ModeGuideBOOTCAMP2010L.pdf

International Technology Engineering Association. (2007). *Standards of Technological Literacy.* Reston, VA: ITEEA. Retrieved from https://www.iteea.org/42511.aspx

Ireland, D. T., Freeman, K. E., Winston-Proctor, C. E., DeLaine, K. D., Lowe, S. M., & Woodson, K. (2018). (Un)hidden figures: A synthesis of research examining the intersectional experiences of Black women and girls in STEM education. *Review of Research in Education, 42*(1), 226–254. doi:10.3102/0091732X18759072

J. Paul Getty Trust. (2011). *Elements of Art.* Los Angeles, CA: The J. Paul Getty Museum. Retrieved from https://www.getty.edu/education/teachers/building_lessons/elements_art.pdf

Jamrisko, M., Miller, L. J., & Lu, W. (2019). These Are the World's Most Innovative Countries. *Bloomberg News.* Retrieved from https://www.bloomberg.com/news/articles/2019-01-22/germany-nearly-catches-korea-as-innovation-champ-u-s-rebounds

Jeong, H., & Chi, M. T. (2007). Knowledge convergence and collaborative learning. *Instructional Science, 35*(4), 287–315. doi:10.100711251-006-9008-z

Johnson, P. A. (1999). Problem-based, cooperative learning in the engineering classroom. *Journal of Professional Issues in Engineering Education and Practice, 125*(1), 8–11. doi:10.1061/(ASCE)1052-3928(1999)125:1(8)

Jonassen, D. H. (2000). Toward a design theory of problem solving. *Educational Technology Research and Development, 48*(4), 63–85. doi:10.1007/BF02300500

Jonassen, D. H. (2003). *Learning to solve problems with technology* (2nd ed.). Upper Saddle River, NJ: Merrill.

Jonassen, D. H., Peck, K. L., & Wilson, B. G. (2001). *Learning with technology: A constructivist perspective.* Upper Saddle River, NJ: Merrill.

Kafai, Y. (2016). From computational thinking to computational participation in K--12 education. *Communications of the ACM, 59*(8), 26–27. doi:10.1145/2955114

Kafai, Y., & Resnick, M. (1996). *Constructionism in practice: designing, thinking, and learning in a digital world.* Mahwah, NJ: Lawrence Erlbaum Associates.

Kálly, É. (2012). Learning strategies and metacognitive awareness as predictors of academic achievement in a sample of Romanian second-year students. *Cognition, Brain, Behavior: An Interdisciplinary Journal, 16*(3), 369–385.

Kaplancali, U. T., & Demirkol, Z. (2017). Teaching coding to children: A methodology for kids 5+. *International Journal of Elementary Education, 6*(4), 32–37. doi:10.11648/j.ijeedu.20170604.11

Kasza, P., & Slater, T. F. (2017). A Survey of Best Practices and Key Learning Objectives for Successful Secondary School STEM Academy Settings. *Contemporary Issues in Education Research, 10*(1), 53–66.

Katehi, L., Pearson, G., & Feder, M. (2009). *Engineering in K-12 education: Understanding the status and improving the prospects.* Washington, DC: National Academy Press.

Khalifa, M. A., Gooden, M. A., & Davis, J. E. (2016). Culturally Responsive School Leadership: A Synthesis of the Literature. *Review of Educational Research, 86*(4), 1272–1311. doi:10.3102/0034654316630383

Kier, M. W., Blanchard, M. R., Osborne, J. W., & Albert, J. L. (2014). The development of the STEM career interest survey (STEM-CIS). *Research in Science Education, 44*(3), 461–481. doi:10.100711165-013-9389-3

Kincheloe, J. (2001). *Getting beyond the facts: Teaching social studies/social sciences in the twenty-first century.* New York: Peter Lang.

Kincheloe, J. (2005). *Critical pedagogy.* New York: Peter Lang.

Kincheloe, J. L., Mckinley, E., Lim, M., & Calabrese Barton, A. (2006). Forum: A conversation on 'sense of place' in science learning. *Cultural Studies of Science Education, 1*(1), 143–160. doi:10.100711422-005-9003-8

King, N. S., & Pringle, R. M. (2019). Black girls speak STEM: Counterstories of informal and formal learning experiences. *Journal of Research in Science Teaching, 56*(5), 539–569. doi:10.1002/tea.21513

Kissling, M. T. (2012). A living curriculum of place(s). *Journal of Curriculum Theorizing, 28*(3), 109–127.

Kitchen, J., & Russell, T. (Eds.). (2012). *Canadian perspectives on the self-study of teacher education practices.* Ottawa, Canada: Canadian Association of Teacher Education.

Klenowski, V. (1995). Student self-evaluation processes: Empowering students in learner-centered contexts. *Assessment in Education: Principles, Policy & Practice, 2*, 145–163. doi:10.1080/0969594950020203

Knapp, C. E. (1996). Just beyond the classroom: Community adventures for interdisciplinary learning. Charleston, WV: ERIC Clearinghouse on Rural Education and Small Schools. (ERIC Document Reproduction Service NO. ED 388485)

Korthagen, F., Loughran, J., & Russell, T. (2006). Developing fundamental principles for teacher education programs and practices. *Teaching and Teacher Education, 22*(8), 1020–1041. doi:10.1016/j.tate.2006.04.022

Kosnik, C., Freese, A., Samaras, A., & Beck, C. (Eds.). (2006). Making a difference in teacher education through self-study: Studies of personal, professional, and program renewal. Dordrecht: Springer Academic Publishers.

Kovach, M. (2009). *Indigenous methodologies: Characteristics, conversations, and contexts.* Toronto: University of Toronto Press.

Kulnieks, A., Longboat, D., & Young, K. (Eds.). (2013). *Contemporary studies in environmental and indigenous pedagogies: A curricula of stories and place*. Rotterdam, The Netherlands: Sense Publishers. doi:10.1007/978-94-6209-293-8

LaBoskey, V. K. (2004). The methodology of self-study and its theoretical underpinnings. In J. J. Loughran, M. L. Hamilton, V. K. LaBoskey, & T. Russell (Eds.), *International handbook of self-study of teaching and teacher education practices* (pp. 817–869). Dordrecht: Kluwer. doi:10.1007/978-1-4020-6545-3_21

Lackney, J. A. (2008). Teacher environmental competence in elementary school environments. *Children, Youth and Environments, 18*(2), 133–159.

Lam, T. C. M., & Bengo, P. (2003). A comparison of three retrospective self-reporting methods of measuring change in instructional practice. *The American Journal of Evaluation, 24*(1), 65–80. doi:10.1177/109821400302400106

Lave, J., & Wenger, E. (1991). *Situated learning: Legitimate peripheral participation.* Cambridge, UK: Cambridge University Press. doi:10.1017/CBO9780511815355

Lawrence Arts Center. (2015, May 9). *About us: History*. Retrieved from http://lawrenceartscenter.org

Lawrence Arts Center. (2018). *2017-2018 Annual Report*. Retrieved from http://lawrenceartcenter.org

Lawrence Arts Center. (n.d.). *ARTspace: Explore. Imagine. Invent.* Lawrence, KS: Lawrence Arts Center.

LearningR. (2019). Retrieved from https://remakelearning.org/

Lee, C. (2007). Boundary negotiating artifacts: Unbinding the routine of boundary objects and embracing chaos in collaborative work. *Computer Supported Cooperative Work, 16*(3), 307–339. doi:10.100710606-007-9044-5

Lee, H.-J. (2005). Developing a professional development program model based on teachers' needs. *Professional Educator*, *27*(1-2), 39–49.

Lee, O., Quinn, H., & Valdés, G. (2013). Science and language for English language learners in relation to Next Generation Science Standards and with implications for Common Core State Standards for English language arts and mathematics. *Educational Researcher*, *42*(4), 223–233. doi:10.3102/0013189X13480524

Lent, R. W., Brown, S. D., & Hackett, G. (1994). Toward a unifying social cognitive theory of career and academic interest, choice, and performance. *Journal of Vocational Behavior*, *45*(1), 79–122. doi:10.1006/jvbe.1994.1027

Lesh, R. A., & Zawojewski, J. (2007). Problem solving and modeling. In F. Lester (Ed.), *Second Handbook of Research on Mathematics Teaching and Learning* (pp. 763–804). Charlotte, NC: Information Age Publishing, Inc.

Liao, C., Motter, J. L., & Patton, R. M. (2016). Learning 21st-century skills through STEAM digital artmaking. *Art Education*, *69*(4), 29–35. doi:10.1080/00043125.2016.1176492

Lichtenberg, J., Wright, M., & Woock, C. (2008). *Ready to Innovate: Are Art Educators and Executives Aligned on the Creative Readiness of the US Workforce?* The Conference Board, Inc.

Lindeman, K., Jabot, M., & Berkley, M. (2014). The Role of STEM (or STEAM) in the early childhood setting: Learning across the early childhood curriculum. *Advances in Early Education and Day Care*, *17*, 95–114. doi:10.1108/S0270-4021(2013)0000017009

Lipman, P. (2013). *The new political economy of urban education: Neoliberalism, race, and the right to the city*. Taylor & Francis. doi:10.4324/9780203821800

Lippman, P. C. (2010). *Can the physical environment have an impact on the learning environment?* Academic Press.

Little Bear, L. (2000). Jagged worldviews colliding. In M. Battiste (Ed.), *Reclaiming Indigenous voice and vision* (pp. 78–85). Vancouver, Canada: University of British Columbia Press.

Long, C. (2012). Teach your students to fail better with design thinking. *Learning and Leading with Technology*, *39*(5), 16–20.

Lortie, D. (1975). *Schoolteacher: A sociological study*. Chicago: University of Chicago Press.

Loughran, J. J. (2002). Effective reflective practice: In search of meaning in learning about teaching. *Journal of Teacher Education, 53*(1), 33 43.

Lowenfeld, V. (1939). *The nature of creative activity*. New York: Harcourt Brace.

Luka, I. (2014). Design Thinking in Pedagogy. *Journal of Education Culture and Society.*, *2*, 63–74. doi:10.15503/jecs20142.63.74

Maeda, J. (2013). Stem+ art= steam. *The STEAM Journal*, *1*(1), 34. doi:10.5642team.201301.34

Malone, J. (1996). *Steppin'on the blues: The visible rhythms of African American dance* (Vol. 21). University of Illinois Press.

Margot, K. C., & Kettler, T. (2019). Teachers' perception of STEM integration and education: A systematic literature review. *International Journal of STEM Education*, *6*(2). doi:10.118640594-018-0151-2

Marra, R. M., Rodgers, K. A., Shen, D., & Bogue, B. (2009). Women engineering students and Self-Efficacy: A Multi-Year, Multi-Institution study of women engineering student Self-Efficacy. *Journal of Engineering Education*, *98*(1), 27–38. doi:10.1002/j.2168-9830.2009.tb01003.x

Martinez, S. L., & Stager, G. (2013). Invent to learn: "making", tinkering, and engineering in the classroom. Torrance, CA: Constructing Modern Knowledge Press.

Martinez, J. (2017). *The search for method in STEAM education*. London, UK: Palgrave Macmillian. doi:10.1007/978-3-319-55822-6

Mathisen, G. E., & Bronnick, K. S. (2009). Creative self-efficacy: An intervention study. *International Journal of Educational Research*, *48*(1), 21–29. doi:10.1016/j.ijer.2009.02.009

Mayer, R. (2014). Cognitive Theory of Multimedia Learning. In R. Mayer (Ed.), The Cambridge Handbook of Multimedia Learning (pp. 43-71). Cambridge, UK: Cambridge University Press. doi:10.1017/CBO9781139547369.005

Mayer, R. (2008). Applying the science of learning: Evidence-based principles for the design of multimedia instruction. *Cognition and Instruction*, *19*, 177–213. PMID:19014238

Mayer, R. E., & Moreno, R. (2003). Nine ways to reduce cognitive load in multimedia learning. *Educational Psychologist*, *38*(1), 43–52. doi:10.1207/S15326985EP3801_6

Mazuzan, G. T. (1994). *The National Science Foundation: A brief history*. Office of Administrative and Public Affairs, the National Science Foundation. Retrieved from https://www.nsf.gov/about/history/nsf50/nsf8816.jsp

McCarthy-Brown, N. (2017). *Dance pedagogy for a diverse world: Culturally relevant teaching in theory, research and practice*. McFarland.

McCarthy-Brown, N. (2017). *Dance pedagogy for a diverse world: Culturally relevant teaching in theory, research, and practice*. Jefferson, NC: McFarland & Company.

McCaughey, T., Budden, D. M., Sanfilippo, P. G., Gooden, G. E., Fan, L., Fenwick, E., ... Hewitt, A. W. (2019). A need for better understanding is the major determinant for public perceptions of human gene editing. *Human Gene Therapy*, *30*(1), 36–43. doi:10.1089/hum.2018.033 PMID:29926763

McNeill, K. L., Katsh-Singer, R., & Pelletier, P. (2015). Assessing science practices – Moving your class along a continuum. *Science Scope*, *39*(4), 21–28. doi:10.2505/4/ss15_039_04_21

Mensah, F. M., & Jackson, I. (2018). Whiteness as property in science teacher education. *Teachers College Record, 120*(1), 1–38.

Michaels, S., & O'Connor, C. (2012). *Talk science primer*. Cambridge, MA: Technical Education Research Centers; Retrieved from http://inquiryproject.terc.edu/shared/pd/TalkScience_Primer.PDF

Miller, G. (1956). The magical number seven, plus or minus two: Some limits on our capacity for processing information. *Psychological Review, 63*(2), 81–97. doi:10.1037/h0043158 PMID:13310704

Mitchell, R. (2012). Simondon, bioart and the milieu of biotechnology. *INFLeXions, 5,* 68-110. Retrieved from http://www.inflexions.org/n5_mitchellhtml.html

Mohammed Jawad, H. (2018). *Exploring how integrating art & animation in teaching text-based programming affects high school students' interest in computer science* (Doctoral dissertation). Retrieved from Eastern Michigan University Digital Commons@EMU: https://commons.emich.edu/theses/916

Moll, L. C., Amanti, C., & Gonzalez, N. (2005). *Funds of knowledge: Theorizing practices in households and classrooms*. Mahwah, NJ: Lawrence Erlbaum Associates.

Moneta, G. B. (2012). On the measurement and conceptualization of flow. In S. Engeser & A. Shipe-Tiska (Eds.), *Historical Lines and an Overview of Current Research on flow* (pp. 23–50). New York, NY: Springer. doi:10.1007/978-1-4614-2359-1_2

Monn, M. A., Weaver, J. C., Zhang, T., Aizenberg, J., & Kesari, H. (2015). New functional insights into the internal architecture of the laminated anchor spicules of Euplectella aspergillum. *Proceedings of the National Academy of Sciences of the United States of America, 112*(16), 4976–4981. doi:10.1073/pnas.1415502112 PMID:25848003

Montessori, M. (1912). *The Montessori Method* (A. E. George, Trans.). New York: Frederick A. Stokes Company.

Morales-Chicas, J., Castillo, M., Bernal, I., Ramos, P., & Guzman, B. L. (2019). Computing with Relevance and Purpose: A Review of Culturally Relevant Education in Computing. *International Journal of Multicultural Education, 21*(1), 125–155. doi:10.18251/ijme.v21i1.1745

Morris, M. W. (2014). *Black stats: African Americans by the numbers in the twenty-first century*. New York, NY: The New Press.

Morrison, J. (2006). *Attributes of STEM Education*. TIES STEM Education Monograph Series. Retrieved from http://www. tiesteach.org/monographs.aspx

Morrison, J., Roth McDuffie, A., & French, B. (2015). Identifying key components of teaching and learning in a STEM school. *School Science and Mathematics, 115*(5), 244–255. doi:10.1111sm.12126

Mulcahy, D., Cleveland, B., & Aberton, H. (2015). Learning spaces and pedagogic change: Envisioned, enacted and experienced. *Pedagogy, Culture & Society*, *23*(4), 575–595. doi:10.1080/14681366.2015.1055128

Munby, H., & Russell, T. (1994). The authority of experience in learning to teach: Messages from a physics methods class. *Journal of Teacher Education*, *45*(2), 86–95. doi:10.1177/0022487194045002002

Murnane, R. J., & Levy, F. (1996). *Teaching the New Basic Skills: Principles for Educating Children to Thrive in a Changing Economy*. New York: Free Press.

Murray, J. (2019). Routes to STEM: Nurturing Science, Technology, Engineering, and Mathematics in early years education. *International Journal of Early Years Education*, *3*(1), 219–221. doi:10.1080/09669760.2019.1653508

MYCEETYA. (2008). *Melbourne Declaration on Education Goals for Young Australians*. Melbourne, Australia: Ministerial Council on Education, Employment, Training and Youth Affairs.

Nakagami-Yamaguchi, E., Fujinaga, K., & Batard, A. (2016). The effect of an animation movie for inpatient fall prevention: A pilot study in an acute hospital. *Safety in Health*, *2*(3). doi:10.118640886-016-0014-9

National Academies of Sciences, Engineering, and Medicine. (2018). Design, Selection, and Implementation of Instructional Materials for the Next Generation Science Standards: Proceedings of a Workshop. Washington, DC: The National Academies Press. doi:10.17226/25001

National Center for Science and Engineering Statistics. (n.d.). *Women, Minorities, and Persons with Disabilities in Science and Engineering*. Retrieved July 19, 2019, from https://ncses.nsf.gov/pubs/nsf19304/

National Coalition for Core Arts Standards. (2012). The 2014 National Core Arts Standards. New York, NY: The National Coalition for Core Arts Standards. Retrieved from https://www.nationalartsstandards.org/content/national-core-arts-standards

National Coalition for Core Arts Standards. (2014). *National core arts standards*. Dover, DE: State Education Agency Directors of Arts Education. Retrieved from https://www.nationalartsstandards.org/

National Governors Association. (2010). *Common core state standards*. Washington, DC: Council of Chief State School Officers & National Governors Association Center for Best Practices.

National Research Council (NRC). (2007). *Taking science to school: Learning and teaching science in grades K–8*. Washington, DC: National Academies Press.

National Research Council (NRC). (2012). *A Framework for K-12 science education: Practices, crosscutting concepts, and core ideas*. Washington, DC: The National Academies Press.

National Research Council. (2011). *Successful K-12 STEM Education: Identifying Effective Approaches in Science, Technology, Engineering and Mathematics*. Washington, DC: The National Academic Press.

National Research Council. (2013). *Next Generation Science Standards: For States, By States*. Washington, DC: The National Academies Press. doi:10.17226/18290

National Science Board. (2010). *Preparing the Next Generation of STEM Innovators: Identifying and Developing the Nation's Human Capital*. Retrieved from https://www.nsf.gov/nsb/publications/2010/nsb1033.pdf

National Science Foundation. (2017). *Convergence Research at NSF.* Convergence Press Release (17- 082). Retrieved from https://www.nsf.gov/od/oia/convergence/index.jsp

National Science Teachers Association (NSTA). (n.d.). About the Next Generation Science Standards. Retrieved May 21, 2019, from https://ngss.nsta.org/About.aspx

Negretti, R. (2012). Metacognition in student academic writing: A longitudinal study of metacognitive awareness and its relation to task perception, self-regulation, and evaluation of performance. *Written Communication*, *29*(2), 142–179. doi:10.1177/0741088312438529

Nelson, P., Martin, S. S., & Baldwin, V. G. (1998). Drawing skills and science concepts in young children: A study of relationships. *Studies in Art Education*, *39*(3), 262–269. doi:10.2307/1320368

New Zealand Ministry of Education. (2014). *The ministry of education's statement of intent 2014-2018*. Wellington, New Zealand: New Zealand Ministry of Education.

Newberry, D. M. (2012). Nanotechnology: A platform for education change. *Nanotechnology (IEEE- NANO), 12th IEEE Conference*, 20-23.

NGSS Lead States. (2013). *Next generation science standards: For states, by states*. Washington, DC: The National Academies Press.

NGSS Lead States. (2013). *Next Generation Science Standards: For states, by states*. Washington, DC: The National Academies Press.

Oblinger, D. G. (2006). Space as a change agent. *Learning Spaces*, *1*, 1–2.

OECD. (2013). *Education at a Glance 2013: OECD Indicators*. OECD Publishing. doi:10.1787/eag-2013-

OECD. (2015). *Education at a Glance 2015: OECD Indicators*. Paris: OECD Publishing. doi:10.1787/eag-2015-

Ogborn, J., Kress, G., Martins, I., & McGillicuddy, K. (1996). *Explaining science in the classroom*. Buckingham, UK: Open University Press.

Oleson, A. K., Hora, M. T., & Benbow, R. J. (2014). *STEM: How a Poorly Defined Acronym Is Shaping Education and Workforce Development Policy in the United States*. WCER Working Paper No. 2014-2. Wisconsin Center for Education Research. Retrieved from http://eric.ed.gov/?id=ED556481

OpenSciEd. (n.d.). OpenSciEd. Retrieved from https://www.openscied.org/openscied/?

Oregon State University. (2011). Retrieved from http://oregonstate.edu/instruct/coursedev/models/id/taxonomy/

Orr, D. W. (1994). *Earth in mind: On education, environment, and the human prospect*. Washington, DC: Island Press.

Orsmond, P., Merry, S., & Reiling, K. (2000). The use of student derived marking criteria in peer and self-assessment. *Assessment & Evaluation in Higher Education*, *25*(1), 23–38. doi:10.1080/02602930050025006

Panadero, E., Andrade, H., & Brookhart, S. M. (2018). Fusing self-regulated learning and formative assessment: A roadmap of where we are, how we got here, and where we are going. *Australian Educational Researcher*, *45*(1), 13–31. doi:10.100713384-018-0258-y

Papanastasiou, G., Drigas, A., Skianis, C., Lytras, M., & Papanastasiou, E. (2018). Virtual and augmented reality effects on K12, higher and tertiary education students' -twentyfirst- century skills. *Virtual Reality*.

Papert, S. (1986). Constructionism: a new opportunity for elementary science education. Massachusetts Institute of Technology, Media Laboratory, Epistemology and Learning Group: National Science Foundation.

Papert, S., & Freire, P. (1995) *Vídeo: O Futuro da escola e o impacto dos novos meios de comunicação no modelo de escola atual.* Márcia Moreno e Marco Aurélio Del Rosso. São Paulo: TV PUC de São Paulo com apoio do Jornal da Tarde e Agência Estado. 1 DVD (115 min): sonoro, colorido. DVD.

Papert, S. (1980). *Mindstorms: Children, computers, and powerful ideas*. New York: BasicBooks.

Papert, S. (1991a). Situating Constructionism. In I. Harel & S. Papert (Eds.), *Constructionism* (pp. 1–11). Norwood, NJ: Ablex publishing Corporation.

Papert, S. (1991b). Perestroika and Epistemological Politics. In I. Harel & S. Papert (Eds.), *Constructionism* (pp. 13–27). Norwood, NJ: Ablex publishing Corporation.

Papert, S., & Solomon, C. (1971). *Twenty Things To Do with a Computer*. Massachusetts Institute of Technology: Artificial Intelligence Laboratory.

Parsad, B., Spiegelman, M., & Coopersmith, J. (2012). *Arts education in public elementary and secondary schools, 1999-2000 and 2009-10*. Washington, DC: National Center for Education Statistics, Institute of Education Sciences, U.S. Department of Education.

Parthasarathy, S. (2017). Grassroots innovation systems for the post-carbon world: Promoting economic democracy, environmental sustainability, and the public interest. *Brooklyn Law Review*, *82*, 761–787.

Penetito, W. (2009). Place-based education: Catering for curriculum, culture and community. *New Zealand Annual Review of Education, 18*, 5–29.

Peppler, K., & Bender, S. (2013). Maker movement spreads innovation one project at a time. *Phi Delta Kappan*, *95*(3), 22–27. doi:10.1177/003172171309500306

Perignat, E., & Katz-Buonincontro, J. (2019). STEAM in practice and research: An integrative literature review. *Thinking Skills and Creativity*, *31*, 31–43. doi:10.1016/j.tsc.2018.10.002

Piaget, J. (1964). Development and Learning. In R. Ripple & V. Rockcastle (Eds.), *Piaget Rediscovered*. Ithaca, NY: Cornel University.

Pinkard, N. (2005). How the perceived masculinity and/or femininity of software applications influences students' software preferences. *Journal of Educational Computing Research*, *32*(1), 57–78. doi:10.2190/3LEE-MLCE-NK0Y-RUEP

Pinkard, N., Erete, S., Martin, C. K., & McKinney de Royston, M. (2017). Digital Youth Divas: Exploring narrative-driven curriculum to spark middle school girls' interest in computational activities. *Journal of the Learning Sciences*, *26*(3), 477–516. doi:10.1080/10508406.2017.1307199

Pinnegar, S. (1998). Introduction to Part II: Methodological perspectives. In M. L. Hamilton (Ed.), *Reconceptualizing teaching practice: Self-study in teacher education* (pp. 31–33). London: Falmer Press.

Pintrich, P. R., & De Groot, E. V. (1990). Motivational and Self-Regulated Learning Components of Classroom Academic Performance. *Journal of Educational Psychology*, *82*(1), 33–40. doi:10.1037/0022-0663.82.1.33

Plattner, H. (Ed.). (2010). *An introduction to design thinking process guide*. Stanford, CA: Institute of Design at Stanford. Retrieved from https://dschool-old.stanford.edu/sandbox/groups/designresources/wiki/36873/attachments/74b3d/ModeGuideBOOTCAMP2010L.pdf

Popova, M. (2012). What is art? A few famous definitions, from antiquity to today. *The Atlantic*. Retrieved from https://www.theatlantic.com/entertainment/archive/2012/06/what-is-art-a-few-famous-definitions-from-antiquity-to-today/258871/

Quigley, C. F., Herro, D., & Jamil, F. M. (2017). Developing a conceptual model of STEAM teaching practices. *School Science and Mathematics*, *117*(1-2), 1–12. doi:10.1111sm.12201

Raca, M., & Dillenbourg, P. (2015). Camera-based estimation of students' attention in class. Lausanne. *EPFL, 2015*. doi:10.5075/epfl-thesis-6745

Raelin, J. A. (2007). Toward an epistemology of practice. *Academy of Management Learning & Education*, *6*(4), 495–519. doi:10.5465/amle.2007.27694950

Raffan, J. (1993). The experience of place: Exploring land as teacher. *Journal of Experiential Education, 16*(1), 39–45. doi:10.1177/105382599301600109

Raffan, J. (1995). Experiential education and teacher education. *Journal of Experiential Education, 18*(3), 117–119. doi:10.1177/105382599501800301

Ravitz, J. L., Becker, H. J., & Wong, Y. (2000). *Constructivist-compatible beliefs and practices among U.S. teachers. Teaching, Learning, and Computing: 1998 National Survey Report #4.* Irvine, CA: Center for Research on Information Technology and Organizations.

Reider, D., Knestis, K., & Malyn-Smith, J. (2016). Workforce Education Models for K-12 STEM Education Programs: Reflections on, and Implications for, the NSF ITEST Program. *Journal of Science Education and Technology, 25*(6), 847–858. doi:10.100710956-016-9632-6

Reiser, B. J. (2013, September). What professional development strategies are needed for successful implementation of the next generation science standards. Paper presented at the International Research Symposium on Science Assessment, Washington, DC.

Relph, E. (1992). Modernity and the reclamation of place. In D. Seamon (Ed.), *Dwelling, Seeing, and Designing: Toward a Phenomenological Ecology* (pp. 25–40). New York: State University of New York Press.

Resnick, M. & Rosenbaum, E. (2013). *Designing for tinkerability. Design, Make, Play: growing the next generation of STEM Innovators*, 163-181. Doi:10.4324/9780203108352

Ricketts, A. (2014). Preservice elementary teachers' ideas about scientific practices. *Science and Education, 23*(10), 2119–2135. doi:10.1007/s11191-014-9709-7

Roosevelt, F. D. (1944). *Letter to Dr. V. Bush from the White House November 17, 1944.* Retrieved from https://www.nsf.gov/od/lpa/nsf50/vbush1945.htm#letter

Root-Bernstein, R., & Root-Bernstein, M. (2003). Intuitive tools for innovative thinking. *The international handbook on innovation*, 377-387.

Rousseau, J.-J. (1961). *Emile; Or, Education.* J.M. Dent & Sons.

Rubin, J., & Chisnell, D. (2008). *Handbook of usability testing: How to plan, design, and conduct effective tests* (2nd ed.). Indianapolis, IN: Wiley.

Rushkoff, D. (2011). *Program or be programmed – ten commands for a digital age*. Berkeley, CA: Soft Skull Press.

Sanderson, S. C., Suckiel, S. A., Zweig, M., Bottinger, E. P., Jabs, E. W., & Richardson, L. D. (n.d.). Development and preliminary evaluation of an online educational video about whole-genome sequencing for research participants, patients, and the general public. *Genetics in Medicine, 18*(5), 501–512. doi:10.1038/gim.2015.118

Sandoval, W. (2014). Conjecture Mapping: An Approach to Systematic Educational Design Research. *Journal of the Learning Sciences, 23*(1), 18–36. doi:10.1080/10508406.2013.778204

Sandoval, W., & Bell, P. (2004). Design-based research methods for studying learning in context: Introduction. *Educational Psychologist, 39*(4), 199–201. doi:10.120715326985ep3904_1

Saurabh, S., & Gautam, S. (2019). Computers and mobile phones are the top two devices for viewing videos. *Computers & Education, 128*, 145–158. doi:10.1016/j.compedu.2018.09.003

Savage, E., & Sterry, L. (Eds.). (1990). *A conceptual framework for technology education.* Reston, VA: International Technology Education Association.

Schank, P., Wise, A., Stanford, T., & Rosenquist, A. (2009). *Can High School Students Learn Nanoscience? An Evaluation of the Viability and Impact of the NanoSense Curriculum.* SRI International.

Schell, J. (2018, March). *Design thinking has a pedagogy problem.* Future20 Session presented at the South By Southwest EDU Conference, Austin, TX.

Schön, D. A. (1987). *Educating the reflective practitioner: Toward a new design for teaching and learning in the professions.* San Francisco, CA: Jossey-Bass.

Schraw, G., & Dennison, R. S. (1994). *Metacognitive awareness inventory.* doi:10.1037/t21885-000

Schwab, K. (2016). *The fourth industrial revolution.* New York, NY: Crown Business.

Schwarz, C., Passmore, C., & Reiser, B. J. (2017). *Helping students make sense of the world using next generation science and engineering practices.* Arlington, VA: NSTA Press, National Science Teachers Association.

Scott, K. A., Sheridan, K. M., & Clark, K. (2015). Culturally responsive computing: A theory revisited. *Learning, Media and Technology, 40*(4), 412–436. doi:10.1080/17439884.2014.924966

Scott, K. A., & White, M. A. (2013). COMPUGIRLS' standpoint: Culturally responsive computing and its effect on girls of color. *Urban Education, 48*(5), 657–681. doi:10.1177/0042085913491219

Segall, A. (2002). *Disturbing practice: Reading teacher education as text.* New York: Peter Lang.

Shernoff, D. J. (2013). *Optimal Learning Environments to Promote Student Engagement.* New York: Springer. doi:10.1007/978-1-4614-7089-2

Shively, K., Stith, K. M., & Rubenstein, L. D. (2018). Measuring What Matters: Assessing Creativity, Critical Thinking, and the Design Process. *Gifted Child Today, 41*(3), 149–158. doi:10.1177/1076217518768361

Shoufan, A. (2019). Estimating the cognitive value of YouTube's educational videos: A learning analytics approach. *Computers in Human Behavior, 92*, 450–458. doi:10.1016/j.chb.2018.03.036

Silk, Y. (2015). *STEAM Grant Program Evaluation.* Pittsburgh, PA: Allegheny Intermediate Unit.

Simon, R. M., Wagner, A., & Killion, B. (2017). Gender and choosing a STEM major in college: Femininity, masculinity, chilly climate, and occupational values. *Journal of Research in Science Teaching, 54*(3), 299–323. doi:10.1002/tea.21345

Smalley, S., Hainline, M. S., & Sands, K. (2019). School-based agricultural education teachers' perceived professional development needs associated with teaching, classroom management, and technical agriculture. *Journal of Agricultural Education, 60*(2), 85–98. doi:10.5032/jae.2019.02085

Smith, G. A. (2007). Place-based education: Breaking through the constraining regularities of public school. *Environmental Education Research, 13*(2), 189–207. doi:10.1080/13504620701285180

Sobel, D. (2004). *Place-Based Education: Connecting Classrooms and Communities*. Great Barrington, MA: The Orion Society.

Steele, F. (1980). Defining and developing environmental competence. *Advances in Experiential Social Processes, 2*, 225-244.

Stevenson, M., Bower, M., Fallon, G., Forbes, A., & Hatzigianni, M. (2019). By design: Professional learning ecologies to develop primary school teachers' makerspaces pedagogical capabilities. *British Journal of Educational Technology, 50*(3), 1260–1274. doi:10.1111/bjet.12743

Story, M. F. (2001). Principles of universal design. *Universal Design Handbook*.

Story, M. F., Mueller, J. L., & Mace, R. L. (1998). *The universal design file: Designing for people of all ages and abilities*. Academic Press.

Stovall, M. (2015). African American cultural technology: The lindy hop, the king of pop, and the factory worker's experience. *Transforming Anthropology, 23*(1), 1–13. doi:10.1111/traa.12042

Strauss, A., & Corbin, J. (1998). *Basics of qualitative research: Grounded theory procedures and techniques* (2nd ed.). Thousand Oaks, CA: Sage.

Sweller, J. (1988). Cognitive load during problem solving: Effects on learning. *Cognitive Science, 12*(2), 257–285. doi:10.120715516709cog1202_4

Sweller, J. (1989). Cognitive technology: Some procedures for facilitating learning and problem-solving in mathematics and science. *Journal of Educational Psychology, 81*(4), 457–466. doi:10.1037/0022-0663.81.4.457

Sweller, J., Van Merrienboer, J. J., & Paas, F. G. (1998). Cognitive architecture and instructional design. *Educational Psychology Review, 10*(3), 251–296. doi:10.1023/A:1022193728205

Szeszak, S., Man, R., Love, A., Langmack, G., Wharrad, H., & Dineen, R. A. (2016). Animated educational video to prepare children for MRI without sedation: evaluation of the appeal and value. *Pediatric Radiology, 46*(12), 1744. doi:10.100700247-016-3661-4

Tanner, C. K. (2008). Explaining relationships among student outcomes and the school's physical environment. *Journal of Advanced Academics, 19*(3), 444–471. doi:10.4219/jaa-2008-812

Tarr, J. E., Reys, R. E., Chavez, O., Shih, J., Confrey, J., & Silver, E. A. (2006). *A longitudinal study of mathematics curriculum and student achievement*. Paper presented at the NCTM research presession, St. Louis, MO.

Tate, W. (2001). Science education as a civil right: Urban schools and opportunity-to-learn considerations. *Journal of Research in Science Teaching*, *38*(9), 1015–1028. doi:10.1002/tea.1045

Taylor, A. (2009). *Linking architecture and education: Sustainable design for learning environments*. UNM Press.

Technical Education Research Centers. (2012a). Goals for productive discussions and nine talk moves. The Inquiry Project: Bridging research & practice supported by the National Science Foundation.

Technical Education Research Centers. (2012b). Reflection tool: Are students progressing toward scientific understanding? The Inquiry Project: Bridging research & practice supported by the National Science Foundation.

TEDEd. (n.d.). In *Facebook*. Retrieved July 22, 2019, from https://www.facebook.com/TEDEducation/posts/are-bats-really-blind-not-exactly-bats-use-a-special-process-called-echolocation/2218872224792638/

Teitel, L. (1998). Separations, divorces, and open marriages in professional development school partnerships. *Journal of Teacher Education*, *49*(2), 85–96. doi:10.1177/0022487198049002002

Ten Hove, P., & van der Meij, H. (2015). Like it or Not. What Characterizes YouTube's More Popular Instructional Videos? *Technical Communication (Washington)*, *62*(1), 48–62.

The College Board. (2012). *The Arts and the Common Core: A Review of Connections Between the Common Core State Standards and the National Core Arts Standards Conceptual Framework*. New York, NY: The College Board.

The Institute for Arts Integration and STEAM. (2019, December 2). *What is STEAM Education*? Retrieved from https://educationcloset.com/what-is-steam-education-in-k-12-schools/#stemvssteam

Theisens, H., Benavides, F., & Dumont, H. (2008). *OECD work on future educational environments*. Academic Press.

Theobald, P., & Curtiss, J. (2000). Communities as curricula. *Forum for Applied Research and Public Policy*, *15*(1), 106–111.

Thomas, H. (2010). Learning spaces, learning environments and the dis 'placement'of learning. *British Journal of Educational Technology*, *41*(3), 502–511. doi:10.1111/j.1467-8535.2009.00974.x

Thomas, K. (2013). Changing mathematics teaching practices and improving student outcomes through collaborative evaluation. *Teacher Education and Practice*, *26*(4), 779–796.

Thompson, K., Kanasa, H., & Chapman, S. (2019). Tools to Facilitate Teacher and Student Collaboration in Assessment. In Lund, K., Niccolai, G. P., Lavoué, E., Gweon, C. H., & Baker, M. (Eds.), A Wide Lens: Combining Embodied, Enactive, Extended, and Embedded Learning in Collaborative Settings, 13th International Conference on Computer Supported Collaborative Learning (CSCL) 2019, Volume 2 (pp. 967-968). Lyon, France: International Society of the Learning Sciences.

Thompson, K., Gouvea, J., & Habron, G. (2016). *A design approach to understanding the activity of learners in interdisciplinary settings: Environment and Diversity*. Presented at the *International Conference of the Learning Sciences*, Singapore.

Tierney, P., & Farmer, S. M. (2002). Creative self-efficacy: Its potential antecedents and relationship to creative performance. *Academy of Management Journal*, *45*(6), 1137–1148.

Timms, M. J., Moyle, K., Weldon, P. R., & Mitchell, P. (2018). *Challenges in STEM learning in Australian schools: Literature and policy review*. Academic Press.

Tousignant, M., & DesMarchais, J. E. (2002). Accuracy of student self-assessment ability compared to their own performance in a problem-based learning medical program: A correlation study. *Advances in Health Sciences Education: Theory and Practice*, *7*(1), 19–27. doi:10.1023/A:1014516206120 PMID:11912331

Towler, L., & Broadfoot, P. (1992). Self-assessment in the primary school. *Educational Review*, *44*(2), 137–151. doi:10.1080/0013191920440203

Towns, M. H. (2010). Where are the women of color? Data on African American, Hispanic, and Native American faculty in STEM. *Journal of College Science Teaching*, *39*(4), 8.

Trahan, K., Romero, S. M., Ramos, R. A., Zollars, J., & Tananis, C. (2019). Making success: What does large-scale integration of making into a middle and high school look like? *Improving Schools*, *22*(2), 144–157. doi:10.1177/1365480219835324

Truth and Reconciliation Commission of Canada. (2012). *Truth and reconciliation commission: Calls to action*. Retrieved May 12, 2016 from http://www.trc.ca/websites/trcinstitution/File/2015/Findings/Calls_to_Action_English2.pdf

Tuhiwai Smith, L. (2012). *Decolonizing methodologies: Research and Indigenous peoples* (2nd ed.). London, UK: Zed Books.

Tushnet, N., Millsap, M.,Abdullah-Welsh, N., Brigham, N., Cooley, E., Elliott, J., & Rosenblum, S. (2000). *Final Report on the Evaluation of the National Science Foundation's Instructional Materials Development Program.* Academic Press.

Tversky, B., Morrison, J. B. & Betancourt. (2002). Animation: can it facilitate? *International Journal of Human - Computer Studies, 57*(4), 247–262. doi:10.1006/ijhc.2002.1017

U.S. Congress, Office of Technological Assessment. (1984). *Commercial biotechnology: An international analysis*. Washington, DC: Author. Retrieved from http://ota.fas.org/reports/8407.pdf

U.S. Congress, Office of Technological Assessment. (1988). *New developments in biotechnology: U.S. investment in biotechnology—Special report*. Washington, DC: U.S. Government Printing Office. Retrieved from http://ota.fas.org/reports/8840.pdf

U.S. Congress, Office of Technological Assessment. *(1991). Biotechnology in a global economy.* Washington, DC: U.S. Government Printing Office. Retrieved from http://ota.fas.org/reports/9110.pdf

United Nations. (2008). *United Nations declaration on the rights of Indigenous peoples*. Retrieved May 12, 2016 from http://www.un.org/esa/socdev/unpfii/documents/DRIPS_en.pdf

Upitis, R. (2004). School architecture and complexity. *Complicity: An International Journal of Complexity and Education, 1*(1).

van Eijck, M. W., & Roth, W.-M. (2009). Towards a chronotopic theory of "place" in place-based education. *Cultural Studies of Science Education, 5*(4), 869–898. doi:10.100711422-010-9278-2

Vasquez, J., Sneider, C., & Comer, M. (2013). *STEM Lesson Essentials: Integrating Science, Technology, Engineering and Mathematics*. Heinemann.

Vossoughi, S., Hooper, P. K., & Escude, M. (2016). "Making" through the lens of culture and power: Towards transformative visions for education equity. *Harvard Educational Review, 86*(2), 206–232. doi:10.17763/0017-8055.86.2.206

Vygotsky, L. (1978). *Mind in Society*. London: Harvard University Press.

Wang, H. H., Moore, T. J., Roehrig, G. H., & Park, M. S. (2011). STEM integration: Teacher perceptions and practice. *Journal of Pre-College Engineering Education Research, 1*(2), 1-13.

Wattchow, B., & Brown, M. (2011). *A pedagogy of place: Outdoor education for a changing world.* Clayton, Australia: Monash University Publishing.

Weintrop, D., & Wilensky, U. (2015, June). To block or not to block, that is the question: students' perceptions of blocks-based programming. In *Proceedings of the 14th International Conference on Interaction Design and Children* (pp. 199-208). ACM. 10.1145/2771839.2771860

Weiss, I. R., & Pasley, J. D. (2004). What is high-quality instruction? *Educational Leadership, 61*(5), 24–28.

Weiss, I. R., Pasley, J. D., Smith, P. S., Banilower, E. R., & Heck, D. J. (2003). *Looking inside the classroom: A study of K-12 mathematics and science education in the United States*. Chapel Hill, NC: Horizon Research. Available www.horizon-research.com/insidetheclassroom/reports/highlights/highlights.pdf

Wells, J. G. (1994). Establishing a taxonometric structure for the study of biotechnology in secondary school technology education. *Journal of Technology Education, 6*(1), 58–75. doi:10.21061/jte.v6i1.a.5

Wells, J., White, K., & Dunham, T. (2000). *Technology education biotechnology curriculum*. Morgantown, WV: Biosens.

Wenger, E. (1998). *Communities of practice*. Cambridge, UK: Cambridge University Press. doi:10.1017/CBO9780511803932

Wenglinsky, H. (2002). How schools matter: The link between teacher classroom practices and student academic performance. *Educational Policy Analysis Archives, 10*(12). Retrieved [2/27/2006] from http://eppa.asu.edu/epaa/v10n12/

Wideen, M., Mayer-Smith, J., & Moon, B. (1998). A critical analysis of the research on learning to teach: Making the case for an ecological perspective on inquiry. *Review of Educational Research, 68*(2), 130–178. doi:10.3102/00346543068002130

Wildcat, M., McDonald, M., Irlbacher-Fox, S., & Coulthard, G. (2014). Learning from the land: Indigenous land-based pedagogy and decolonization. *Decolonization, 3*(3), i–xv.

Wilson, S. (2007). Guest editorial: What is an Indigenist research paradigm? *Canadian Journal of Native Education, 30*(2), 193–195.

Women, Minorities, and Persons with Disabilities in Science and Engineering: 2019 | NSF - National Science Foundation. (2019). Retrieved 23 November 2019, from https://ncses.nsf.gov/pubs/nsf19304/digest/employment#employment-status

Women, Minorities, and Persons with Disabilities in Science and Engineering: 2019 NSF - National Science Foundation. (n.d.). Retrieved July 19, 2019, from https://ncses.nsf.gov/pubs/nsf19304/

Wood, D., Bruner, J. S., & Ross, G. (1976). The role of tutoring in problem solving. *Journal of Child Psychology and Psychiatry, and Allied Disciplines, 17*(2), 89–100. doi:10.1111/j.1469-7610.1976.tb00381.x PMID:932126

Woodhouse, J. L., & Knapp, C. E. (2000). *Place-based curriculum and instruction: Outdoor and environmental education approaches*. Charleston, WV: ERIC Clearinghouse on Rural Education and Small Schools.

Yisa, N. C., & Ojiaku, F. C. (2016). Effectiveness of Computer Animation on a Progressive Learning Achievement of Secondary School Biology Students in Niger State, Nigeria. *International Journal of Education and Evaluation, 2*(4), 72-81.

Yosso, T. J. (2005). Whose culture has capital? A critical race theory discussion of community cultural wealth. *Race, Ethnicity and Education, 8*(1), 69–91. doi:10.1080/1361332052000341006

Young, J. L., Ero-Tolliver, I., Young, J. R., & Ford, D. Y. (2017). Maximizing opportunities to enroll in advanced high school science courses: Examining the scientific dispositions of Black girls. *Journal of Urban Learning, Teaching, and Research, 13*, 174–183.

Yue, C., Kim, J., Ogawa, R., Stark, E., & Kim, S. (2013). Applying the cognitive theory of multimedia learning: An analysis of medical animations. *Medical Education, 47*(4), 375–387. doi:10.1111/medu.12090 PMID:23488757

Zeichner, K., & Tabatchnik, B. (1981). Are the effects of university teacher education "washed out" by school experiences? *Journal of Teacher Education, 32*(3), 7–11. doi:10.1177/002248718103200302

Zhao, Y. (2016). From deficiency to strength: Shifting the mindset about education inequality. *The Journal of Social Issues, 72*(4), 720–739. doi:10.1111/josi.12191

Zhao, Y., Wehmeyer, M., Basham, J., & Hansen, D. (2019). Tackling the wicked problem of measuring what matters: Framing the questions. *ECNU Review of Education, 2*(3), 262–278. doi:10.1177/2096531119878965

Zollars, J. (2018) *Flow Theory and Engagement: Observing Engagement through the Lens of Flow in a Middle School Integrated Maker Space* (Doctoral Dissertation). University of Pittsburgh.

About the Contributors

Kelli Thomas, Ph.D., is Associate Dean for Teacher Education and Undergraduate Programs in the School of Education and Human Sciences at the University of Kansas and Associate Professor of Mathematics Education in the Department of Curriculum and Teaching. Dr. Thomas's professional work has focused on mathematics teaching, learning, and assessment in schools for more than 25 years. She has been involved in educator preparation and professional development of educators in university and K-12 settings. She began her career in education teaching in public school settings including the Department of Defense Dependent Schools (DoDDS) in Germany. Her research agenda focuses on teaching, learning, and assessment of mathematics with a focus on teacher development and classroom practice. She also has expertise in program evaluation. Her work is published in journals such as Journal of Educational Research, School Science and Mathematics, Teaching Children Mathematics, Action in Teacher Education, American Journal of Evaluation, New Directions for Evaluation, and Teacher Education & Practice.

Douglas Huffman, teaches courses in elementary education and also has expertise in assessment and program evaluation. Dr. Huffman previously served as the associated editor for the Journal of Research in Science Teaching. Additionally, he has publications in numerous journals, such as the American Journal of Evaluation, Journal of Research in Science Teaching School Science & Mathematics and the Journal of Science Teacher Education.

* * *

Oludurotimi Adetunji, Ph.D., is the Associate Dean for Undergraduate Research and Inclusive Science and an adjunct assistant professor of physics at Brown University, where he leads the SciToons Creation Group. He has given over 60 talks at national and international conferences, and he is the author/co-author of over 75 articles, conference presentations, and reports. His B.S. in Physics and Computer Science is from Fisk University, and his Ph.D. in Physics is from The Ohio State

University. He also holds an MLE Certificate from the Institute for Management and Leadership Education at the Harvard Graduate School of Education.

Ayana Allen-Handy, Ph.D., is an Assistant Professor of Urban Education in the School of Education at Drexel University. Her research interests include examinations of the social contexts of education through ethnographic and critical participatory action research (CPAR) methods towards empowering youth, teacher, and community member capacity and agency to address critical issues of educational equity and social justice in urban schools and communities. She is the co-editor of three books, Autoethnography as a Lighthouse: Illuminating Race, Research, and the Politics of Schooling (Information Age); Black Female Teachers: Diversifying the United States' Teacher Workforce (Emerald); and Global Perspectives on Issues and Solutions in Urban Education (Information Age). She has published articles in various journals such as the Journal of Negro Education, Urban Education, Teachers College Record, International Journal of Qualitative Studies in Education, & Equity and Excellence in Education. She received a B.A. from the University of North Carolina at Chapel Hill, a M.Ed. from the University of St. Thomas, a Ph.D. in Curriculum & Instruction from Texas A&M University, and she completed a Post-Doctoral Fellowship at The Urban Education Collaborative at UNC Charlotte.

Neal Barbour cut his teeth as a visual artist in the Seattle area. Neal has a wide range of experience as a videographer, screenprinter, ceramicist, and digital installation artist, working in a variety of locations and settings including Chile, Germany, and across the west coast. After ten years of itinerant artistry, Neal moved to Portland, Oregon to obtain a Masters in Art Education from Portland University. After teaching in the public school system for seven years, he began working at the Lawrence Arts Center as Director of Youth Education. He has shown work across the country in many solo and group shows and continues to show work locally with emphasis in ceramics and printmaking, although much of his devotion now funnels into education where he strives to create relevant educational opportunities for youth.

James D. Basham, Ph.D., is an Associate Professor in the Department of Special Education at the University of Kansas. Dr. Basham is also the Senior Director for Learning & Innovation at CAST and founder of the Universal Design for Learning - Implementation and Research Network (UDL-IRN). His research is focused on the implementation of UDL, STEM education, learner-centered design, innovation, and technology in human learning. He is well-published, given hundreds of talks, and serves on various boards for journals, companies, and organizations.

Lizette Burks, is the secondary science education director for the Arkansas Academy of Educational Equity graduate program at the University of Arkansas. Previously, she directed K-12 science programs as the state science supervisor for the Kansas State Department of Education. She also served as a science teacher in Texas and Kansas at the high school and middle school levels. Dr. Burks serves on committees with the Council of State Science Supervisors and the National Science Education Leadership Association. Her research interests include studying the pedagogical shifts that emerge alongside science education reform and how pre-service and in-service teachers can more effectively implement instruction that moves towards more equitable science teaching and learning. Dr. Burks earned her B.S. in biology from Texas Tech University, her M.Ed. in Teaching from Emporia State University, and her Ed.D. in Curriculum and Instruction from the University of Kansas.

Susan Chapman is a lecturer in the School of Early Childhood and Inclusive Education at the Queensland University of Technology. She has worked professionally as an actor and musician and has experience in teaching the Arts in primary, secondary and tertiary sectors. Her publications focus on integrated interdisciplinary learning using an Arts Immersion approach. Susan has completed a Research Fellowship in STEAM at Griffith University, and conducts workshops on developing engaging academic presentation strategies.

Becky Eason received her doctorate in Higher Education from the University of Kansas. She is the President of WordCraft LLC, which has as its mission to provide excellent, individualized strategic services to arts organizations and other non-profits to encourage their viability and growth. WordCraft is currently providing project facilitation and evaluation services to the US Department of Education, The NAMM Foundation, the Kansas Department of Transportation, and Harvesters Community Food Network, among others. Prior to founding her own company, Eason was the Associate Director of the Center for Public Partnerships & Research at the University of Kansas, where she served as Principal Investigator/ Project Director on numerous multi-million dollar grants. Her specialties include strategic planning, group facilitation, wellness coaching, and project evaluation. Becky is an experienced presenter who has traveled around the country working with companies and nonprofits to maximize their potential for growth and success. Becky has also completed Levels I and II Certification from the Kansas Leadership Center, Health & Wellness Courses from MentorCoach, and is an Associate Certified Coach with the International Coach Federation.

Rachel Louise Geesa, EdD, is an Assistant Professor of Educational Leadership at Ball State University. Dr. Geesa is an experienced art and gifted resources educator and school administrator in PreK-12 schools to include Department of Defense Education Activity (DoDEA) overseas schools. She has additional skills working with students, teachers, and administrators to create and execute integrative STEM-based learning experiences. Dr. Geesa earned a Bachelor of Arts degree in art education, art history, fine arts, and photography from Purdue University and a Master of Art Education degree from Indiana University. Additionally, Dr. Geesa holds a Doctor of Education in educational administration and supervision degree, Master of Arts degree in educational psychology, a Master of Arts in Education degree in educational administration and supervision, and a Specialist in Education degree in school superintendency from Ball State University.

Cassandra L. Hunt is a Doctoral Candidate in the Department of Special Education at the University of Kansas. Ms. Hunt was formerly a special education teacher having taught from K-12, in self-contained and co-teaching settings, and in rural and urban school districts during her career. Her research interests include UDL implementation, measurement of UDL implementation, and UDL implementation for student with emotional and behavioral disorders.

Valerie Ifill, M.F.A., is a dance educator, researcher and performer focused on the intersections of dance and the community and making dance more accessible. Assistant Professor of Dance at the University of Akron, Valerie is invested in university-community dance initiatives as well as using the Africanist perspective to support university dance curriculum. Valerie earned her Master of Fine Arts degree in Dance from the University of Oregon, completed the Independent Study Program at The Alvin Ailey American Dance Theater, and earned her Bachelor of Business Administration degree with a Dance minor from Kent State University. Valerie has founded and directed university-community dance programs at Drexel University through Dornsife Center for Neighborhood Partnerships in Philadelphia, PA, and at Johnson C. Smith University in Charlotte, NC.

Haidee Jackson, M.Ed., is a doctoral student in the Department of Special Education at the University of Kansas. Her research is focused around the incorporation of UDL and student-centered instructional practices in schools, space design, technology, and innovation in education.

Harry Kanasa has taught in primary, secondary and tertiary settings over his 22 year teaching career. Starting his career as a mathematics, science and biology teacher, he coordinated Senior Biology before becoming a middle school Head of mathematics and science. He has also taught gifted and talented mathematics and science topics to primary school students. He currently teaches mathematics and science education to pre-service mathematics and science teachers. His current research interests is STEAM inquiry and how it can benefit inquiry-based and problem-based learning.

Roger Levine earned a BS in Biology from Rensselaer Polytechnic Institute, a master's degree in Psychology from Tufts University, and a Ph.D. in Behavioral Endocrinology from Stanford University. He was employed by research/consulting firms for over 30 years, specializing in survey research, program evaluation research, education research, and health services research. He was the Project Director for NSF's Opportunities for Enhancing Diversity in the Geosciences Program, providing technical evaluation assistance to dozens of grantees' evaluation efforts and developing procedures to enable an overall evaluation of the program's effectiveness. His resume includes 55 peer-reviewed journal publications and nearly 100 presentations.

Kevin O'Connor is an Associate Professor and Chair for the Department of Education, at Mount Royal University, Canada. He has taught in elementary and secondary schools and was educational administrator for 15 years. Much of his current research and publications are based on the synthesis of multi-sensory pedagogy and interdisciplinary curriculum through the integration of experiential and place-based learning, science field studies and Indigenous education. Dr. O'Connor is the National Chair of the Canadian Association of Teacher Education CATE-Self-Study in Teacher Education Practices SSTEP group, NASA-GLOBE Canada Country Coordinator, serves as an editor and reviewer for numerous education research journals worldwide and sits as an Advisory Board member on the Canadian Research Institute for Social Policy CRISP, NASA-GLOBE International Science Education Group, European Scientific Institute ESI and The Paulo and Nita Freire International Center for Critical Pedagogy.

Renata de Almeida Ramos is a doctoral student in Social and Comparative Analysis in Education at the University of Pittsburgh. Her research interests are teacher autonomy and pedagogical change, educational policies, and critical pedagogy. Her professional experience includes teaching (elementary and tertiary levels), program evaluation, and educational research in Brazil and the United States of America. She received her M.S.Ed. in Program Evaluation from the Duquesne University in 2014 and her B.S. in Pedagogy from the Pontifícia Universidade Católica de São Paulo (Brazil) in 2000.

Michelle L. Rogers, PhD, is an associate professor in the College of Computing and Informatics at Drexel University. For 15 years, she has used human factors engineering methods and socio-technical systems theory to study the impact of health information technology (HIT) on clinical workflow and usability of technology with and by medically under-served populations. Most recently, Dr. Rogers has been investigating the usability and utility of patient portals and electronic medical records. Internationally, her investigations are focused on HIT in Uganda health systems. In addition, she is collaborating with faculty from industrial design, dance and education to understand how making, arts, and coding can assist in making the realization of a career in STEM fields achievable.

Stephanie Romero received her BA from Arizona State University and both her MA and EdD from the University of Pittsburgh. She worked in public schools as a teacher for 15 years, holding teaching, supervisor of curriculum and principal certifications. Most recently she founded and runs a nonprofit, Awaken Pittsburgh, working to transform our local educational systems through mindfulness practices and programs.

Raja Schaar is Program Director and Assistant Professor of Product Design at Drexel University. Raja is an industrial designer, museum exhibition designer, and STEAM education researcher. Her work focuses on decolonizing STEM/STEAM education, maternal health, and biologically inspired design and sustainability. She graduated from GA Tech with a BS Industrial Design in 2001, and she completed her MA in Art Education in 2003 from the School of the Art Institute of Chicago. Prior to joining Drexel's faculty, Raja taught design in GA Tech's Biomedical Engineering and Industrial Design schools and designed museum and visitor services exhibitions all over the Southeast.

Gladys Sterenberg is a Professor in the Department of Education at Mount Royal University. Her program of research is focused on relational ethics and can be described as encompassing three interrelated research interests: mathematics education, Indigenous ways of knowing, and overlapping communities of practice within teacher education.

Krista Stith, PhD, is the Director of the Center for Gifted Studies and Talent Development at Ball State University. Her primary responsibilities include serving stakeholders in gifted education and promoting inter-, multi-, and transdisciplinary experiences in STEM education for all youth through teacher education. she has served as a member of the Indiana STEM Taskforce to strategize STEM education initiatives across the state. Dr. Stith has co-designed multiple Integrative STEM education courses at the undergraduate and graduate levels and is the coordinators of the Project Lead The Way (PLTW) partnership between Ball State and PLTW. Her research interests are in design thinking and STEM-literacy outcomes. Dr. Stith graduated from Virginia Tech with a Doctorate of Curriculum and Instruction with a focus in Integrative STEM education and a Masters in Agricultural Extension & Education.

Cynthia Tananis is an Associate Professor in Administrative and Policy Studies at the University of Pittsburgh and has been an educator for 40 years in K-12 andHigher Education settings. She was the Founder and Director of the Collaborative for Evaluation and Assessment Capacity (CEAC) for 12 years at Pitt.

Wei Tang is a Graduate Evaluator at Collaborative for Evaluation and Assessment Capacity, School of Education, University of Pittsburgh.

Kate Thompson is an Associate Professor of Digital Pedagogies in the Faculty of Education at Queensland University of Technology. Kate teaches all things digital and technology related, and her research tends to be with teams of researchers, using innovative analytic or data collection techniques, about connecting people and knowledge to answer some big questions about learning and teaching. The underlying focus of Dr Thompson's research is learning sciences, specifically collaborative learning and discovering patterns of learner interaction that could be used by an instructor to indicate progress through a task. Her research in this area influences how educators design for the use of a range of digital technologies as the tool around which collaboration occurs in increasingly complex face-to-face and online collaborative environments. Kate's research has been applied in collaborative learning and learning situations that include school students (primary and secondary), undergraduate and postgraduate students. A recent focus has been in interdisciplinary collaboration with environmental science undergraduate and graduate students in the USA. In Australia, Dr Thompson leads teams working with STEAM preservice teachers, teachers and students.

Keith Trahan, Ph.D., is Visiting Faculty and Interim Director of the Collaborative for Evaluation and Assessment Capacity at the University of Pittsburg, Keith has served as lead evaluator and research on a variety of programs in the areas of K-12 math and science reform and school leadership, IHE STEM curriculum, instruction, and learning, IHE international education, and community based human services. He holds a B.A. in government and a B.A. in sociology from McNeese State University and an M.A.T. from Charleston Southern University. Before joining CEAC, Keith worked as a support enforcement specialist for the State of Louisiana, project director at the Canon Street YMCA in Charleston, SC, and teacher and coach for the Charleston County School District. Keith has earned his Doctor of Philosophy in the administrative and policy studies at the University of Pittsburgh.

Norman Vaughan's teaching background includes graduate and undergraduate courses in educational technology, K–12 education in northern Canada, technical training in the petroleum industry, and English as a Second Language in Japan. He is the Co-founder of the Blended Online Design Network (BOLD), a member of the Community of Inquiry Research Group, the Associate Editor of the International Journal of Mobile and Blended Learning and he is on the Editorial Boards of the International Journal of Excellence in e-Learning, Canadian Journal of Learning and Technology, the International Journal of E-Learning & Distance Education, the Journal on Centres for Teaching & Learning, and the Learning Communities Journal.

Monique Woodard is a second year PhD student at Drexel University in Educational Leadership Development and Learning Technologies. Her research focuses on developing creative processes in Black girls as they learn how to code and design virtual reality environments.

Jeffrey Zollars received his Ed.D., Superintendent's Letter of Eligiblity, and M.A. in school leadership and curriculum supervision through the University of Pittsburgh. He served as a classroom educator for 15 years and as a school district curriculum director for 10 years. Currently, his primary research interest includes Flow Theory and student engagement as observed in semi-structured learning environments found in Maker classrooms.

Index

H

I

J

L

M

N

O

P

www.ingramcontent.com/pod-product-compliance
Lightning Source LLC
LaVergne TN
LVHW061550290926
849162LV00004B/968